W0071796

Research Reports ESPRIT

Subseries PDT (Product Data Technology)

Project 5109 · Neutral Interfaces for Robotics (NIRO)

Edited in cooperation with the European Commission and the Product Data Technology Advisory Group (PDTAG)

I. Bey D. Ball H. Bruhm T. Clausen W. Jakob
O. Knudsen E. G. Schlechtendahl T. Sørensen
(Eds.)

Neutral Interfaces in Design, Simulation, and Programming for Robotics

Springer-Verlag

Berlin Heidelberg New York
London Paris Tokyo
Hong Kong Barcelona
Budapest

Contact Editor

Ingward Bey
Kernforschungszentrum Karlsruhe GmbH
Postfach 3640
D-76021 Karlsruhe

ESPRIT Project 2614/5109 "Neutral Interfaces for Robotics (NIRO)" belongs to the Subprogramme "Computer Integrated Manufacturing" of ESPRIT, the European Specific Programme for Research and Development in Information Technology supported by the European Commission.

Project 2614/5109 aims to develop neutral interfaces and neutral data formats for the exchange of geometrical, technological and programming functions and data between CAD, robot planning and robot control. The user-oriented robot programming language is treated as an interface between the programmer and the programming system. These developments are based on formal techniques for specifying the interfaces that are relevant for the robotic application area. The main aim is to develop processors, software tools and methods that will enable multivendor systems to be put together. An important topic of the project is to demonstrate the feasibility of the approach by constructing prototypes to prove the intercommunication capabilities between systems built from multivendor modules. The project follows the approach of ESPRIT project 322 and extends its scope into the robotics domain.

CR Subject Classification (1991): D.3.4, H.2.3

ISBN-13: 978-3-540-57531-3   e-ISBN-13: 978-3-642-85057-8

DOI: 10.1007/978-3-642-85057-8

CIP data applied for

Publication No. EUR 15675 EN of the European Commission, Dissemination of Scientific and Technical Knowledge Unit, Directorate-General Information Technologies and Industries, and Telecommunications, Luxembourg.

Neither the European Commission nor any person acting on behalf of the Commission is responsible for the use which might be made of the following information.

Typesetting: Camera-ready by authors
SPIN: 10132142   45/3140 – 543210 – Printed on acid-free paper

Foreword

Product Data Technology encompasses the information related to all stages in the product life cycle from product design via production planning, production processes, production control, etc. through the delivery and operational stages of the technical product. Product Data Technology takes a coherent, unified view of the information captured in this whole life cycle and provides methodologies to support this integrated perspective.

The Product Data Technology Advisory Group (PDTAG), is a Special Interest Group supported by the CIME Division within DG III of the European Commission. Founded in 1991, the PDTAG has encouraged the formation of a new subseries on Product Data Technology within the existing series of Research Reports ESPRIT. This subseries will provide a depository for the important contributions made by ESPRIT projects to the evolving area of Product Data Technology, particularly also those based on the STEP (ISO 10303). PDTAG is grateful to Springer Publishers for establishing this subseries which will serve to report recent international developments in Product Data Technology.

The current volume describes the results of the NIRO Project (ESPRIT 2614/5109), dealing with neutral interfaces for robotics. This project has made important contributions to international standardization for the software interfaces linking systems for computer-aided design, for industrial robot programming and robot simulation. These systems cover a key part of the CIM process chain. NIRO has had important influence on international standards for kinematics (Part 105 of STEP) and on the robot programming languages IRL and ICR: This volume presents NIRO's contributions to these standards and its own results in developing STEP processors as well as four demonstration pilots of neutrally interfaced systems for CAD supported offline robot programming and robot simulation. These results are very suitable to demonstrate the current potential and future promise of Product Data Technology.

Berlin, November 1993   H. Nowacki

Preface

The standardisation of neutral interfaces for product data exchange is a complex, long-lasting, but very important task to bring information in a fast, secure and efficient way from system to system, company to company, and from person to person.

In this book a further step in this direction is reported. The project NIRO (ESPRIT 2614/5109) contributed to the development of the international STandard for the Exchange of Product model data (STEP initiative at ISO TC184SC4).

This volume covers descriptions of the specification work for the neutral interfaces STEP (extensions with kinematics), ICR (interface to robot controllers), and IRL/PLR for robot off-line programming. Processors and interpreters developed for several commercial CAD, programming, simulation, and control systems are described as well as the application of the results and the tests of the processors in four experiments in an industrial CIM environment. They cover applications in welding big ship parts, assembly of car motors, and filament winding of aircraft parts.

An overview of project results on the international standardisation process is also given.

Acknowledgements

It is a great pleasure for me to present the results of the ESPRIT project 2614/5109 NIRO together with the partners within the project to a greater interested audience via this book in the Springer Series "Research Reports ESPRIT", a series which was started in 1986 with the former ESPRIT project 322 "CAD Interfaces" (CAD*I). I would like to thank all people involved in the project for their spirit of cooperation and their help and effort during the project but especially all the authors and chapter editors who contributed to the presentation and the publication of the project results in this book.

Also I want to thank Mr. P. Maebe from the European Commission and the four reviewers of the NIRO project, Messrs. P. Coiffet, G. Destefanis, G.A. Parker, and E. Warman for their fruitful hints from the beginning until the end of the work and their technical and administrative support.

I would like to express a lot of thanks to Mrs. Ursula Frey: She has acted as the spirit of the project, the organizing master of meetings, reports and actions, and the central point of the NIRO communication network, and contributed with her efforts and enthusiasm in large measure to the positive results of the project which are reported here.

Karlsruhe, October 1993   Ingward Bey

Partner Companies

BYG Systems Ltd, William Lee Building,
  Highfield Science Park, Nottingham NG7 2RQ, UK

CASA, Manufacturing Engineering, Construcciones Aeronauticas SA
  E-28900 Getafe-Madrid

DIS Industrial Consultants a|s, Granskoven 10, DK-2600 Glostrup

Danmarks Tekniske Hojskole, Instituttet for Styreteknik,
  Bygning 424, DK-2800 Lyngby

DISEL, Avenida de Burgos, no 8-B, Edificio Génesis, E-28036 Madrid

FIAT, Systemi e Telecomunicazioni, Corso Marconi 10, I-10125 Torino

Kernforschungszentrum Karlsruhe GmbH, IAI/PFT,
  Postfach 36 40, D-76021 Karlsruhe

Odense Steel Shipyard Ltd, P.O. Box 176, DK-5100 Odense

PSI Gesellschaft für Prozeßsteuerungs- und Informationssysteme mbH,
  Heilbronner Str. 31, D-10711 Berlin

Reis GmbH & Co Maschinenfabrik, Postfach 11 01 61, D-63777 Obernburg

Contributors

J.P. Allen, BYG
E. Baena, CASA
D. Ball, BYG
G. Barker, BYG
A. Bautista, DISEL
M. Bettinardi, COMAU
I. Bey, KfK
U. Brockel, Reis
H. Bruhm, Reis
C. Busk, DIS
E. Cabaleiro, CASA
G. Cantello, COMAU
B. Carlsen, DIS
T.G. Clausen, DTH
C. D'Elia, SESAM
T. Elter, Reis
U. Frey, KfK
E. de Frutos, CASA
U. Gengenbach, KfK
L. Gil, CASA
A. Gonzalez, DISEL
S. Haas, KfK
H.U. Hohn, KfK
L. Huggler, DIS
W. Jakob, KfK
J.C.S. Jensen, DTH
O. Knudsen
B. Kottke, PSI
U. Kroszynski, DTH
E. Kroth, Reis

H.C. Larsen, OSS
H.-P. Lorenz, KfK
A. Ludwig, KfK
J. Lundberg, DIS
R. Lutz, KfK
R. Marshall, BYG
W. May, Reis
F. Naccari, FIAT
L.F. Nielsen, DTH
J.M. Nuñez, DISEL
M. Odifreddi, SESAM
M. Pasini, FIAT
M. Pfeiffer, Reis
J. Reim, KfK
G. Reitmayer, Reis
T. Rienmüller, Reis
F. Ruiz, DISEL
F. Rusina, SESAM
D. Sanchez-Brunete, CASA
E.G. Schlechtendahl, KfK
H. Schneider, Reis
C.E. Skjoelstrup, OSS
T. Sørensen, DTH
P. Sorenti, BYG
B. Spatz, Reis
J. Triginer, DISEL
E. Trostmann, DTH
S. Trostmann, DTH
A. Zaragoza, CASA

Table of Contents

1. **Introduction** 1
   I. Bey

1.1 The Need for Standardized Interfaces for Electronic Data Transfer 1
1.2 Short Summary of Interface Development 2
1.3 ESPRIT Project Nr. 2614/5109 NIRO 4

2. **State of the Art upon Project Start** 6
   I. Bey

2.1 CAD Interfaces 8
2.2 Interfaces in Robotics 9
     2.2.1 Robot Programming 11
2.3 Standardization 12

3. **Aims of the NIRO Project** 15
   I. Bey

4. **Results Related to the STEP Standard** 17

   Chapter Editor: E.G. Schlechtendahl

4.1 The NIRO Contributions to STEP 18
     E.G. Schlechtendahl
     4.1.1 The STEP Kinematics Model 18
     4.1.2 Physical File Structure 20
     4.1.3 STEP Data Access Interface 21
     4.1.4 Specification 21

4.2 The NIRO Specification for STEP Kinematics Models 22
     U. Kroszynski

5. **Development of STEP Processors** 25

   Chapter Editor: T. Sørensen

5.1 From CAD Systems to STEP 27
     5.1.1 GRASP 27
          P. Sorenti
     5.1.2 BRAVO3 31
          A. Ludwig and J. Reim
     5.1.3 CADDS 5 37
          C. D'Elia

5.1.4 CATIA (Kinematics) 41
   T. Sørensen and U. Kroszynski
5.1.5 KISMET... 45
   A. Ludwig

5.2 From STEP to Robot Programming and Simulation................... 56
5.2.1 GRASP.. 56
   P. Sorenti
5.2.2 CATIA Robotics.. 59
   R. Lutz
5.2.3 KISMET.. 64
   H.-P. Lorenz and S. Haas
5.2.4 Scanner/Parser for CATIA and KISMET................................. 69
   H.-P. Lorenz and S. Haas

5.3 Test of Processors and Consistency of the Specification 74
   T. Sørensen and U. Kroszynski
5.3.1 Test Suites, Library of Test Models...................................... 74
5.3.2 Evaluation of Test Results. Experiences................................ 80

6.   **Results Related to IRL and ICR** 85

   Chapter Editor: W. Jakob

6.1 Industrial Robot Language (IRL)... 85
   U. Schmiedecke
6.1.1 Present Status.. 85
6.1.2 The Language Features... 86
6.1.3 General Purpose Language Features 86
6.1.4 Robot Specific Features... 90
6.1.5 Recommendations for Further Enhancements 96
6.1.6 NIRO Contributions to IRL... 96
6.1.7 Contextfree Grammar of IRL.. 97

6.2 Intermediate Code for Robots (ICR) 106
   T. Clausen
6.2.1 Introduction to ICR.. 107
6.2.2 Technical Aspects of ICR 113
6.2.3 NIRO Contributions to ICR 120
6.2.4 Experiences with ICR.. 121
6.2.5 Present Status 123
6.2.6 Recommendations for Further Enhancements 123

7. Development of Processors for IRL and ICR 126

Chapter Editor: D. Ball

7.1 Processors for IRL 126
    7.1.1 IRL Preprocessor for GRASP 126
    P. Sorenti
    7.1.2 IRL to ICR Compiler 131
    U. Schmiedecke

7.2 Preprocessors for ICR 139
    7.2.1 ICR Preprocessor for GRASP 139
    P. Sorenti

7.3 ICR Interpreters 144
    7.3.1 Overview 144
    W. Jakob and J. Reim
    7.3.2 Robot Independent Part 147
    J. Reim
    7.3.3 Robot Dependent Part for the KISMET Simulation System 148
    J. Reim
    7.3.4 Robot Dependent Part for the ABB IRB 2000 Robot 149
    T. Clausen
    7.3.5 The Robot Dependent Part for the ABB IRB 60 Robot 154
    A. Bautista
    7.3.6 Robot Dependent Part for Reis ROBOTstar-IV 157
    H. Bruhm

7.4 ICR Transformation to Robot Vendor Code 160
    7.4.1 Overview 160
    C. Busk
    7.4.2 ICR Transformation to the Hirobo Robot 162
    C. Busk
    7.4.3 ICR Transformation to PDL2 (the Comau Robot) 164
    M. Odifreddi and C. D'Elia
    7.4.4 ICR Transformation to Reis Robot Language 174
    H. Bruhm

7.5 Test of Processors 177
    A. Bautista

8.   Application of Developments and Results................................... 180

Chapter Editor: H. Bruhm

8.1   The Rationale.. 180
      H. Bruhm

8.2   The Demonstration at CASA/DISEL....................................... 180
      A. Bautista
      8.2.1 Criteria for Task Selection.. 180
      8.2.2 Task Description 181
      8.2.3 Information Flow Structure....................................... 184
      8.2.4 Implementation Details.. 186
      8.2.5 Results... 189

8.3   The FIAT Demonstration. Experiences Gained 190
      C. D'Elia

8.4   Demonstration at Odense Steel Shipyard 197
      C. Busk and O. Knudsen
      8.4.1 Criteria for Task Selection.. 197
      8.4.2 Task Description 198
      8.4.3 Information Flow Structure....................................... 199
      8.4.4 Implementation. 200
      8.4.5 Experiences and Recommendations 204

8.5   The Reis Demonstration 205
      H. Bruhm
      8.5.1 Criteria for Task Selection.. 205
      8.5.2 Task Description 205
      8.5.3 Information Flow Structure....................................... 210
      J. Reim
      8.5.4 Implementation Details.. 210
      J. Reim
      8.5.5 Results... 213
      H. Bruhm, Reis and J. Reim

8.6   Other Applications Outside the NIRO Project........................ 214
      S. Trostmann, E. Trostmann, and L.F. Nielsen
      8.6.1 Introduction.. 214
      8.6.2. Architecture and Information Flow.............................. 215
      8.6.3 Models 216
      8.6.4 Modules... 217
      8.6.5 The Ropsim Model Interface..................................... 221
      8.6.6 Example 221
      8.6.7 Conclusion 224
      Acknowledgement... 225

9.   **Conclusions of the Project** 226

9.1   Interface Between CAD, Robot Programming
      and Simulation Systems 226
      E. Trostmann
      9.1.1 General 226
      9.1.2 Contributions to STEP standard 228

9.2   Interface Between Robot Off-Line Programming
      and Control Systems 229
      H. Bruhm

10.   **Outlook** 231
      O. Knudsen

References 239

Appendix 1: Specification for
      Kinematics and Robotics CAD Data Exchange 243

Appendix 2: Example IRL Output for GRASP Preprocessor 313

Appendix 3: Example ICR Output for GRASP Preprocessor 317

Appendix 4: Test Example:
      Movement File and Related ICR Code File 323

1. Introduction

I. Bey

1.1 The Need for Standardized Interfaces for Electronic Data Transfer

The integration of markets and companies and people acting in them is an actual historical process which is based mainly on better opportunities for easy communication. Information technology is a driving and supporting factor in this evolution. However, besides the consideration of cultural, language and other "originalities" when linking together specific persons, companies, or markets, a lot of technical problems must be solved to get a smooth data and information transfer from one participant to another in the big game of communication all around the world.

As cultures, languages grew up separately in different areas of our globe, in the same way very different systems for data processing and transfer were developed and introduced in societies, enterprises, in divisions, parts of factories, etc. The task is now how to find a way to bring these dissimilar computer systems to communicate with each other. The answer is: standardized interfaces.

Of course, this task is impossible to solve in a general way without standardised interfaces. The overall number of systems involved is too big to make an individual interface for each pair. Vendors of IT and communication systems may have limited interest in standard interfaces according to their market penetration; users of such systems on the other hand have the tendency to highly promote this approach to avoid dependencies on one single source of equipment and software.

Computer Integrated Manufacturing, or CIM, is the computer supported integration of the information flow within a company or between the company with its customers and suppliers or between several companies acting together in a big project. Thus implemented in the correct way, CIM is a strategic approach providing more flexibility, shorter throughput time and better quality. An important part of this is the choice of the right interfaces to transfer information between people and systems.

In manufacturing of mechanical products, information flow is mainly concerned with technical data. From the idea of a specific product, until its delivery to the customer (via the whole process of design, planning, manufacturing, assembly, final quality control, distribution) and its further life time including service and maintenance and perhaps recycling, a lot of product oriented information is generated, processed, stored, changed, adapted, transferred and finally deleted. The integration of this product data flow via interfaces is the challenging task which has been tackled in a series of ESPRIT projects, of which the project ESPRIT 5109 "NIRO - Neutral Interfaces for Robotics" is a link in this chain. This book summarizes the results of the NIRO project.

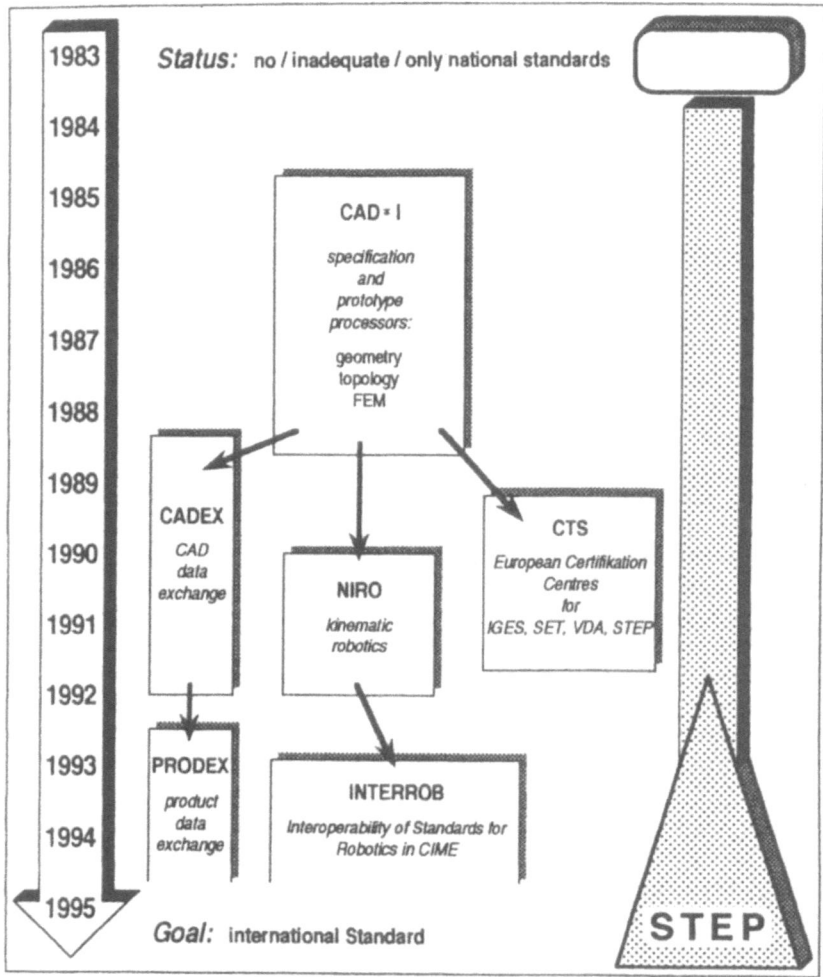

Fig. 1.2-1: Product data exchange: European based projects

1.2 Short Summary of Interface Development [Trostmann 91]

The "history" of standard interface development for mechanical product data exchange started in the late seventies with IGES. The IGES Version 1.0 became a US national standard for the first time in 1981 [ANSI Y 14.26 M]. The following versions during the eighties improved the possibilities of data transfer through a relatively high acceptance of IGES by CAD vendors and CAD users. However, the many deficiencies in IGES prompted the first new approach in this

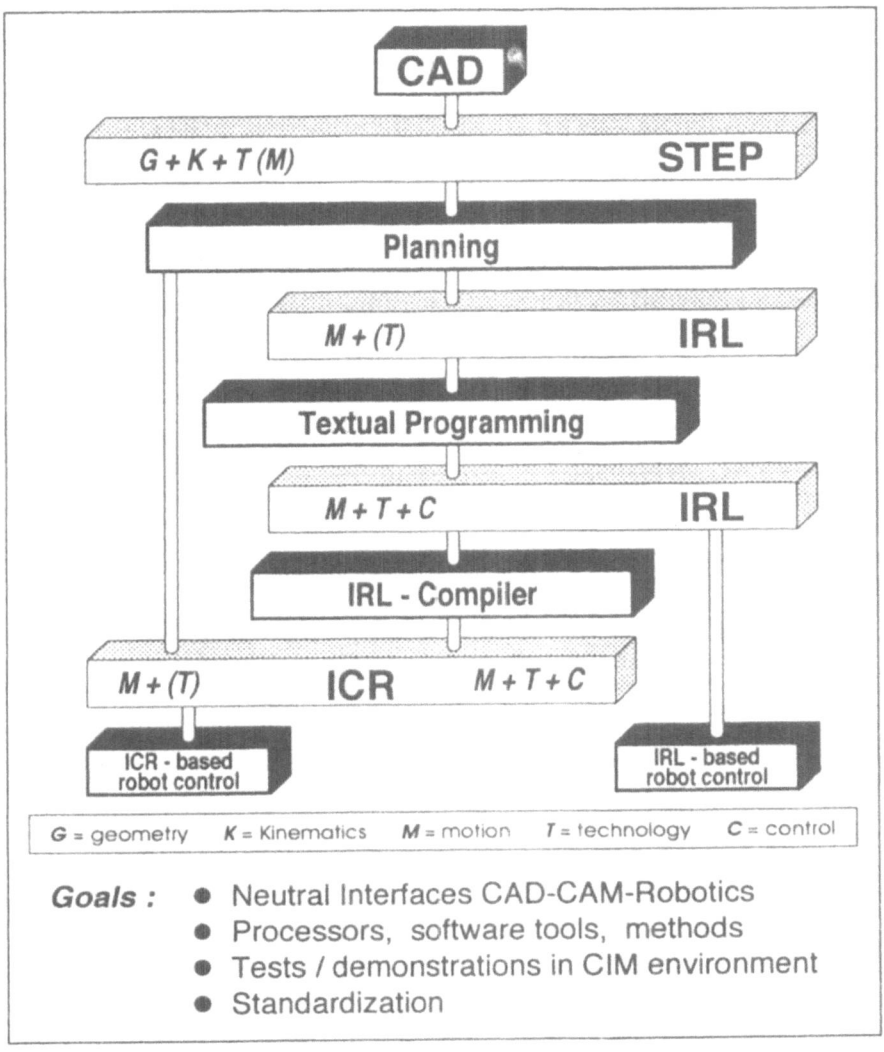

Fig. 1.2-2: Neutral Interfaces for Robotics (NIRO): Project overview

started 1984 and concluded 1989 with a series of new definitions for a neutral interface (including almost full geometry/topology models for mechanical objects) and the corresponding pre- and post-processors for data transfer between CAD, FEM, and simulation systems from different vendors.

Figure 1.2-1 gives an overview over the whole development process in this area of activities funded under ESPRIT since 1983, including the follow-up projects from ESPRIT 322, especially ESPRIT 2195 CADEX and ESPRIT 2614/5109

NIRO (Neutral Interface for Robotics). The main goal of all these efforts was and is to come from a situation of no internationally available standard in the field of product model data transfer to the opposite: a functionally complete, broadly accepted international standard. The now valid international platform for this is the activity STEP (Standard for Exchange of Product Model Data) carried out by and via ISO TC 184 SC4 (ISO: International Standardisation Organisation).

The projects NIRO, which is reported here, and the follow-up project ESPRIT 6457 INTERROB: Interoperability of Standards for Robotics in CIME (Computer Integrated Manufacturing and Engineering) extend the line of model transfer at the CAD level to the shop floor level including robot control, while at the same time the geometry and topology definitions at the CAD level are broadened with kinematics and application related information. This is shown in more detail in Fig. 1.2-2.

1.3 ESPRIT Project Nr. 2614/5109 NIRO

The project NIRO started in December 1989 with 9 partners from 5 EC countries. At the ending stage the following partners - which are the contributors to this report - were involved:

vendors:	Reis GmbH & Co Maschinenfabrik Obernburg, Germany	robots, handling systems
	BYG Systems Ltd Nottingham, U.K	software, software tools
vendors:	PSI - Gesellschaft für Prozeß- steuerungs- und Informations- systeme mbH Berlin, Germany	software, software tools, engineering, training
consultants, engineering companies:	DISEL Madrid, Spain	engineering, EDP system
	DIS Industrial Consultants a|s Glostrup, Denmark	engineering, EDP systems
users:	CASA Manufacturing Engineering Madrid, Spain	aeronautical industry
	Odense Steel Shipyard Ltd Odense, Denmark	ship building

	FIAT Sistemi e Telecomunicazioni Torino, Italy	automotive industry
research institutions:	Kernforschungszentrum Karlsruhe GmbH Karlsruhe, Germany	applied information technologies, standards, prototype software
	Danmarks Tekniske Hojskole Instituttet for Styreteknik Lyngby, Denmark	applied information technologies, prototype software

The project has been industry driven and its aims are the development of neutral interfaces (neutral data formats, processors and software tools) for the exchange of geometrical, technological, kinematic and programming functions and data between CAD/CAM systems, simulation and planning programmes, and robot control. This will help users to become independent from single vendor CIM solutions and small and medium size European vendors to get better market opportunities for CIM components and systems.

The results of the project are combined into several prototype solutions to demonstrate the practicality of the approach in an industrial CIM environment, e.g. for welding, parts manufacturing, assembly processes in the automotive and aerospace industry and ship building (see Fig. 1.3-3). Here are shown the different systems used at the partners' sites:

- at the CAD level: the systems CATIA, BRAVO 3, CADDS.5 and HICADEC
- at the simulation level: the systems GRASP and KISMET
- at the real-time level: the robot control systems for the robots COMAU PMAST-25, ASEA IRB 60 and IRB 2000, REIS RV15 and HIROBO.

For these systems the corresponding prototype interface processors for STEP and ICR (Intermediate Code for Robotics) had been developed.

In this book the results of the 3-years work of the above mentioned consortium are reported . It begins with a short overview on the state of the Art and the goals when the project was started.

Chapters 4-8 cover the results in detail: Chapter 4 the neutral file specification work, especially on Kinematics and the corresponding contributions to STEP, followed by a description of the prototype processors for the extended STEP definitions (Chapter 5 and 6). The same way, specification results and processors are described for the ICR and IRL (Industrial Robot Language) interfaces (Chapter 7). Chapter 8 contains the description of the four demonstration pilot systems defined and developed to show how information flow can be easily performed via neutral interfaces in an industrial CIME environment.

An outlook on the role of interfaces for CIME in the future concludes this book. Some important details are covered extensively as an appendix.

2. State of the Art upon Project Start

I. Bey

Most partner companies have experienced the need for extended integration and automation of their manufacturing operations. Having for many years been used to introducing new and improved methods for processing materials they now more and more experience that there is a pressing need for integration at all levels of manufacturing from the identification of business objectives through design and planning, to real-time control of machining and assembly at the shop floor and further to the shipping of finished products to customers.

It is now widely believed and accepted that manufacturing companies who achieve large-scale integration, but not necessarily full automation, of their operations will be competitive to a much higher degree than those who only integrate and automate in separate areas.

The most difficult problem to overcome in advancing the concept of integrated manufacturing systems is the integration process itself. Generally speaking integration is in this context understood as linking different system components together into a complete system, which fulfils the objectives and specifications of the integrated manufacturing system. The basic function of the system components in such a system is to transform an input to a desired output by certain mechanisms (transfer functions).The components are controlled by constraints specifying defined goals, performances, plans etc. to be fulfilled.

A typical problem example is found in an enterprise where designs generated by a CAD-system cannot readily be transferred to a production department using CAM-systems because of data communication incompatibilities and lack of compatible product and engineering data base systems.

A main target of research in the domain of computer integrated manufacturing systems addresses the properties, conditions, possibilities and means by which system components can be integrated and how the integrated system can be modified, updated and controlled.

However, very often the integration strategy in manufacturing is misunderstood. Integration in this context does not mean centralizing and merging all kinds of sub-systems but integration should allow for the building of integrated manufacturing systems of a type, size, order, degree of automation and implementation speed that fits into the practical needs of a given user company - whether it is a small, medium-sized or large company.

A development like the one described above demands the standardization of the interfaces between the individual sub-systems in a CIM environment. Such a standardization is a prerequisite in the development of the future CIM systems, namely for replacing existing individual components in the CIM system by more effective ones as they appear on the market. This leads to the application of multi-vendor systems. Furthermore, with standard interfaces, the so-called "islands of automation" can be avoided in the manufacturing process.

Also standardization and modularity will allow small as well as large vendor companies to market specialized CIM systems and sub-systems (modules) on a competitive international market.

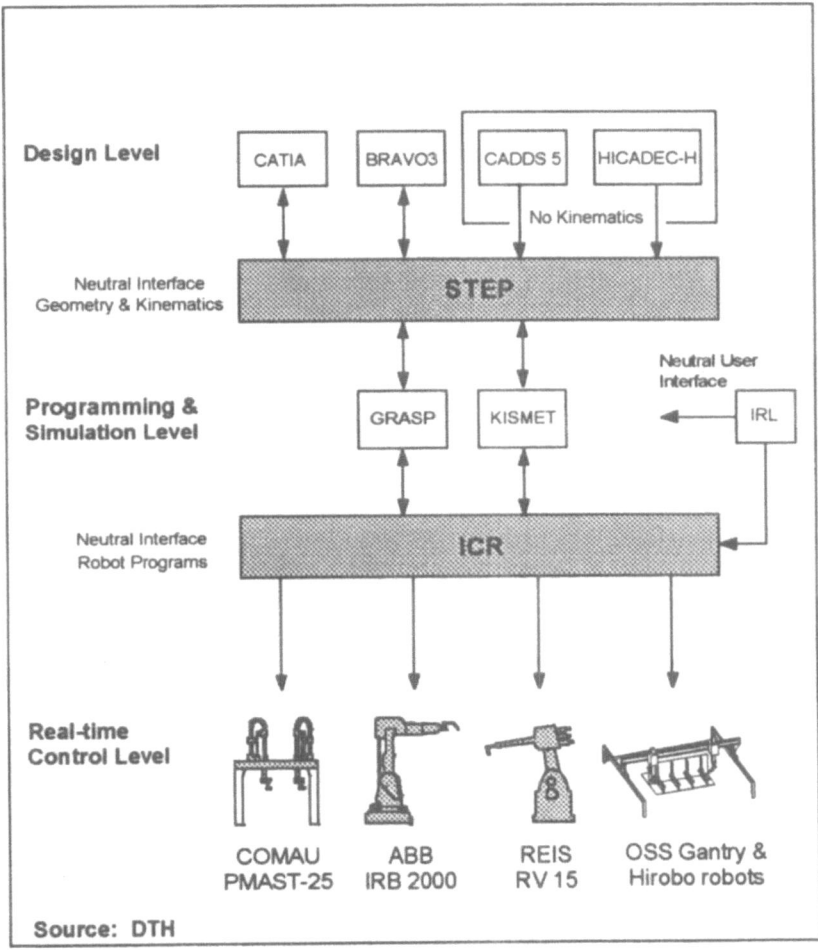

Fig. 1.3-3: Schematics of the NIRO implementation goals

There are three main aspects or fields of interest concerning the NIRO project. A first main aspect to the NIRO work is related to the data of CAD systems and their interchange with other systems during design and production planning. There is a need of interfaces between CAD and robot programming, planning and simulation. The second point concerns the robotics interface between programming and control. The third part corresponds to the international standardization for neutral interfaces in robotics and is described at the end. The respective state of the art is now discussed in the following section.

2.1 CAD Interfaces

The situation when starting the NIRO project was that many commercial CAD-systems have one or more interfaces through which other programs or applications can access the geometrical data in the CAD data base. Since 1980 a considerable interest has been shown concerning standardization of these interfaces with the intention of automatic transfer of CAD data between CAD systems of different origin and of CAD data between CAD systems and dedicated application systems from different vendors.

The need for such communication occurs at all levels within a factory, in different departments in a company and between different companies, specially between subcontractors and their main contractors.

Typical examples of dedicated computer aided application systems, where automatic CAD data transfer is of great interest, are:
- Kinematic Simulation and Collision Checking
- Dynamic Optimization of Part Geometry
- Process Planning
- CNC Programming
- Robot Programming.

Many national and international projects and activities have been initiated during the eighties with the aim of specifying interfaces in the scope of CIM: A major effort of this sort was undertaken by the ESPRIT project 322 CAD interfaces (CAD*I), which dealt with the development of standard format interfaces in CAD/CAM. These formats are tried out for a set of independent commercial CAD systems. Results from ESPRIT project CAD*I have been utilized in this project [Raflik, Schlechtendahl, Thomas]. Also the results of the CAD*I project and of the PDES project in the US have played a major role in formulating the ISO STEP proposal on CAD geometry now under way [ISO/89].

The project partners had so far reached specific achievements in the area of CAD data exchange when starting the project. BYG had already developed special interfaces for its GRASP system to receive CAD data from various CAD systems in IGES and several native formats. DISEL was involved in CAD*I and participated in the CADEX project and hence had experience in neutral file interfaces. DTH and KfK have played a leading role as interface processor developers in the CAD*I project. KfK has been involved in the STEP development since 1985. The industrial partners CASA, DIS, FIAT and OSS contribute their industrial needs in interfacing CAD to robotics and supplement from their experience in solving this problem in a special (non-neutral) way.

2.2 Interfaces in Robotics

Computer Integrated Manufacturing (CIM) comprises product design, production planning, production control, production processes, quality control, production equipment, and plant facilities. The creation of the product starts in the design department, where its function, physical design, and manufacturing methods and processes are established. The next steps are done with the help of the CAM-modules (planning, programming and simulation), which generate the needed data and/or instructions for performing the production process. The real-time control units for machines, transport systems and robots are executing the generated programs.

Of course it is possible to build so-called monolithic systems without neutral or with unpublished interfaces between the CAD and CAM modules. Such systems are designed and implemented for the special hardware and robots of one robot and robot controller producer. It is not possible to apply CAM-modules to other robots and control units without a total re-design and large amount of implementation work. Normally the more developed CAM systems are implemented by a company of large size, because it can offer a higher number of robots/controls to share the development costs. Medium or smaller sized companies cannot develop all CAM-modules on their own, they need to integrate e.g. a programming system developed by other software houses. Therefore especially for the smaller and medium sized robot and controller producers neutral interfaces are of special interest. A neutral interface between the components would give the industrial user the choice of components which are best suited for his specific tasks.

Thus, the CAM modules in planning, programming and simulation have been of interest in the NIRO project. They can be identified in the higher level of the robot specific area in connection with other machines and the whole production process. The work in the project was concentrated on neutral interfaces
- between this CAM-area and CAD
- between this CAM-area and robot control unit
- between the CAM modules themselves.

Therefore, the main effort and interest was directed to the work on processors and software tools in the CAM area.

An important technical aspect for the introduction of neutral interfaces into robotics and for CAM-modules is the quick availability of simulation tools. When a new robot type is developed or a user wants to buy such a robot, it is an advantage to simulate the robot and some robot tasks by a simulation system. The use of neutral interfaces reduces the amount of work to apply the simulation tool to the new robot type. The work to simulate the robot is restricted to the new definition of the robot geometry, kinematics and dynamics, while the robot task description and programming is unchanged.

It is standard practice to perform the design of manufacturing system and the layout of the working environment with CAD techniques. Several CAD systems with kinematic simulation facilities have become available. Some systems provide for translation of the operation sequence into a programme formulated in

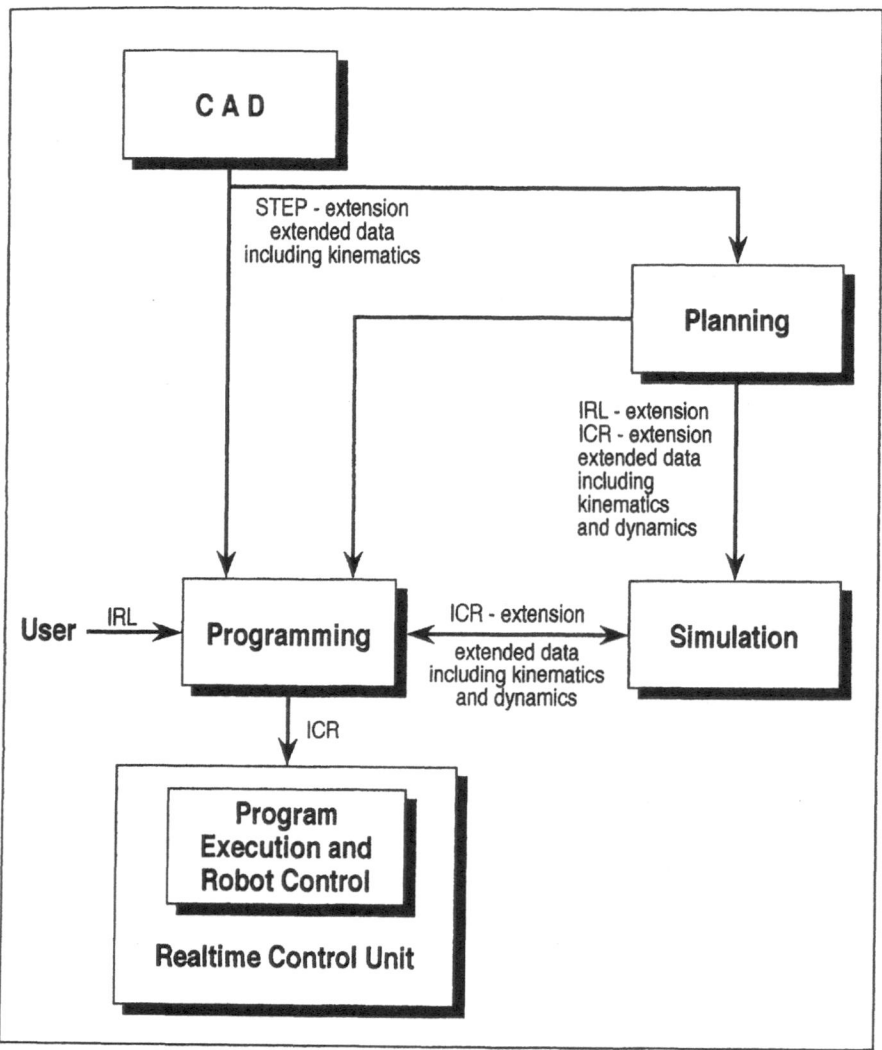

Fig. 2.2-1: Neutral interfaces between components in robotics

one of the off-line robot programming languages. However, systems of this kind usually have highly integrated interfaces and force the user to use the software of a single vendor for all applications. Figure 2.2-1 shows the interfaces between the outer world and the CAM, but not within CAD and the robot control unit itself which includes sensor interfaces and hardware level interfaces.

The interfaces between the different modules on Figure 2.2-1 show the area of main interest for the project. When starting the project there existed only first drafts for common standards for data or program function exchange between these modules. The proof of a practical implementation - a very necessary issue - had not been done so far. Fig. 2.2-1 shows also where extension to the existing draft standards has been scheduled for specification within the project.

The modules for planning, programming and part of robot control has been within the scope of ESPRIT project 623. Hence, the results and interfacing requirements established in project 623 were used as starting point in the NIRO project.

2.2.1 Robot Programming

The programming of robots is a bottle-neck in many companies, whose production is mainly based on small batches of frequently changing products. The most often used methodology for robot programming is based on using the robot itself in a so-called teach-mode and thereby the robot and the equipment to be served by the robot are idling during the programming phase.

More recent methods, based on off-line programming are more efficient since the robot can be used in production while the programming of the next tasks is being prepared. Also the off-line programming allows for the use of advanced computer-assisted tools for modeling, task planning, path generation, simulation and program verification.

The situation is that there are many different robot manufacturers all of them using their own programming systems with special programming languages such as VAL, ARLA, BAPS, LM, SIGLA, HELP, etc..

These languages differ in various aspects concerning syntax program structure and features. For companies using different robots this situation requires

- that every robot system used needs a specially trained programmer and
- that an existing robot program for one robot system must be re-programmed in order to be executed by another robot system even though the robots may have a similar kinematic structure.

A solution to this problem of robot program portability within the CIM domain requires a standardization of the interfaces between robot planning and programming systems on the one side and (1) CAD-systems as well as (2) robot real-time controllers on the other side.

Considering the off-line robot programming system to be a sub-system truly integrated in a CIM system it becomes obvious that already existing pertinent geometry data residing in the solid geometry data base of the CAD system included in the overall CIM system should be automatically transferable to the robot programming system. Again adhering to the "open system" concept as described above this data transfer should be adopted through a standard interface (neutral file format).

Originally, programming of the robot was done with simple methods that had little resemblance to conventional computer programming. Since the 1960s scientists have been working to introduce artificial intelligence (AI) into robot technology. So far, this attempt has reached the testing state and is being further developed in companies. The high level programming languages for robots where developed since the end of the 1970s. Presently, industry is at the beginning of a wider use of high-level programming, graphical tools and textual programming.

Emphasis in the area of robotic applications has so far been on robotics equipment with conventional kinematics and on manufacturing applications. Only recently, less conventional designs and applications have been investigated. These include redundant kinematics and applications for work (maintenance, inspection and rescue operations) in an environment that is either very large, of complex shape (with increased collision danger), or difficult and dangerous to access. Such applications require supervisory control, which means that the performance of the robot and its control must be supervised by an independent system which performs on-line collision analysis and allows a human operator to visually supervise the operation, monitor the sensor states and to intervene with the automatic control if required. Such systems need geometric modelling, collision analysis and geometry display capabilities which have real-time response characteristics. Interfacing such supervisory control systems with CAD is as yet a research and development domain.

Because development in robotics is very rapid and the robot is integrated into the whole system of manufacturing, the need for new standards arose early. The first step of the standardization of the interface between robot programming and control is IRDATA (DIN/90), but it lacks many required functions.

Broad accepted international standards for interfaces between CAD programming, planning and simulation were still missing when starting the NIRO project. There was some work in progress for the standardisation of the ICR interface between programming and robot control. ICR mean "Intermediate Code for Robots" and is a Draft Standard Proposal at ISO (ISO/89-1). Necessary development were identified as following:

- Standard interfaces to CAD levels
- Standard interfaces between software modules on the operational control level like programming languages, simulators and data base systems
- Standard interfaces to robot control levels.

The project partners started their work having some experience in this field. I.e. BYG had developed some interfaces from its GRASP system to various industrial robot controllers. DTH had developed CAD*I and IRDATA interfaces for GRASP. PSI had many years experience in developing off-line programming systems. KfK was and is involved in the standardisation of ICR and IRL and also had several years of experience in off-line programming and interfacing between programming and robot control. Reis is the project's vendor of robotic hardware and controllers. The industrial users CASA, DIS, DISEL, FIAT and OSS contribute from their practical experiences.

2.3 Standardization

For the development of new or extended neutral interfaces there are the following main foundations to the NIRO project:

- ESPRIT project 322 (CAD*I)
- STEP (Standard for the exchange of product model data)
- IRL (Industrial Robot Language)
- ICR (Intermediate Code for Robots)
- IRDATA (Industrial Robot DATA).

When starting with NIRO, these standard formats were not yet in world-wide use. CAD*I was designed for interfacing between CAD systems as well as from CAD systems to other systems on the level of geometric information. CAD*I has influenced the development of the emerging international standard STEP to a great extent, but the CAD*I project did not develop neutral solutions to the transfer of Kinematic information.

STEP (ISO/DP 10303) is the future international standard for product definition data. Its geometry, topology and shape definition parts agree to about 90% with the CAD*I reference model. The same holds for the physical file format. As a consequence of work based on CAD*I results a topical information model for kinematics had been developed by NIRO partners to be included in STEP. This proposal concerned information regarding

- kinematic topology
- geometry association
- kinematic configuration sequences and paths.

This kinematics proposal had been accepted by ISO as a basis for STEP kinematics when starting NIRO.

The teach-in procedure is used in many cases to program the moves of industrial robots. By teach-in the robot itself is used as a tool for programming. But for more complex applications like handling or assembly operations with sensor integration or for external data integration (e.g. CAD data) the pure teach-in method is not sufficient. Therefore a number of high-level programming languages for industrial robots had been developed.

In Germany at national level by DIN, section NAM (Standards for machinery tools) and in parallel in ISO TC184/SC2/WG4 at international level the work of standardizing a robot programming language started in 1988. The first draft of the Industrial Robot Language (IRL) was published in 1989 (DIN/89). The IRL definition is based on the modern concepts for structured programming. Therefore IRL includes the element for modularization, program flow control, procedures, functions, local and global constants and variables, block concept, recursivity and others. Also IRL will provide in a future extension parallel execution of program parts.

At the starting point of NIRO, the definition of the neutral software interface ICR (Intermediate Code for Robots) was in progress and handled by the ISO TC 184/SC 2/WG 4 "Programming Methods and Languages for Manipulating Industrial Robots". The purpose of this recommendation is to establish a standard structure and representation of control information in ICR for industrial robot controllers as well as a standard data link control procedure. The code is used for a data link between external, controller independent industrial robot programming systems and the robot controller system software, equipped with an ICR interface.

ICR consists of records, each representing a function which is executed immediately by the robot control. The functions and their parameters expressed by the code define a superset of robot controller functions. That means, hopefully each robot controller can use a smaller or larger subset of ICR code elements to express its functions. But a robot controller doesn't need to include all ICR code elements! The neutral interface can be regarded as a wide bridge or gate, which is used as a general connection between any robot control and any programming system.

The code consists of sequences of records representing operations in prefix manner. Such operations may be data definitions also. The operations include arithmetic and stack operations, which allow an efficient evaluation of arithmetic expressions with respect to a postfix notation. Parts of code called ICR units can be a user program, a user module or transfer files used for data exchange between various programming and control components. It is also possible to exchange data from one robot control unit to another by transfer files.

One field of application of ICR is the use as an interface between robot simulation systems and robot control. After the definition of robot kinematics and perhaps its dynamics the user wants to see the geometry of robot moves and the danger of collisions. ICR can be the internal interface between simulation calculation and simulation graphics for displaying the calculated move sequences. On the other hand the user can download the simulated programs via ICR to robot control and compare simulation and real execution of his program. The effort for implementing the simulation of new robot types is reduced by the code.

These code elements allow the expression of various robot programming language elements by ICR. The definition of ICR is in progress by TC 184/SC 2/WG 4 and a first draft was issued by the end of 1989. Because ICR is based mainly on IRDATA, which is a German Standard (DIN 66 313, Part 1) basic concepts and main elements of both codes are similar.

3. Aims of the NIRO Project

I. Bey

The main goals of the NIRO project have been threefold:
- to develop new interface specifications for product data transfer in CIM, especially for the CAD, simulation, programming and robotics applications,
- to develop corresponding prototype processors for the above mentioned interfaces, and
- to influence the standardization process, especially in the international context at ISO.

The neutral interfaces (neutral data formats) and associated pre - and postprocessors were developed for the exchange of geometrical, technological and programming functions and data between CAD, CAM-modules (robot planning, programming and simulation) and robot control. Also the user-oriented robot programming language was treated as an interface, namely between the programmer and the programming system. These developments were based on formal techniques for specifying the interfaces that are relevant for the robotic area. In this sense, the project followed the approach of ESPRIT project 322: CAD*I and extended its scope into the robotics domain.

One main aim has been to develop processors, software tools and methods that will enable multi-vendor systems to be composed to the benefit of both users and vendors. An important topic of the project was to demonstrate the feasibility of the approach by prototypes (distributed over the partners) by proving the inter-communication capabilities between systems built from multi-vendor modules. The work was performed in close cooperation with the international development of the relevant standards: STEP, IRL and ICR.

The development of neutral interfaces consists basically of the following steps:
1. Identification of the relevant interfaces and analysis of requirements
2. Identification of the information representing the functions and data for planning, programming and program execution in robotics, which have to flow across the interfaces
3. Formal specification of the data types and statements, their semantics and their syntactical representation
4. Development of processors, testing and enhancement of formats and processors
5. Implementation of the software tools in an industrial environment.
6. Demonstration and documentation
7. Contribution to international standardisation activities.
8. Publication of results, technology transfer actions.

Common concepts, specification techniques and languages were used for these applications. Also results of ESPRIT projects 623 (Operational Control for Robot System Integration into CIME), 688 (AMICE: A European Computer-Integrated Manufacturing Architecture) and 955 (CNMA: Communications Network for Manufacturing Applications) were taken into account. The development results had been demonstrated at several partners' sites and public events. Important part of the project's efforts have been devoted to introducing the neutral interface specifications into international standardisation activities, especially for STEP and standards based on ISO or CEN/CENELEC proposals.

At the beginning of the project it was stated as one important objective, that the neutral interfaces and processors to be established and introduced for standardization and the new developed software tools like compilers, simulators and interpreters will allow for:

- exchange of kinematics and robotic technological information between CAD systems, planning tools, programming systems and simulation
- writing user programs in a powerful, standardised high level robot programming language
- using the robot programming language for real industrial applications on different equipment for several implementations
- exchange of user programs between different robot control units and simulation
- exchange of geometrical and robot specific technological information between robot control units and simulation tools
- introduction of an improved and extended standard for kinematics, robot specific, technological information, robot programming language and intermediate code for robot control units
- simulation and test of user programs for an industrial robot on different simulation and test facilities.

The results gained during the 3-years work period of NIRO along the above stated guide-lines and goals are documented in the following chapters.

4. Results Related to the STEP Standard

E.G. Schlechtendahl

STEP, the unofficial acronym for the International Standard ISO 10303, means *Standard for the Exchange of Product Data*. STEP is being developed by an international community of experts within ISO/TC184/SC4 since 1984. Europe has been actively involved in this development mainly via the contributions of several ESPRIT projects. The first one of these projects which had significant influence on STEP in the time period from 1985 through 1989 was ESPRIT 322, CAD*I (CAD-Interfaces). Project NIRO continued the contributions which started in CAD*I and added a new aspect: kinematics.

At the beginning of the NIRO project the results of the first official ballot on the STEP standard were being received. The ballot version was the so-called Tokyo version of STEP. One result of that ballot was the evidence that STEP could not be produced as one single standard document. During the project years from 1990 through 1992, the philosophy that underlies STEP has undergone several changes: from a single monolithic standard via a set of loosely coupled ISO documents to the present philosophy of a well-harmonised set of documents whose content is properly integrated, which need to pass an intensive qualification and editing procedure before publication. This process turned out to be much more time-consuming than anybody had expected (without questioning the necessity and the amount of detail work involved in this process). As a consequence, while NIRO started with the assumption that all those parts of the standard that related to shape information exchange were already stable and that only the kinematics part had to be added, no single document remained as it was in 1990, even in its essentials. Furthermore, also the working organisation in ISO/TC184/SC4 was completely restructured during these years. While originally a single Working group WG1 covered all aspects of STEP, now a total of eight active Working Groups (WG2 through WG8 plus the Joint Working Group JWG9 for electrical and electronic matters) have been established.

In order to allow practical experience to be gained in NIRO with implementation of the new STEP technology (the technology remained practically constant while the documents were changing) and to allow feedback from NIRO into the STEP work it was necessary to "freeze" i.e. to select certain work documents as a reference during the course of the project as an implementation basis. Consequently, the present implementations in NIRO do not conform to the current and as yet still not final ISO-STEP documents. All implementations, however, are sufficiently close to whatever the final standard version will be so that they can be adapted easily to the "real" standard as soon as this standard becomes officially released.

4.1 The NIRO Contributions to STEP

In the course of the NIRO project, project members were actively involved in the standardisation process for STEP. The main areas of contribution from NIRO to STEP were:

1) the development of the kinematics information model for STEP in ISO 10303-105;
2) the final definition of the physical file format in ISO 10303-21;
3) contributions to the development of the STEP Data Access Interface (SDAI) in ISO 10303-22.
4) the NIRO Specification for STEP Kinematics Models for the transfer of robotics information based on ISO 10303-105 (kinematics).

4.1.1 The STEP Kinematics Model

The technical content of the STEP kinematics information model as defined in ISO 10303-105 has undergone very little change after it was fixed during the early phases of the NIRO project. This technical content was then "frozen" within NIRO to serve as a basis for the implementations (see NIRO Specification for STEP Kinematics Models below). The internal structure of the information model was, however, changed drastically as a consequence of the STEP integration process. The amount of changes also caused a significant delay of the document development. So, while a technically integrated and qualified document is available at the end of the NIRO project, it could not yet be distributed for formal balloting.

As an example of the changes made in the integration process we select the information model for the entity "revolute pair".

Specification as used in NIRO-STEP:

```
ENTITY revolute_pair
SUBTYPE OF (kinematic_lower_pair);
   actual_rotation :
     kinematic_angle_value;
   lower_limit_actual_rotation : OPTIONAL
     kinematic_angle_value;
   upper_limit_actual_rotation : OPTIONAL
     kinematic_angle_value;
WHERE
   number_of_possible_motions = 1;
   number_of_independent_motions = 1;
   actual_pair_parameter [1] = actual_rotation;
   pair_parameter_type [1] = 'kinematic_angle_value';
END_ENTITY;
```

Specification as defined in ISO 10303 Part 105 Version "London 92":

```
ENTITY pair_value;
   applies_to_pair : kinematic_pair;
END_ENTITY;

ENTITY simple_pair_range;
   applies_to_pair : kinematic_pair;
END_ENTITY;

ENTITY revolute_pair_value
SUBTYPE OF (pair_value);
   actual_rotation : plane_angle_measure;
WHERE
   WR1: 'kinematics_schema.revolute_pair' IN
        TYPEOF (SELF\pair_value.applies_to_pair);
END_ENTITY;

ENTITY revolute_pair_range
SUBTYPE OF (simple_pair_range);
   lower_limit_actual_rotation :
     plane_angle_measure;
   upper_limit_actual_rotation :
     plane_angle_measure;
WHERE
   WR1: 'kinematics_schema.revolute_pair' IN
      TYPEOF (SELF\simple_pair_range.applies_to_pair);
END_ENTITY;

ENTITY revolute_pair
SUBTYPE OF (kinematic_pair);
END_ENTITY;
```

It is evident that the size of the kinematics information model (e.g. the total number of specified entities) has increased significantly. Moreover information that before integration was grouped together in a way that appeared intuitively logical from the kinematics point of view has now been spread between several entity types. This results from a number modelling principles imposed on the integration process:

- An entity has to carry a clear semantics otherwise it is superfluous. The subdivision between lower and higher pair via subtypes fell into that category. While this subdivision has its merits from the theoretical point of view, it does not add semantics to the data model and has now been removed.

- Optional attributes like lower_limit_actual_rotation in the revolute_pair_definition have to be avoided. Optional attributes are an indication that "the cardinality was unclear". Hence, pair_range entities were introduced that point to the pair entity. The pair_value entity was also introduced for data dependency reasons leaving the revolute_pair entity without any attributes.
- Information on the shape aspects of the kinematic structure is now provided by the imbedding of the kinematic model in the product property definition of a product model. Shape is also a product property. Hence, shape is no longer directly tied to the kinematic structure but has to be searched in the structure of the product definition. Thus, the ground entity and the link_shape entity were removed from the kinematics specification and replaced by a ground_representation and link_representation entity respectively. These entities are subtypes of the representation entity which is part of the product definition.

The new information model now fits better into the style of the other STEP models that have passed the integration process; but without doubt, the previous information model as it was used in the NIRO implementation was much better suited for efficient development and application of interface processors between CAD and robot programming systems.

4.1.2 Physical File Structure

Work in this area was more or less a continuous series of update processes with which the physical file specification had to follow changes of the EXPRESS language according to ISO-10303-11. The NIRO project did not follow through these changes and, hence, STEP files from NIRO do not conform with the now valid Draught International Standard (DIS) version of ISO-10303-21. All other known implementation projects (such as the ESPRIT project CADEX and several implementations done by companies that are organised in PDES Inc.) are in a similar situation. This deviation from the most recent version of the standard document is much less serious than it might seem. This was proven by successful interchange of solid models on STEP files between the two ESPRIT projects CADEX and NIRO.

The most significant change to the physical file structure which occurred during later phases of the project is not due to a change in document ISO 10303-21, but rather results from changes in the EXPRESS language specification ISO 10303-11 as well as from the new style in information modelling imposed on the STEP information models in the course of the so-called integration process. When applying the mapping rules specified in ISO 10303-21 these changes have a significant impact on the physical file. For those readers who are familiar with the EXPRESS language we give an example: by default the so-called ANDOR supertype clause now applies to many entity types with the consequence that the "external mapping" specified in ISO 10303-21 has to be used for a majority of entities instead of the simpler "internal mapping". From the point of view of the NIRO experience one has to state that the transfer of robotics models via STEP will be less satisfactory according to the present ISO documents. It is hoped that further iterations of the standard which involve industrial practice and more industrial influence will produce a corrective feedback and will allow transfer of kinematics and robotics in the very successful and effective way which was demonstrated by NIRO.

4.1.3 STEP Data Access Interface

SDAI is the acronym for STEP Data Access Interface. SDAI shall become the specification standard for accessing STEP-structured data either in data bases or in primary memory from application programs. The interface is specified as a structured set of service routines (functions procedures) in a way that does not depend on a particular programming language. Specific programming language implementations of this generic SDAI shall be specified in separate standard documents as so-called language bindings. The languages for which bindings are to be developed with high priority are C, C++, and Fortran.

As an initial step towards the SDAI specification a methodology had to be developed. From NIRO, the proposal was made to use a finite state machine approach. Using this approach, the SDAI will be in a well-defined and formally specified state in each instant and every routine will either modify that state based on the state proper and some input data, or will return data depending on that state and some input data. The proposal also said that the EXPRESS language should be used to specify the data structure which represents the SDAI state. This proposal has been accepted by the STEP implementation Working Group and is now the basis for defining SDAI in ISO 10303-22.

4.1.4 Specification

This last contribution - the *NIRO specification for STEP Kinematics Models* - will be described in more detail in Chapter 4.2. While the concept of STEP Application Protocols was not yet known when NIRO started the NIRO Specification for STEP Kinematics Models may be regarded as a predecessor of an Application Protocol for robotics applications.

The NIRO specification was not submitted to the ISO community as this was not intended and since it does not conform to the present layout rules for Application Protocols. The purpose of the specification was to serve as a basis for implementing the NIRO STEP processors. Nonetheless, the experience gained in writing this specification and in implementing it was used in updating the STEP Kinematics Model in ISO 10303-105.

4.2 The NIRO Specification for STEP Kinematics Models

U. Kroszynski

Early in the project, the necessity to define the scope and type of information to be exchanged was identified. This activity resulted in a formal work document to be used by processor implementors. According to the general principles of the project this specification was to fit into the upcoming standard STEP. STEP was, however, still in a very early stage when the specification had to be "frozen". It became necessary to define a reference basis for the NIRO/STEP specification. This reference basis is spelled out in more detail in Appendix A. The areas of product information necessary to describe robotic mechanisms include:

geometry - to represent the shape of robotic arms, of the objects to be handled by the robots, and of the work place scene.

kinematics - to represent the topology and transformations that describe how the different parts of the robots (each having a geometric shape) are linked together and articulated to form mechanisms.

robotics - to describe concepts that are characteristic for industrial robots, like tool attachment point frame, robotic tool, tool centre point frame, etc.

Concerning geometry, a subset of STEP Part 42 was selected, dealing with the description of polyhedron solid models, also called Facetted Boundary Representations. The reason was that all the commercial systems employed by the partners have this modeling capability and that more comprehensive STEP geometry is dealt with in detail in ESPRIT projects (CAD*I, CADEX). Facetted boundary representations constitute a simple yet wide and self-contained family of shapes, for which only point coordinates are needed, with some topology concerning how these points are sequenced in planar contour boundaries or loops. Besides, the STEP "replica" entity was adopted that describes instances of frequently used objects by means of a reference to the object and a placement matrix indicating the new location and orientation in space.

Concerning kinematics, a subset of the proposed STEP Part 105 was adopted, featuring prismatic and revolute lower pairs as usually found in robots. Among the different methods to describe the transformations for the articulations, it was decided to employ matrices as the neutral representation. Otherwise, the entire range of entities described in STEP Part 105 was included. Important features are closed kinematic chains and the possibility of defining the base of a mechanism on an arm of another mechanism (e.g. a robot suspended from a gantry).

Concerning robotics entities, there is no existing baseline in the STEP integrated resources. Robotics entities like the robotics tool with its tool centre point frames, the tool attachment point frame on a robotic mechanism model, or the actuator entity to indicate driving pairs in closed kinematic chains featuring a motor and a gear were defined in NIRO as user-defined entities. These entities may be considered as a starting set for a possible future definition of an industrial robot application protocol.

The inclusion in the NIRO information model was done employing EXPRESS (STEP Part 11) and the implementation on the physical level was by means of the STEP user defined entity, a provision for describing records according to private agreements between sender and receiver.

The NIRO specification exactly defines the mapping and restrictions of the adopted STEP entities as well as their appearance on the physical file. This is necessary not only for the NIRO-specific entities dealing with robotics but generally since the STEP definition in EXPRESS leaves the possibility to implement entities in different ways.

For example, it was determined in NIRO which STEP entities should have a SCOPE section on the physical file, i.e. include other entities in their definition. On the physical level, the rules for mapping EXPRESS onto an ASCII textual file were adopted (STEP, Part 21).

These activities resulted in the following releases of the NIRO/STEP specification for the exchange of robotics models.

L01V01	of July 1990
L01V02	of December 1990
L01V03	of March 1991, which served as basis
	for the first pilot processors.
L02V00	of December 1991 (final version),
	which is presented in Appendix A.

For the benefit of processor implementors informal explanations and representative examples of actual files were included.

The baseline for the L02V00 final version is found in STEP documents which were available at that time. These documents have evolved since then, and further versions will continue to appear until STEP becomes officially released. Although different in detail from the ones adopted in NIRO, the contents and structure of the entities remain ed. unchanged and therefore applicable throughout the project. Those who are familiar with the present (1992) status of STEP will easily recognize the differences when they read the specification. The most notable difference is that and older version of the EXPRESS language is used. The NIRO/STEP specification makes frequent use of the (old) EXPRESS construct MAP. MAP has since been replaced by the language constructs USE and REFERENCE. The reader is advised to interpret MAP as USE.

Table 4.2-1 presents an overview of the STEP entities considered in the NIRO specification.

Table 4.2-1: STEP entities

General

File-identifier	and
File-description	according to STEP requirements in the HEADER section
Index-entry	associates a user name with an entity
Export-list	allowing an entity to be recognized outside its scope

Geometry

Facetted-brep	including in it's scope ...
Cartesian-point	defined by it's three coordinates
Direction	a unit-vector defined by its three components
Poly-loop	a sequence of references to points defining a planar closed polygonal contour
Face	the planar region between an outer loop and inner ones
Closed-shell	the set of all faces constituting a polyhedral solid
Solid-instance	a replica of a polyhedral solid, including in it's scope
Transformation	a matrix indicating the placement of the replica

Placements

axis2-placement	a homogeneous rigid transformation defined as a 4x4 matrix

Kinematics

kinematic-model	containing the description of the ground and of the mechanisms
ground	a co-ordinate system to which mechanisms are related
mechanism	a topological description of an articulated construction containing
kinematic link	the rigid parts of the mechanism
kinematic joint	the topological junction of two links
kinematic pair	the functional specification of a kinematic joint
pair placement structure	the location of kinematic joints on a kinematic link
kinematic structure	the list of kinematic joints which constitute an articulated construction

Robotics

actuator	indicating which articulations in a closed kinematic chain have a motor
tool attachment point	where a fixture for attaching tools to the robot exists
tool	a simple model of a robotic tool with offsets to the tool centre points
mount tool	indication of the fact that a tool is mounted

Rendering

colour table	a specification of the colour codes employed in the model
rgb colour	the red-green-blue weighting of each colour code
colour attribute	the colour code assigned to an object
point direction pair	auxiliary information on the facets of a polyhedron needed for the
render face	assignment of a colour code to each facet

Other NIRO specific entities

solid mass properties	volume, centre of mass, inertia moments of a polyheron
link mass properties	mass, centre of mass, and inertia moments of a robotic arm

5. Development of STEP Processors

Chapter Editor: *T. Sørensen*

The STEP specification [Kroszynski, Sorensen, and Schlechtendahl 1991], addressed in Chapter 4, was the basis for a development of several pre- and post-processor programs within the NIRO project.

These programs are the topic of this chapter as shown in Fig. 5.0-1.

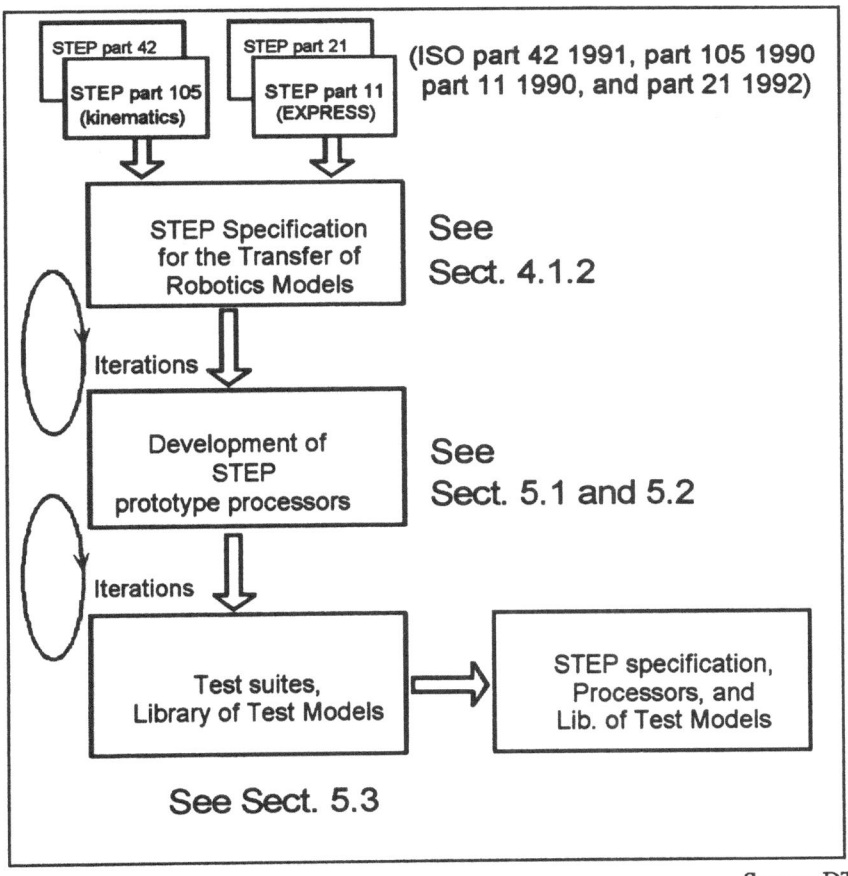

Fig. 5.0-1. The development and test of STEP processors and the validation of the specification are described in Chaps. 4 and 5 as indicated

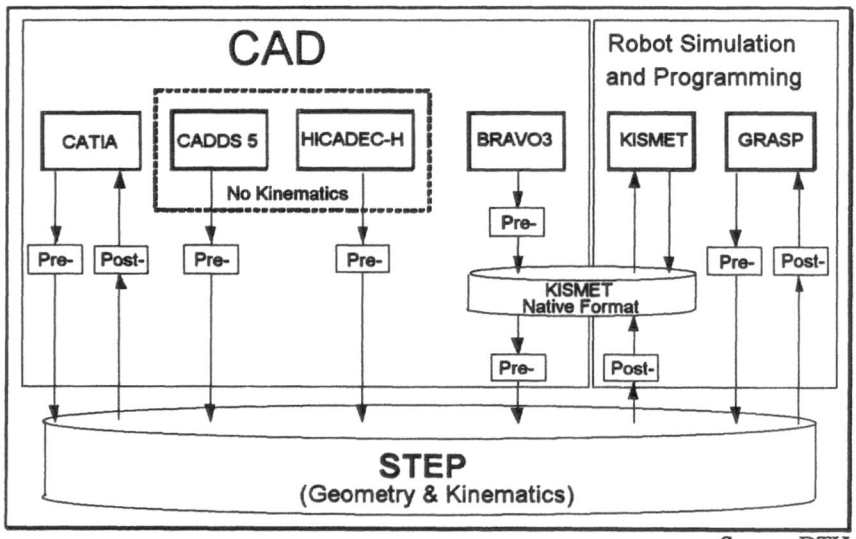

Source: DTH

Fig. 5.0-2: Several pre- and post-processor programs have been developed in the NIRO project on the basis of the STEP specification. These programs, denoted 'pre-' and 'post-' in the figure, are the main topic of this chapter.

A diagram of the developed processors is shown in Fig. 5.0-2.

The function of a pre-processor program is to realise the transfers of information from a CAD or other system to the neutral medium, e.g. convert the information from its native format to the neutral STEP format. Correspondingly, a post-processor program realises the transfer of information from the neutral medium to the receiving system, e.g. CAD.

The development of such processor programs is based on classical and modern compiler techniques. This topic is not addressed here in detail, since the principles of compilers are well documented in the literature, e.g. in [Schlechtendahl 1989] and [Aho and Ullman 1979].

Concerning concepts of kinematics and robotics employed in the text, the reader should refer to such textbooks as [Fu, Gonzalez and Lee 1987, Craig 1989, McCarthy 1990]. For more background please refer to [Kroszynski et al. 1989 and 1991, Trostmann 1990, and Sorensen et al. 1992].

The development of STEP processors from CAD system to STEP is described in Sect. 5.1, and the development of STEP processors from STEP to robot programming and simulation is described in Sect.5.2. Section 5.3 addresses the test of the processors and the consistency of the specification.

5.1 From CAD Systems to STEP

5.1.1 GRASP

P. Sorenti

System description

GRASP is a general simulation tool from BYG Systems Ltd, using 3D graphical modelling for a wide range of applications. It offers a facility for discrete event simulation, (e.g. factory layout design, production planning), as well as the more specific applications of robot modelling and programming. The GRASP software has its own high level textual programming language for defining the geometry of the workplace and the geometry and kinematics of the robot. However, the facility to import geometric data from CAD systems is essential to avoid duplication of work and to ensure consistency. The most efficient way to do this is by using neutral file formats rather than develop processors dedicated to a specific target system.

The STEP processor to the GRASP simulation system (see Sect. 5.2.1) is written as an independent tool which translates the STEP neutral file format into the high level language recognisable by GRASP (GRASP source). A different approach has been adopted for the GRASP to STEP processor in that the processor has been embedded in the simulation system and thus forms an integral part of that system. There are several reasons for this approach, namely:

- the data structure in the GRASP internal database is such that it is straightforward to output in the format required by STEP.
- translating GRASP language into STEP would introduce unnecessary mapping problems and involve additional development of external processor software to perform lexical, syntactical and semantic analysis of the GRASP source file.

Some of the information in the STEP file has no equivalent in the GRASP system. Although currently much of this information is ignored, it will be a requirement that future processors can output such data back to the STEP format. The approach adopted allows data to be read in and stored in the GRASP internal data structure until it is required. The alternative of significantly extending the GRASP source language is cumbersome and inefficient.

The GRASP to STEP processor has thus been developed as an integral part of the GRASP software. It is available on any graphics workstation supporting GRASP, for example SUN SPARCstations, Hewlett Packard and IBM Risc System 6000 workstations. The basic principle is very straightforward since all that is required is a traversal of the internal data structure to identify GRASP entities that can be translated into the GRASP entity schema. Then, for each valid entity found it is then mapped onto the STEP physical file according to the formats described by the appropriate documents. Thus, all that is really required of such a pre-processor from GRASP is the ability to identify common entity types and associations and output them to an external file using a standard library of functions to put them in the correct format.

Method of model output from GRASP to a STEP data file

GRASP allows its internal data structure to be output either in GRASP source file format or STEP neutral file format. All robot structures are output as STEP kinematic mechanisms. All GRASP parameterised solids (CSG - constructive solid geometry) may be expanded internally into a GRASP-specific geometric entity called a General Module and as such will have a one-to-one mapping to the 'FACETTED_BREP' STEP entity.

Geometry output to the STEP file

To illustrate this mapping, consider a solid model, e.g. a box. The basic components of a GRASP General Module are the Vertex, the Edge and the Face. These map to the 'CARTESIAN_POINT', the 'POLYLOOP' and the 'FACE' of the STEP 'FACETTED_BREP', with a closed ring of GRASP edges forming the 'POLYLOOP'. Holes in the GRASP solid model may also be output but appear implicitly in the STEP solid. This is because solids created in GRASP always adhere to the Eulerian solid rule relating the number of faces, edges and vertices to define a closed solid of finite thickness. It is not true however that non-Eulerian solids defined in a STEP file cannot be read into GRASP, it is just that the user is warned that such a solid has been detected.

The output of surfaces (parametric and B-spline) from GRASP to the STEP file is not implemented since they were not included in the scope of the NIRO project and the specification contained therein. This is true also of entities such as parametric and B-spline curves, simple line segments and points (vertices).

Kinematic model output to the STEP file

GRASP is a very powerful tool for modelling a vast range of kinematic structures. These include those with closed chains, robots mounted on other mechanisms, degenerate jointed systems and parallel link configurations. The current implementation of the GRASP to STEP processor can produce STEP files for any kinematic arrangement that can be described within the GRASP system. It does so by using the STEP-NIRO 'MECHANISM' entity.

Kinematic models with simple serial configurations of up to 24 joints can be transferred from GRASP. More complex systems with closed loops and parallel link arrangements may also be catered for as well as those where several individual kinematic entities are connected through common links. The NIRO-STEP schema takes into consideration all these cases that have been implemented accordingly in the GRASP pre-processor.

In addition tools, tool attachment points (TAP) and tool centre points (TCP) in the model are also output to the STEP file as the STEP-NIRO 'TOOL' and 'TOOL_ATTACHMENT_FRAME' entities. The 'TCP' description is embedded within the 'TOOL' entity. The requirement for the 'TOOL' and 'TCP' entities becomes apparent when considering the question not only of the application for which a robot is being used, e.g. drilling, pick and place, welding, etc., but when the method of control is addressed. It is the TCP that the user of a robot (real or simulated) is concerned about. This is the 'point of action' for the system, that is considered a point of reference for a co-ordinate system and controlling device. A method has been devised within the NIRO-STEP schema by which the relevant information may be transferred.

Name references in the STEP file

All geometric and kinematic entities in the GRASP system may have a unique name associated with them, or, in most cases (except for example for a robot) they may be anonymous, i.e. have no name. Where an entity has been given a name in GRASP this name is transferred to the STEP file by the 'INDEX_ENTRY' mechanism.

All (solid) geometric entities in a GRASP model may be output in the STEP neutral file format. There is no capability in the GRASP system to generate STEP entities that have no GRASP equivalent. However, future generations of the processor will allow such information stored as a result of reading in a STEP file to be output without modification.

User defined entities

The NIRO protocol supports the ability to transfer entity types not formally defined in its application schema. This is also true of the STEP interface from which much of the NIRO work takes its base. This means that information specific to a given CAD and/or planning system may be exported between like systems without the loss of data. A system unfamiliar with such a 'user-defined' entity type will discard or ignore the information during the post-processing stage. The GRASP pre-processor could quite simply output examples of such entities. However, this feature has not been implemented since it has found no practical use in the context within which it has been used. The feature has however been tested and implemented in other areas of the NIRO project.

Library of test models

The output of STEP files from GRASP has been validated on a wider platform than simply by internal testing. Contributions have been made to the library of test models (described in Chapter 5.3) which has been used by other partners in the project to verify the action of their STEP post-processors. An entry in the library indicates its adherence to the correct STEP file protocol on the physical file format.

Summary

The (pre)processor to create STEP neutral files from a model in the GRASP planning and simulation system has been successfully tested using the library of models set up within the project. The implementation of the NIRO protocol for robot model transfer from GRASP has proven to be a working solution for the use of neutral files in this field.

The future

The nature of the role of a planning and simulation system in the complete production process from computer aided design to real robot control places greatest emphasis on the transfer of CAD data from the neutral file format to the simulation system rather than vice versa. The integrity, accuracy and consistency of the data transferred to a system such as GRASP can affect on the results of off-line generated robot programs; especially in areas such as collision avoidance and path planning.

Thus further development work will be driven by the requirements of the STEP to GRASP processor but will be incorporated into both processors to ensure compatibility between the two directions of data transfer.

It is envisaged that the inclusion of more advanced STEP entities such as B-spline/Parametric curves and surfaces and other solid geometries will be a significant improvement to the neutral interface pathway between CAD and planning/simulation. This is unlikely to prove a difficult task since the required data types already exist in both the STEP protocol and GRASP's internal data format. It should be simply a question of writing the output routines to the STEP physical file from the parameters stored in the simulation system.

In addition, the adoption of other lower pair (possibly higher pair) kinematic configurations in the STEP-NIRO specification would extend the application of the neutral file beyond common robot and geometry data transfer to the realms of more advanced, and specialised research fields.

Further areas that need to be addressed are those concerned with the dynamics of the mechanisms being transferred. Kinematic information transfer in the scope of this project considers links that form the mechanism 'flesh' to be rigid, weightless bodies. Whereas some progress has been made in defining the dynamics of these rigid bodies in the NIRO-STEP protocol (defining moments of inertia, centre of mass, etc.) considerable work remains. This is chiefly because of the advances in robot technology where the devices are becoming faster and lighter and performance is of increasing importance.

The extent to which technological information is catered for in the STEP file should also be considered. This would include the definition of joint velocities and accelerations, maximum speeds, configuration information (the rules that govern which inverse solution to adopt during Cartesian motion control) and other controller-related information.

5.1.2 BRAVO3

A. Ludwig and J. Reim

System description

Bravo3 is an integrated CAD/CAM system distributed by Schlumberger CAD/CAM Division. In the NIRO project an implementation on a DEC VAX cluster was used at KfK. The cluster is formed by a DEC VAX 8650 computer and a DEC VAX 6310 computer as so-called boot nodes and several Vax stations as clients. In addition, about 25 Schlumberger graphical work stations are connected directly to the boot nodes.

Bravo3 comprises several application programs including:

1. A graphic editor (called AE) for handling all communications and controls and for producing 2D or 3D wire frame geometry and 3D surface geometry;
2. SOLIDS to produce geometric models defined by primitive solid shapes using the CSG (Constructive Solid Geometry) approach;
3. GRAFEM to create finite element meshes;
4. NC to produce files for numerical controlled machines;
5. MECHANISM for kinematic/dynamic analysis;
6. IGES, VDAIS, and VDAFS converters to write and read neutral files according to these interface specifications;
7. APP and IAGL, that are two different programming interfaces to Bravo3 that allow the user to create his own application programs.

The application program, SOLIDS, allows to produce solid geometry interactively at a workstation connected to the system. The CSG construction tree is stored in the Bravo3 database along with a precise specification of the geometry of the primitives. Since an evaluation of the construction tree using precise geometry takes a rather long time, it is performed only in background jobs and the results of this evaluation may be presented at the work stations if wanted. To provide a fast response to the user during the design process, however, SOLIDS uses in addition a foreground modeller that produces polyhedron representations (planar-facet approximations of curved surface sections).

While the constituents of the exact solid model (the CSG primitives and the Boolean operators) can be accessed readily with means provided by Bravo3, this is not possible for the facetted model because the facetted model is intended to be used for visualisation only. Furthermore, the MECHANISM module merely provides an interface to another program that is also run in batch mode. The format of the files output by this interface is not suited for general use.

For these reasons KfK has added (linked) an application program package, called ROBOT, to Bravo3 using the APP programming interface. There are two main goals for this package:

1. It allows the user to extract the shape information from the Bravo3 database, either through a "guided" interactive redefinition of shapes, based on the standard Bravo3 representation, or automatically by evaluating the internal polyhedral approximation of Bravo3 (this latter feature is not yet fully integrated into ROBOT; the program is still executed independently). In the interactive mode, besides polyhedral models also regular shape models (spheres, cylinders, truncated cones, boxes) as well as linear or rotational sweeps and "pipes" (a specific type of sweeps) can be created. Since, however, shape representation in NIRO is restricted to the FACETTED_BREP entity of STEP, these capabilities must not be used if the model is to be exchanged via a STEP file.

2. It provides a simple, but powerful editor to add kinematic and relative placement information to Bravo3 assembly models, to check the kinematics within a Bravo3 session, and to output these data on a suitably formatted file.

All information that has been created interactively or collected with ROBOT is stored primarily in a binary file specific for ROBOT. Then some ROBOT output routines can be used to write the shape, relative placement, or kinematic data on files that are formatted according to the appropriate native file format of KISMET (a description of KISMET is given in Sect. 5.1.5).

Processor overview

In the framework of the NIRO project, pre-processing of Bravo3 models is divided into four principal steps (see Fig. 5.1.2-1):

1. extract the shape of all solid models which constitute the entire model, either interactively using ROBOT or automatically from the corresponding Bravo3 data base;

2. add all relative placement and kinematic information needed using ROBOT (as Bravo3 proper provides only shape information and parts of the relative placement data);

3. merge shape, relative placement, and kinematic data to one ROBOT model which is then written to an equivalent KISMET file structure (see Sect. 5.1.5);

4. convert this file structure to a STEP file using the KISMET-STEP pre-processor (see Sect. 5.1.5).

If the creation and extraction of data are performed entirely by means of the interactive ROBOT modules, then also the merging step may be performed within ROBOT. If on the other hand the automatic extraction procedure is used, shape information along with static relative placement information is obtained. Any kinematic models have to be defined and written separately into the STEP file. The points (1-4) addressed above are discussed in further details in the following text.

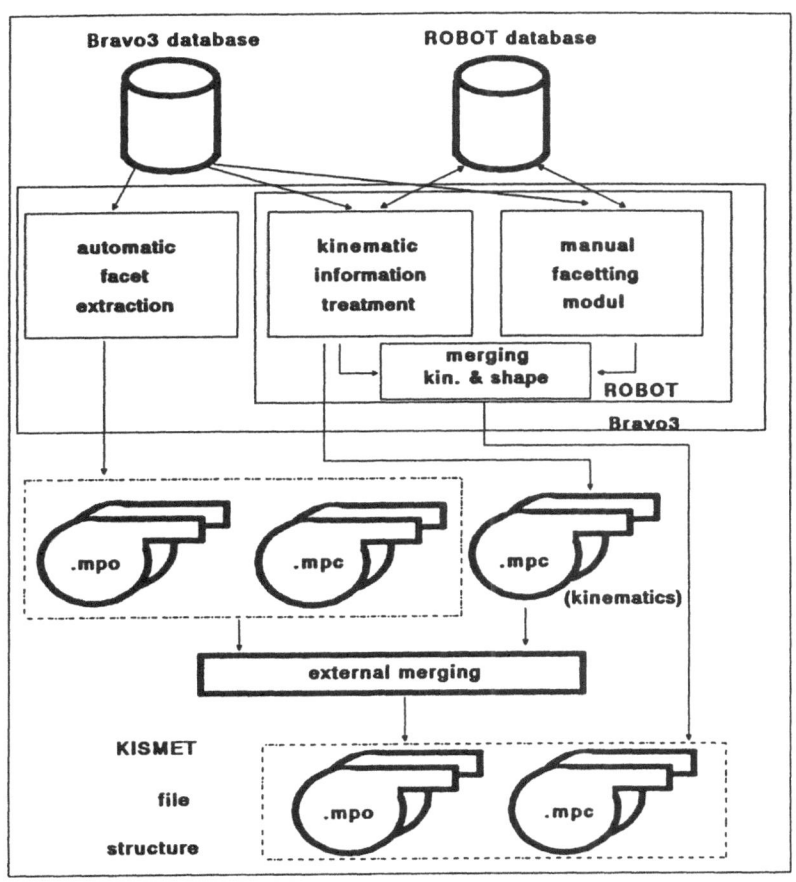

Source: KfK

Fig. 5.1.2-1: Extraction of Bravo3 models to KISMET model structures

Interactive facetting approach

This short description cannot outline all features of the procedure. Nevertheless, it illustrates that this interactive faceting approach demands an effort similar to the original design work. It has been developed as a work-around, because any access to the facetted models of Bravo3 is not documented and not supported by Schlumberger.

The creation of other types of shape primitives that are common to ROBOT and KISMET (box, sphere, cylinder, cone, sweeps and "pipes") is also supported by ROBOT and is performed much more easily. These types, however, cannot be used in the framework of NIRO.

The interactive facetting approach is a synonym for creating a polyhedral model using the respective module of ROBOT. Its only mandatory requirement is that a Bravo3 model cell is open; no solid model need to be there, and even an empty cell without geometric models (wire frame or solid) would be sufficient. In such a case the vertices of the polyhedron would be defined "on the fly", using the normal point-defining facilities of Bravo3. Usually, however there exists a solid model that is then shown by the Bravo3 foreground facetted modeller as a faceted model, and may be used as a "work base".

The following is a rather compact description of the interactive process for creating a polyhedron. It starts with the determination of the first vertex of the first loop (an outer loop) on the first face. The term "determination" means that either a point can be "picked" from a model stored in the database, or it can be arbitrarily input by one of the various means provided by the Bravo3 editor. Then the next vertices of this loop are determined, until some vertex is no longer co-linear with the first and the second one. Now the plane of the new face is computed automatically, and its outer normal is shown to the user, who has the opportunity to reject this plane and to start the loop again. If he accepts it, he is asked to determine the remaining vertices of the loop (which will be projected onto the face plane, if necessary), until the program detects that the loop has been closed, or until the user explicitly demands to close the loop. Now the edges of the face are computed and stored in the Bravo3 database as wire frame lines that are marked suitably.

The other faces of the polyhedron are created by a similar procedure. The main difference is that the first edge of an outer loop must be "picked" out of those edges of other loops that have been used only once up to now. In this way, the connectivity of the faces is enforced. Any inner loops have to be defined immediately after the outer loop enclosing them; their vertices are projected onto the plane of the respective outer loop, and their edges are checked not to cross the edges of the outer loop or the edges of any previously defined inner loops. Due to the marking of the edge wire frame lines, ROBOT detects when the shell of the new polyhedron is closed, and allows the polyhedron model to be stored to the ROBOT database.

Automatic facet extraction and optimisation

The automatic facet extraction is based on the fact that the Bravo3 editor and its modules are themselves written in APP, which is actually PL/I enhanced by a huge set of data structures (only partially documented) and another set of routines working on these structures. The routines are all well documented, and all the data structures are delivered to an APP customer within an include library. Thus, indeed, an APP user can find all data structures stored in a Bravo3 database. In this way, the internal representation of Bravo3 faceted solid models have been explored.

The steps of the automatic facet extraction procedure are distributed in two programs. The first one is a PL/I program using the APP interface and working in the Bravo3 environment. It scans a Bravo3 SOLIDS ASSEMBLY cell for SOLIDS PART cell references, stores the placement information and enters the referenced PART cells. Then it searches for the facet data structures and evaluates them. After some checks and corrective algorithms (described later) have been applied, the evaluated data are stored into an intermediate file in binary format, together with the reference tree of the ASSEMBLY cell.

The second program is started outside the Bravo3 environment. It reads the intermediate file and formats the data into the KISMET file structure.

There are some problems with the extracted facet data. Since Bravo3 uses them exclusively for the visualisation of solid models, there is no need for sophisticated checks on these data, and thus they do not always conform to the STEP rules for polyhedron representation. Typical shortcomings arise from the Boolean operations that have been applied to the faceted representations of primitives:

- After Union operations, sometimes some faces are instanced twice;
- not all faces in the interior of the solid body have been removed;
- due to numerical inaccuracies, sometimes corresponding vertices of neighbouring faces do not coincide, thus causing a gap between the faces or an overlap.

Other shortcomings originate in an internal optimisation strategy used by Bravo3 that removes vertices between co-linear edges. This causes:

- some edges to appear twice and
- some edges are referred to by only one face rather than two.

These shortcomings lad to the implementation of some corrective algorithms working on the extracted facet data. In the very first step, the data are checked for inconsistencies, i.e., for the existence of edges referred to only once. Then, the polyhedron is optimised in the following fashion:

1. Search for double edges and remove loops that are instanced exactly twice. Signal an error abort, if the doubles cannot be removed (fortunately all polyhedrons extracted from Bravo3 group their double edges in double loops).
2. Put all edges that are referred to by only one face into a list.
3. Group these edges to closed loops that are as short as possible. Again signal an error abort, if unpaired edges remain on the list (this only occurs when the Bravo3 faceted modeller itself sent an error message while creating the polyhedron).
4. If these loops describe a face, add them to the polyhedron. If some of the edges are co-linear with other edges, divide the longer edges at vertices of the other loop.

5. Verify the modifications of the polyhedron and rewrite it to the intermediate data structure.

As indicated above, the extractor and corrective programs are written in PL/I and embedded into the Bravo3 environment via APP. They have a modular structure and use highly optimised algorithms that provide for linear increase of computing time with increasing model size (extractor: vertex sorting with hashing; optimiser: indexed list to get edges by their first vertex). Also the KISMET file formatter is written in PL/I, using some existing routines.

The automatic extraction has been tested and verified by the successful transfer of several models from Bravo3 to KISMET, including the "solitaire" set-up of the Reis demo for NIRO (see Sect. 8.5).

Extraction of kinematic information

KISMET divides its models into several sub models that, in turn, arrange some partial models and define kinematic or static relative placement relationships among them. The sub models correspond to the configuration files in the native file format of KISMET (see Sect. 5.1.5). To supply a model to KISMET from a CAD system, one has either to extract the corresponding information from the data structures stored by the CAD system or to define and store them in appropriate data structures provided by an application program like ROBOT.

Kinematic information cannot be extracted from a Bravo3 database since Bravo3 does not deal with kinematics, rather, it has to be created in an interactive session where both Bravo3 and ROBOT, are executed concurrently. For each partial model that will become part of a kinematic sub model, a Bravo3 "cell" must exist which contains either the true shape model of this partial model or a suitable substitute (e.g., a wire frame model). Then the kinematics module of ROBOT is invoked. It uses the cell referencing mechanisms provided by Bravo3 and guides the user to add the necessary information in at least two steps:

- Enter each cell belonging to a partial model, one after the other. In each cell, define "joint frames" for each joint to which the partial model may be connected as a link, at the root of the kinematic tree as well as at any of its leaves. Where a non-kinematic partial model may be attached, a "reference frame" has to be defined. Such a frame (which is a Cartesian axis system triplet) is defined by determining its origin and the directions for two of the three axes.
- Enter another cell that will serve as an assembly cell for the kinematic sub model (under Bravo3 SOLIDS, it must be a SOLIDS ASSEMBLY cell). This cell may already contain references to the partial models of the sub model, arranged in the "zero state" (where all variable joint parameters are set to zero). In this case, the user "picks" each partial model, starting at the root and progressing to the leaves, and starting again at any branch. For each partial model he denotes its kinematic predecessor frame in one of those partial models already held in the chain (these are presented in a menu). After the user has confirmed his choice, the partial model is added

to the kinematic chain. The user may also decide to open an empty assembly cell. Then the user has to select the partial model from a menu. For the base model, the user determines the placement. After the chain has been started, the user again denotes a suitable frame in one of the partial models already included. Now ROBOT inserts automatically a cell reference to the cell of the partial model, placing it according to the frame data. Some minor corrections can be done at this stage, and after confirmation by the user the partial model is added to the kinematic chain.

In a third step, the user may define special dependencies between the degrees of freedom of the various joints in a kinematic model. This definition is needed for simulation purposes within ROBOT as well as within KISMET. In the framework of NIRO, however, this feature is irrelevant, as it is not mapped to the STEP file.

The resulting kinematic models that are stored now in the ROBOT database are equivalent to those required by the KISMET native format. Therefore, they may easily be written out to files in this format.

ROBOT provides a very similar procedure for the definition of static relative placement relationships. Most of this information may be obtained from a Bravo3 assembly cell, also by the automatic extraction procedure described in the previous section. Kinematic data, however, cannot be extracted automatically since they are unknown to the proper Bravo3 models.

Besides this, also hierarchical dependencies between sub models, as used in KISMET (see section 5.1.5), may be defined using ROBOT. For this purpose, only the parent partial model has to be chosen within a kinematic model or a static relative placement model. Then the user determines one of the other existing sub models as the "son". Since the partial models of the "son" may be themselves "parents" of other sub models, the nesting depth of this model tree is virtually unlimited. Any cyclic references in this tree are prevented by checks that are performed automatically by ROBOT.

Another ROBOT module allows to simulate motion of a kinematic model defined with the ROBOT Kinematics module. Although far from being a real-time simulator system like KISMET it proved to be a very helpful tool for checking the correctness of kinematics prior to their transfer to other systems. It is also helpful in testing new designs of complicated mechanisms that have to be examined regarding the design goals.

5.1.3 CADDS 5

C. D'Elia

System description

CADDS 5 is a family of stand-alone personal productivity tools for computer-aided design, developed by COMPUTER VISION at Bedford MASS, (US).

Among these products is the CVware Solid Modelling software that has been used in the NIRO project. CVware Solid Modelling offers advanced representations of geometry, including full support of Non-Uniform Rational B-Spline (NURBS) curves and surfaces, and provides powerful solid modelling operations (Boolean, sculpting, filleting, sweeping along free-form curves, and converting wire frames into solids). It represents all solids using an exact boundary representation (B-rep).

The pre-processor developed within the NIRO project allows to export solid models from the CADDS 5 data base to STEP files.

The pre-processor supports only faceted B-Rep format. Kinematics is not provided.

CADDS 5, as used in the NIRO project runs on a Prime station model 52 based on a SUN Sparc workstation. The CADDS 5 modules used are: Base 2D, 3D, Wire frame, NURBS (for surface development) and the Solid modeller.

Inside CADDS 5 the information is organised in a data base of parts. A part is a basic object in the CADDS 5 environment, and it is created during a series of interactive graphic sessions. Each part has its own associated data base for creation, editing, or analysis. A part may refer to other parts, but only one part per task may be active at one time.

The CADDS 5 data base management routines interface with the File Manager to store each part's data base on disk. Each part is referenced by a name that is a concatenation of all catalogue levels used in the name.

The CADDS 5 part data base contains a part description distributed on several UNIX files. The part itself is made up of several entities. An entity can be compared to an object to which geometric or non-geometric attributes or properties are assigned in a hierarchical fashion. Various entities relate to each other in ways that convey specific associative information.

Processor description

The pre-processor developed by FIAT/SESAM in the NIRO project allows for the extraction of product data from CADDS 5 to STEP. The processor is written in FORTRAN 77 and it makes use of Computer Vision CVDORS utilities to access the CADDS 5 data base. CVDORS utilities features a comprehensive set of functions and procedures for C, C++ and FORTRAN 77 environments. These features allow for entity reading and modifying although the knowledge of the CADDS 5 data base is not required.

The interaction occurs outside the CADDS 5 environment. The user that runs the processor is requested to enter:

a) The name of the CADDS 5 part name, the file name of the output STEP neutral file and a description text to appear in the header section of the STEP file

b) Additional parameters to control processing

Processor architecture

The software architecture can be described as in the Figure 5.1.3-1 below.

The pre-processor is structured as a main routine calling sub-routines from a library of procedures. According to architectural guidelines for processor development of the ESPRIT project 2195, CADEX, processor software architecture is structured in two sections, pre-processor front-end and pre-processor back-end, sharing an internal data structure.

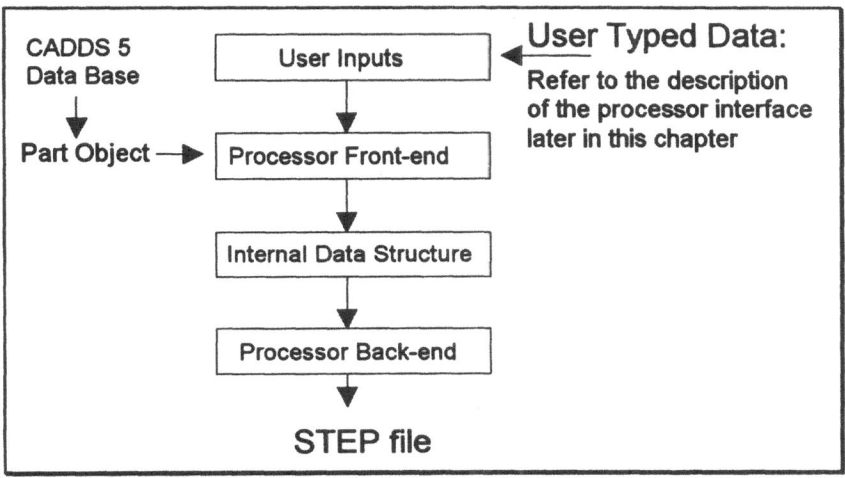

Figure 5.1.3-1 Processor software architecture

Pre-processor front end

The pre-processor is started outside the CADDS 5 graphic session. It first requires the name of the solid model (part name) to be exported and some other information (STEP output-file name, additional information for the header section, etc.) . The pre-processor creates a temporary structure by means of the CVDORS utilities and then opens the part file. The part_name must contain the path name to allow processor to find the corresponding data base.

The pre-processor then loops to find all polyhedrons belonging to the part. For each polyhedron it extracts solids and related entities.

Loop information is not provided directly by the CVDORS tool kit utilities thus the processor front-end obtains point sequences from the face boundary.

At the end of a processing step if no diagnostic errors have been detected the processing may continue.

The entities that can be exported include: solid, solid instance and solid mass properties. A colour code is assigned to each solid in order to differentiate them.

Pre-processor back-end

The pre-processor back-end produces the STEP file. It runs on Internal Data Structure (IDS) entities processing an entity at time.

The pre-processor outputs the records formatting them following the NIRO description .

Processor interface

When running processor the operator is prompted for information necessary to open and access the CADDS 5 data base and to create the STEP file as required. The processor interface looks as follows:

```
STEP Pre-processor for CADDS 5 (L02V00)
          Corrado D'Elia - January 1992
CADDS 5 under PRIME  - SPARC model 52
FORTRAN 77 compiler/linker
SESAM - Software e Sistemi per l'Automazione Manifatturiera
          Corso Svizzera 185   10149 Torino ITALY
          Tlf.  (+39) 11 74261   Fax (+39) 11 7492043


Enter CADDS 5 part name
Enter output file name   (default part_name)
Enter number of digits for reals
Enter specification version L02V00
Enter Author
Enter description text
```

The processor starts its activity and signals when a step has been completed successfully:

```
.. .Please wait
1..Model accessed
2..Solids finished
4..Files released
   Good-bye
```

If an error occurs the processor stops, displays the error, and writes it to the STEP output file.

Remarks on processor testing

Testing activity has been accomplished by exporting in neutral format the model of a COMAU MAST-3 robot and importing it to other systems via STEP (refer to 5.3).

The testing activity has, for instance, verified the importance of correct orientation of the loops when the receiving system uses this information for solid generation.

A general remark concerns the use of double versus single precision real number arithmetic when creating solid models and the number of digits used for co-ordinates in neutral format to avoid geometrical errors (such as derivations from planarity, etc.).

The resulting STEP files have been imported and tested on the following CAD and simulation systems: KISMET, GRASP and CATIA.

5.1.4 CATIA (Kinematics)

T. Sørensen and U. Kroszynski

System description

CATIA is a CAD system developed by Dassault Systems in France and distributed world-wide by IBM. It features 2D and 3D geometry modules as well as diverse application modules. It has been implemented under various platforms, namely the MVS and VM/CMS operating systems for IBM mainframes as well as under UNIX/AIX for IBM's RISC System/6000 series.

The CATIA software is written in FORTRAN 4, FORTRAN 77, C, and ASSEMBLER.

CATIA supports the following four layers:

1. Base and interfaces with the CATIA base module, data/communication management library functions and data base access routines libraries (CATGEO, CATMSP);
2. Geometry modelling, calculation and simulation applications comprising drafting, 3D-design, advanced surfaces, finite elements, kinematics and solid modelling capabilities;
3. Manufacturing with application-dependent modules for numerical control machines and robotics;
4. Architecture and plant design with application-dependent modules for piping and tubing, structural design, schematics and building design.

Modules related to robotics

CATIA kinematics- and CATIA robotics-modules are two specific modules for dealing with 2D/3D kinematics and robotics tasks. CATIA robotics is discussed in chapter 5.2.2. CATIA kinematics, used in conjunction with CATIA Base, CATIA 3D Design, and CATIA Solids Geometry, allows the user to create models and simulate the operation of kinematic mechanisms, and to analyse the results of the simulation. It includes the following two functions: *KINEMAT* and *KINEMUSE*.

The *KINEMAT* function is divided into two parts depending on the selected working mode: 2D SPACE or 3D SPACE. It is used to:

a) Define the joints, stops, fixed part and commands.
b) Analyse the kinematic mechanism to control the joint definition.
c) Erase joints from a mechanism.
d) Erase, copy or rename a kinematic mechanism.
e) Transform a kinematic family and unite two kinematic mechanisms.

The *KINEMUSE* function is used to:

a) Simulate the operation of a kinematic mechanism and generate traces as well as numerical outputs if required.
b) Detect automatically a collision between two parts.
c) Study the velocity range and acceleration of the parts.
d) Define and modify command laws.
e) Modify the joints, impose stops, fixed parts or commands and analyse the kinematic mechanism.

Interface capabilities related to NIRO

Access to the CATIA data base is possible through a FORTRAN subroutine package called CATGEO. The CATIA system supports the access and management of both the polyhedron (faceted B-rep) and the CSG representation of solid geometry. For NIRO only the faceted B-rep representation is used. The CATIA system supports also the access and management of approximately twenty different types of higher- and lower order kinematic pairs (prismatic, revolute, gear, screw, etc.). For NIRO only prismatic and revolute kinematic pairs are used.

The *user interface* allows to create CSG models, interactively, with primitives such as Box, Cylinder, Cone, Rotational volumes by sweeping closed contours around an axis, and extrusions by sweeping planar closed contours along a distance (also a curve). These primitives are "simple" leaves of the CSG tree.

Solids can also be generated by the conversion of a surface model into a solid model, i.e. by merging the surfaces of the surface-model. Such solids derived from surface models will form more "complex" leaves of the CSG tree.

The binary CSG tree with simple and complex leaves is stored as the definition of the solid in the database. An "evaluated" approximate representation in the form of a polyhedron is also stored and used for computations and visualisation.

This strategy permits to change the discretisation of curved surfaces (e.g. a cylinder as a prism of 6 lateral faces to 12 lateral faces) as well as the parameters of the leaves (e.g. the cylinder radius or height). Primitives can also be repositioned, re-oriented in space, removed from the binary tree, replaced by other primitives, etc. When such modifications take place, the polyhedral model has to be re-evaluated.

Program design of the STEP pre-processor for CATIA

Since in the pilot program development, not extremely large models occur, it was selected to hold all information in large arrays (memory + disk handled by the operating system) rather than using direct access files to store the intermediate information.

Some FORTRAN routines are based on C/UNIX/AIX and therefore provisions are given for dynamics storage utilities, a feature not normally allowed in FORTRAN programming. This is a great advantage since the size of the models (e.g. nr. of points, etc.) are not known beforehand. Because of the dynamic memory facilities the arrays and tables are allocated, extended, or freed as convenient.

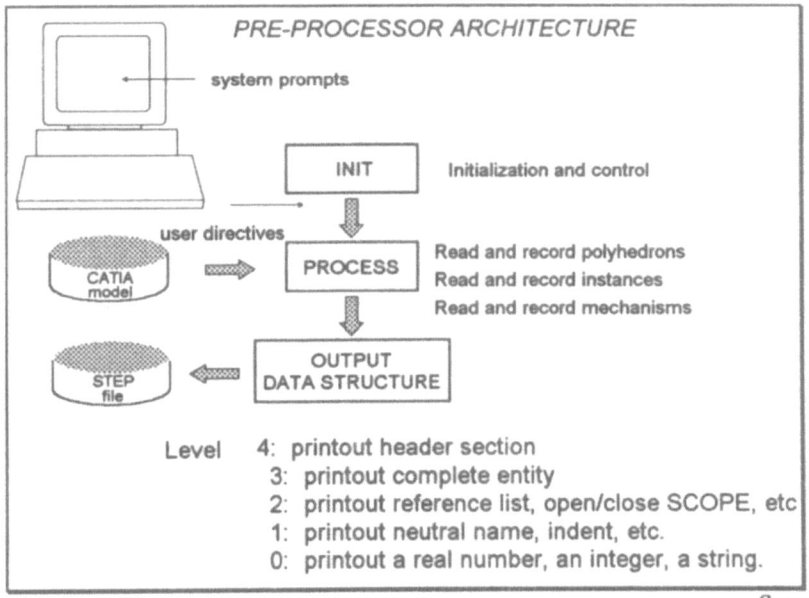

Source: DTH

Figure 5.1.4-1: The architecture of the CATIA pre-processor

The user interaction occurs outside the CATIA environment. When running the pre-processor the user is requested to enter:

- The name of the output STEP neutral file,
- nr. of significant digits to be used for real numbers,

- CATIA user name (Password, group),
- title of CATIA model to be retrieved,
- authors name, and
- a description text to appear in the header section of the STEP file.

The architecture of the processor can be described as in the Figure 5.1.4-1. A comprehensive description of this architecture can be found in [Schlechtendahl 1989]. A similar description is not given here, so please refer to [Schlechtendahl 1989] for further details on the processor architecture.

Processor interface

When running processor the operator is prompted for information necessary to open and access the CATIA data base and to create the output STEP file as required. The processor interface looks as follows (default values are in parentheses):

```
STEP Pre-processor for CATIA (L01V03) (L02V00)

(Here follows a description of Hard- and software versions currently running, the
author and the (DTH) organisation)

Enter output file name  ...................(niro)..>
Enter number of digits for reals ..............(4)..>
Enter specification version 1-L01V03 2-L02V00  (2)..>
Enter your CATIA user name ..................(uri)..>
Enter CATIA member name  ........(CATGEO : SAMPLE)..>
Enter Author  ................(Uri I. Kroszynski)..>
Enter description text  ...............(Test file)..>
```

The processor starts its activity and signals when a step has been completed successfully:

```
    ... Please wait ...
1. Model accessed  ...
2. Solids finished ...
3. Kinematics copy ...
4. Files released  ...
    ... Good-bye   ...
```

If an error occurs the processor stops, displays the error on the screen, and writes it to the STEP output file.

The processor supports the entire NIRO information scheme (see Appendix A) with the exception of INDEX-ENTRY and the user-defined entities RGB_COLOUR_TABLE, RGB-COLOUR, POINT_DIRECTION_PAIR, RENDER_FACE, and LINK_MASS_PROPERTIES. Some modeling conventions have to be followed in order to produce the TAP frame TOOL and MOUNT entities.

5.1.5 KISMET

A. Ludwig

System description

KISMET (Kinematic Simulation, Monitoring and Off-Line Programming Environment for Telerobotics) is a graphical system whose characteristics are briefly described in chapter 5.2.3 in the context of post-processing.

In the native format of KISMET, a kinematic model is described by a hierarchical structure of "configuration files" and "geometry files" Each geometry file contains one solid model; the configuration files give topological, kinematical, and relative placement information, attached to several partial models (called "abstract frames") in the file. Some of these partial models may include one or more "work frames" indicating gripping positions, tool attachment points, or tool centre points.

There are three types of configuration files ("BASE_file", "FRAME_file", and "ROB_file"). A BASE_file may appear only at the root of a model tree and provides for a model base co-ordinate system to which all transformations in the model are related, either immediately or via transformation chains. The partial models are static or quasi-static models with no internal degrees of freedom. Only a ROB_file can contain kinematic structures, while a FRAME_file is similar to the BASE_file, except that it contains no model base co-ordinate system and may appear anywhere in the model tree.

The file structure is built up by references to other configuration files and to geometry files within the partial models of configuration files; only the root configuration file is never referred to, and the geometry files do not reference other models. Hence, each solid model within a kinematic model is an instance of an independent solid model template. The "hierarchy" in this file structure evolves from the fact that each partial model of a configuration file may reference another configuration file as its "son". Such a "son" can either replace the solid instances of the "parent" model if desired (e.g., to represent a model of interest at a higher degree of detailing), or it defines a mechanism that is mounted to the "parent" model. The parent model may, in turn, be part of the environment or of a link within another mechanism.

The "abstract frames" are divided into two categories. The first category, called "MOD_abs frames", represents model components without internal degrees of freedom; this is the only category that appears in files of type BASE_file and FRAME_file, but it is found also in a ROB_file. These frames define a local co-ordinate system by means of six transformation parameters and denote a "predecessor" to which this transformation is related. All other transformations given in the frame must be related to the local co-ordinate system. The role of the predecessor may be played either by the model base co-ordinate system (in a BASE_file only), by the local co-ordinate system of the "abstract frame" referencing the file under consideration as its "son", or by the local co-ordinate system of any other "abstract frame" preceding the actual frame in the same file. In the last case, the predecessor is denoted by the name of its frame.

The second category, allowed in a ROB_file only, comprises "KIN_abs frames". Its model components exhibit exactly one degree of freedom - either a prismatic or a rotational motion about a joint axis. Hence, this ROB_file represents a kinematic link. This axis is assumed to coincide with the z-axis of the local co-ordinate system of the "predecessor" (which is denoted in the same way as before). The local co-ordinate system of a "KIN_abs frame" is given by four so-called DH-parameters (DH = Denavit-Hardenberg, see [Kühnapfel 1991A]. Additionally, the lower and upper limits as well as an initial value for the "free parameter" (which defines the actual position along the axis) are supplied. The z-axis of the local co-ordinate system may coincide with the co-ordinate system that is chained to the present link, and a MOD_abs frame must be inserted which denotes the present KIN_abs frame as its predecessor and has a local co-ordinate system according to the succeeding joint. In the same way, branches of kinematic chains can be modelled.

The internal format of all KISMET model files is described in detail in [Kühnapfel 1991B]. Format descriptions for the most relevant file types are also given in [Kühnapfel 1991A].

Processor overview

The KISMET-STEP pre-processor is written in C and was developed on a DEC MicroVAX III and on a DEC VAX 8350 using ANSI C language under VMS. It is operational also under UNIX/IRIX on Silicon Graphic workstations.

In KISMET, a model is entered by naming a "simulation file" that references a couple of other files most of which contain KISMET-specific control information irrelevant for the pre-processor. One of the files referenced, however, is the root of the model tree to be transferred. The pre-processor ignores the simulation file and takes the model root file immediately as its primary input.

The pre-processor has to translate solid and kinematic models given in the native format of KISMET into equivalent models formatted according to the NIRO specification [Kroszynski, Sorensen, and Schlechtendahl 1991]. It can be run in two modes. In the first mode, it accepts only one polyhedron model or one extended polyhedron (another KISMET feature, where for each vertex of each face of the polyhedron a virtual normal direction is given, allowing improved rendering of curved surfaces compared to the faceted representation). The polyhedron model is transferred to a STEP file as a 'FACETED_BREP' entity that, in this mode, will be the only top entity on the file.

The second mode of running the pre-processor works on model trees as explained above including kinematic models. The model tree is read in two passes. During the first pass, all solid models referenced by the configuration files are written to the STEP file provided they are given as correct polyhedron models. Solid models that are instantiated more than once are transferred to the STEP file only once, while for each reference a 'SOLID_INSTANCE' entity is created. These 'SOLID_INSTANCE' entities are also evaluated and written to the STEP file during this pass. The base co-ordinate systems having their transformations referred to, will be written to the STEP file later, but, of course, they must be known implicitly at this stage to get proper transformations for the instances. These transformations may be different in the STEP environment as compared to the KISMET environment. Therefore, when the "scope" of a new base system is entered, a stack of relative placements is always incremented. The stack maintains the relative placements to the actual base system.

In the second pass the topological and kinematic information is read and stored into internal data structures that are then evaluated and written out as a 'KINEMATIC_MODEL' entity, with as many 'MECHANISM' entities in its scope as mechanism models (i.e., ROB_files) are contained in the model tree. If there is no ROB_file in the tree, no 'KINEMATIC_MODEL' is created. Non-kinematic hierarchically dependent models are associated with their parent models - either a link of a mechanism or the ground representation. Thus, the hierarchical structure of the input model is flattened, and some parts may be represented more than once with different shapes, although they exist only once if the original KISMET model has not been cleaned accordingly.

The architecture of the pre-processor is sketched schematically in Fig. 5.1.5-1. Besides the KISMET model tree file, the program uses as input a control file, which provides some path specifications for model files and STEP files and optionally some header information and numerical control parameters (which otherwise have to be input interactively at the start of each pre-processor run). The output consists of the resulting STEP file accompanied by an error log file.

Mapping of KISMET data to STEP entities

The native formats of KISMET are very compact, often storing in a single attribute information which in STEP has to be distributed among many entities. Therefore, the conversion is not a simple one-to-one entity mapping; rather, it may depend on the context. One example of such context-dependent mapping has already been mentioned, namely the transformation values of 'SOLID_INSTANCE's. Or, when an "abstract frame" is part of a FRAME_file, its 'SOLID_INSTANCE's (if any) contribute to the geometric representation of the 'GROUND' entity, provided that the "abstract frame" is not denoted as a tool (see below), and provided that none of the parents of the FRAME_file up to the root file is contained in a ROB_file. In the latter case, the 'SOLID_INSTANCE's would contribute to the geometric representation of a 'KINEMATIC_LINK' entity, instead.

Mapping of shape models (geometry files)

The mapping of a polyhedral model of KISMET to a STEP 'FACETED_BREP' is straight-forward. The KISMET "list of points" is converted into a sequence of

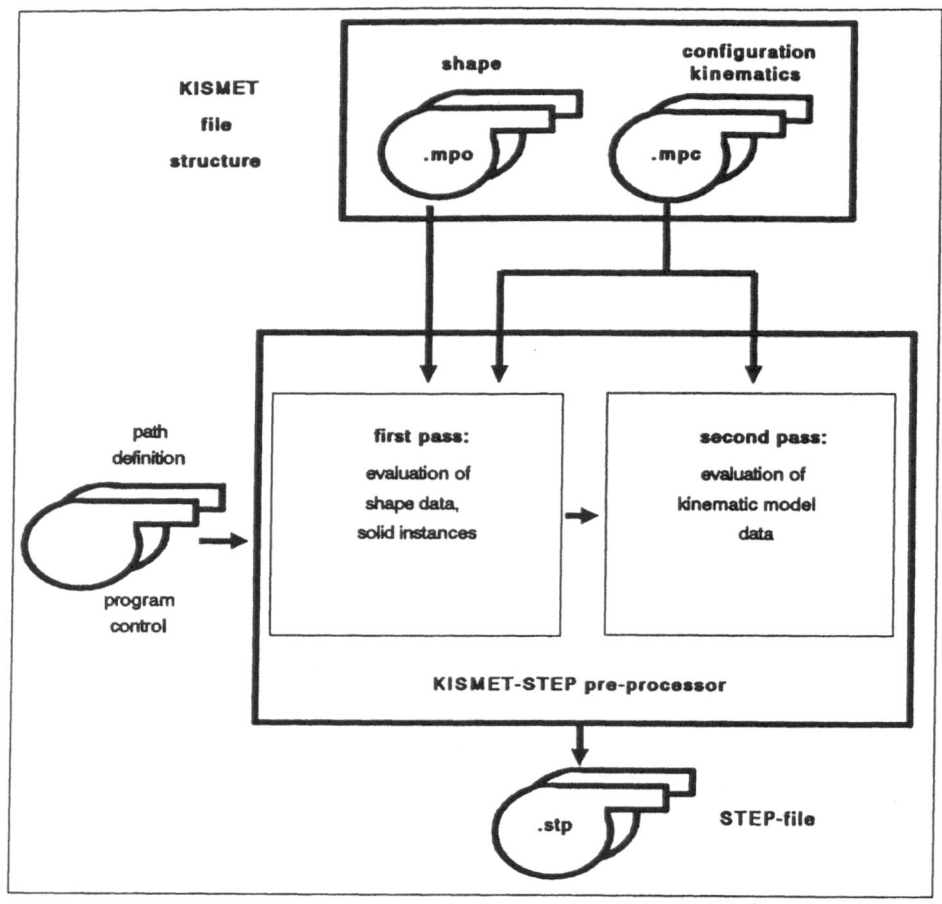

Fig. 5.1.5-1: KISMET STEP pre-processor environment

'CARTESIAN_POINT' entities. Each entry in the "vertices_per_face_list" causes a 'POLY_LOOP' entity to be created. The point references to be inserted can be evaluated from corresponding entries in the "vertex_index_list". Whether a 'POLY_LOOP' is used as the outer loop of a face or as an inner loop depends on the value of the corresponding entry in the "facet_colour_list" (KISMET POLYHEDRON primitive; value unequal -2: outer loop; value equal -2: inner loop in the face of the previous outer loop) or on the sign of the entry in the "vertices_per_face_list" (KISMET EXTENDED_POLYHEDRON primitive; positive value: outer loop; negative value: inner loop in the face of the previous outer loop). For each outer loop a 'FACE' entity is written to the STEP file. It contains in its "bounds" attribute first the outer loop, followed by references to all its inner loops. These 'FACE' entities are referenced now in the "shell_boundary" attribute of a 'CLOSED_SHELL' entity, which, in turn, is referenced by the "outer" attribute of the 'FACETED_BREP'.

The "material_index" of a KISMET Polyhedron is mapped to the "colour_code" attribute of a '!COLOUR_ATTRIBUTE' entity, pointing to the 'FACETED_BREP' in its "entity_to_be_coloured" attribute (the exclamation mark indicates that this is a NIRO-specific entity). If a "facet_colour_list" exists in the KISMET files with non-negative values of its entries, the corresponding entries are mapped to the "colour" attribute of a '!RENDER_FACE' entity that refers to the respective 'FACE' instance in its "face" attribute. In the case of a KISMET POLYHEDRON primitive the attribute "point_direction_pair" of this entity is an empty set. EXTENDED_POLYHEDRON primitives, however, contain additionally a "normal_list" of direction vectors and a "normal_index_list" that corresponds to the "vertex_index_list" and the entries of which point to entries in the "normal_list". Although these additional lists are used only to improve the surface rendering in visualisation it may be desirable to keep this information. Therefore, the "normal_list" is mapped to a sequence of 'DIRECTION' entities, and the corresponding entries of the "vertex_index_list" and of the "normal_index_list" are evaluated to '!POINT_DIRECTION_PAIR' entities. All of these entities that belong to one face, are now collected in the "point_direction_pair" attribute of the '!RENDER_FACE' entity, which is created for each 'FACE' entity in this case (if there is no "facet_colour_list" in the file or if its respective entry does not determine a colour, the "colour" attribute is mapped from "material_index" which is available in any case).

Finally, for each polyhedron model an 'INDEX_ENTRY' entity is created which assigns the name of the KISMET geometry file (without extension) to the 'FACETED_BREP'.

Mapping of kinematic models (configuration files)

The mapping of the KISMET configuration files to STEP is much more complicated. First, any root file (no matter which type) is mapped to a 'KINEMATIC_MODEL' entity provided only that there is at least one ROB_file in the model tree. (If the model tree contains no ROB_file, the STEP file will contain only 'FACETED_BREP's and 'SOLID_INSTANCE's but no kinematics.) If the root file is of type BASE_file (the normal case) its model base co-ordinate system (given by six parameters) is mapped to an 'AXIS2_PLACEMENT' that is used as the "ground_frame" attribute of the 'GROUND' entity. If, however, the root file is of type FRAME_file or ROB_file, a unit transformation is assumed for the missing base co-ordinate system. Since the ground frame is likely to be referenced from outside the 'GROUND' scope it is included into an 'EXPORT_LIST' entity. Any BASE_file found in the tree apart from the root will cause the pre-processor to issue an error message and to stop.

Each ROB_file in the tree will be mapped to a 'MECHANISM' entity in the scope of the 'KINEMATIC_MODEL'; details are given below. FRAME_files appearing outside the root will not be mapped explicitly.

If the root file is not of type ROB_file, any shape information referenced by the "abstract frames" of the root file will contribute to the "geometric_representation" attribute of the 'GROUND' entity. This is done by means of 'SOLID_INSTANCE' entities, whose 'TRANSFORMATION' results from the evaluation of the transformation chain starting at the base co-ordinate system and ending at the geometry reference. The same holds for the geometry references in any hierarchically dependent FRAME_file, if none of its "parent files" up to the root is a ROB_file. There is one exception to this mapping rule: it concerns "abstract frames" denoted as "tools" (see below). Apart from this exception any "work frames" contained in the "abstract frames" of these files will not be mapped.

Mapping of mechanisms

The equivalent to a STEP mechanism is the ROB_file in KISMET.

The situation is different for the contents of ROB_files. As already mentioned above, the ROB_file itself is mapped to a 'MECHANISM' entity. Its attribute "reference_frame" depends on the position of the ROB_file in the tree. If none of its parent files is another ROB_file, the ground frame will be addressed; otherwise, an "additional_frame" must have been created at a link of the 'MECHANISM' that represents the parent ROB_file and be included there in an 'EXPORT_LIST'. In both cases, any intervening transformations must be taken into account when evaluating the "actual_position" attribute of the new 'MECHANISM' entity.

In the scope of this entity, first a default 'AXIS2_PLACEMENT' is inserted, representing a unit transformation. This is due to the different modelling approaches for kinematic chains in KISMET and STEP. KISMET is using a Denavit-Hartenberg like notation (see [Kühnapfel 1991A]) which gives a transformation from the axis of the root-sided joint of a link to any position connected to the link. This position is used by the pre-processor as the (only implicitly given) "link frame"; frequently it is coinciding with the axis position of another joint, which then will be placed in the STEP file using the default 'AXIS2_PLACEMENT'.

Mapping of rigid parts and rigid connections from KISMET

MOD_abs frames in the KISMET terminology when they appear in a ROB_file correspond roughly to the base of a mechanism or to a rigid extension of a kinematic link. While the base is a STEP concept, the rigid extension of a link has to be treated specially.

Usually the first "abstract frame" in a ROB_file is a "MOD_abs frame" relating its local co-ordinate system to the parent system. In this case, the parameters defining the local co-ordinate system are mapped to another 'AXIS2_PLACEMENT' that defines the link frame of the "base link" and will be referenced as the "actual_position" attribute of the 'MECHANISM'. If the first "abstract file" is already a "KIN_abs frame", a virtual "MOD_abs frame" is inserted with a unit transformation, but without any references to geometry files. In both cases, of course, the base link frame has to be evaluated in such a way that any intermediate transformations in the relation to the reference frame are reflected. In any case, the "base link" will be represented as a 'KINEMATIC_LINK' entity that will be referenced as the "base" attribute by the 'MECHANISM' entity.

Any other "MOD_abs frames" in the ROB_file have to be dealt with differently. Their transformation parameters are also mapped to an 'AXIS2_PLACEMENT', but this has to reflect the transition from the relevant link frame to the local co-ordinate system of the MOD_abs. The meaning of the relevant link frame is evaluated from the "predecessor" statement in the MOD_abs:

- If the transformation parameters are also related to the parent system, then the MOD_abs becomes a part of the base link, and the 'AXIS2_PLACEMENT' gives the transition from the base link frame to its local co-ordinate system (evaluated by appropriate matrix inversions and multiplication's).
- If the MOD_abs denotes the first MOD_abs or any other MOD_abs related to the first MOD_abs as its predecessor, then it becomes also part of the base link. The 'AXIS2_PLACEMENT' has to reflect the transformation chain from the base link frame to the local co-ordinate system (some matrix multiplication's have to be performed).
- If the MOD_abs is related to a KIN_abs (immediately or through intervening MOD_abs frames) it becomes a part of the link constituted by that KIN_abs. The 'AXIS2_PLACEMENT' is evaluated according to the transformation chain from that link frame to the local co-ordinate system.

Geometry references within a MOD_abs are mapped to 'SOLID_INSTANCE' entities, where the transformation matrix defined by the parameters in the file has to be pre-multiplied by the matrix corresponding to the respective 'AXIS2_PLACEMENT' frame (except the first MOD_ABS). Geometry references become part of the "geometric_representation" attribute of the respective 'KINEMATIC_LINK' entity.

Again an exception applies for MOD_abs frames that are denoted as tools.

Mapping of link/joint connections from KISMET

Since KISMET is based on the DH-approach to defining kinematics the information about a joint and one associated link is not separate (as in STEP) but more closely integrated. It is kept in so called "KIN_abs Frames" in a ROB_file.

For each "KIN_abs frame" in a ROB_file, the following is done:

- Evaluate the DH-parameters to a transformation matrix describing the transition from the root-sided joint of the link to the link frame. Invert this matrix and map the inverted matrix to an 'AXIS2_PLACEMENT', which will be used to give the joint position with respect to the link frame.

- Depending on the "kin_type" stated in the KIN_abs, create a 'REVOLUTE_PAIR' or a 'PRISMATIC_PAIR' entity that uses the initial position and the lower and upper limits from the KIN_abs as its attributes (after some unit conversions).
- Create two 'PAIR_PLACEMENT_STRUCTURE' entities both of which reference the pair just created in their attribute "pair". In the first entity, the attribute "placement" has to reference that 'AXIS2_PLACEMENT' that gives the pair placement with respect to the link frame of the preceding link. If the "predecessor" in the KIN_abs addresses either the first MOD_abs (constituting the base link) or another KIN_abs, the default unit transformation is used here which has previously been inserted at the beginning of the 'MECHANISM' scope. In all other cases, the 'AXIS2_PLACEMENT' created for the preceding MOD_abs has to be addressed. The second entity addresses that placement that has just been evaluated from the DH-parameters of the present KIN_abs.
- Create a 'KINEMATIC_LINK' entity. In its attribute "pairs", all pair entities that belong to this link are referenced (according to the root-sided joint defined by the KIN_abs as well as to all immediately succeeding joints. These are defined by any KIN_abs denoting the present KIN_abs or a MOD_abs dependent on it as its predecessor). Whether the "additional_frames" attribute is omitted or not, depends on the presence of "work_frames" associated with the link. The "geometric_representation" attribute collects all 'SOLID_INSTANCE's according to the geometry references in the present KIN_abs and in the MOD_abs frames dependent on it.
- Create a 'KINEMATIC_JOINT' entity which denotes the preceding link entity (root-sided) as its "first_link", the link entity just created as its "second_link", and the root-sided pair entity as its "pair" attribute".

Following these steps, any simple or branched open kinematic chain may be converted from the KISMET ROB_file into the scope of a STEP 'MECHANISM' entity. However, KISMET allows for modelling some special kinematic meshes, too. These are planar meshes that connect two neighbouring links in the "primary chain" (which is modelled as usual) with a "kinematic by-pass" consisting of other two links. The two links in the primary chain must be connected by a 'REVOLUTE_PAIR'; the joint in the by-pass may be of revolute or prismatic type. Each end of the by-pass is connected to one of the links in the primary chain by means of revolute joints whose axes must not coincide with any other joint axis participating in the mesh. Such meshes may be cascaded, thus allowing for rather complex kinematic structures.

Formally, the KISMET model of such a mesh looks like an open branched chain where the by-pass forms one branch. However, the two KIN_abs frames defining the by-pass are marked appropriately, and the second one denotes a MOD_abs frame depending on the second link in the primary chain explicitly as its "successor". Thus, the closing joint of the mesh is defined implicitly. In the STEP model, it has to be mapped explicitly to a 'REVOLUTE_PAIR' entity with almost arbitrary attributes (we use 0.0 for the "actual_rotation", -2p for "lower_limit_actual_rotation", and 2p for "upper_limit_actual_rotation"). This pair is placed by two 'PAIR_PLACEMENT_STRUCTURE' entities, the first one referencing the above mentioned unit transformation entity (pair placement with respect to the link frame of the second link in the by-pass), the second one addressing that frame entity corresponding to the local co-ordinate system of the MOD_abs denoted as by-pass successor. Finally, a 'KINEMATIC_JOINT' entity is created ("first_link": second by-pass link; "second_link": second link in the primary chain; "pair": the closing 'REVOLUTE_PAIR' just created).

Now, after all "abstract frames" in the ROB_file are dealt with, a 'KINEMATIC_STRUCTURE' entity is created which references all 'KINEMATIC_JOINT' entities of the present 'MECHANISM' in its "joints" attribute. Then all additional frames and '!TOOL_ATTACHMENT_FRAME' entities eventually created are collected in an 'EXPORT_LIST' entity, as they are likely to be referred to from outside the scope of the 'MECHANISM'. Now this scope is closed, the 'MECHANISM' keyword is written, and its attributes are added as stated before. The "kinematic_structure" attribute addresses the corresponding entity just created. Finally, the "ROB_name" associated with the kinematic mechanism in the ROB_file is attached to the 'MECHANISM' entity by means of an 'INDEX_ENTRY' entity.

Mapping of "work frames"

Work frames in KISMET represent gripping positions, tool attachment points, or tool centre points. Only the latter ones are included in the NIRO/STEP specification.

As mentioned above, any "abstract frame" in any configuration file of the KISMET model tree may include the definition of any number of "work frames" that may be used for various purposes. The data in such a work frame definition relevant to the present pre-processor are the "WFR_type" denoting the intended usage of the work frame, and "trans_par", a transformation given by six parameters and defining a frame (i.e., a Cartesian axis co-ordinate system and its origin position) with respect to the local origin of the "abstract frame" containing the definition. The other work frame data are not relevant here and are ignored by the pre-processor.

The mapping of such work frames to the STEP file depends on both the value of "WFR_type", and the position of the work frame definition in the model tree. If any one of the work frames defined in an "abstract frame" is marked by its WFR_type as a "tool base frame", the whole "abstract frame" is separated from the remaining model and mapped to a '!TOOL' entity as described below. This happens independently from the type of the containing configuration file and from the position in the model tree. If none of the work frames indicates this usage, the position in the tree and the file type become relevant and influence the conversion into the STEP representation.

Again, "abstract files" in a ROB_file or in a FRAME_file dependent on a ROB_file are treated differently. All work frame definitions found here, except those belonging to a tool, are mapped as additional frames within the 'MECHANISM' scope, where the 'AXIS2_PLACEMENT' corresponds to "trans_par", pre-multiplied by the transformation denoting the transition from the respective link frame to the local co-ordinate system of the defining "abstract frame". The intended usage is indicated by an 'INDEX_ENTRY' entity pointing to this 'AXIS2_PLACEMENT', which is also referenced in the "additional_frames" attribute of the respective 'KINEMATIC_LINK'. If WFR_type demands it, this placement is also referenced by a '!TOOL_ATTACHMENT_FRAME' entity. As mentioned above, all 'AXIS2_PLACEMENT' entities used as additional frames, and all '!TOOL_ATTACHMENT_FRAME' entities are included in an 'EXPORT_LIST'.

Creating '!TOOL' entities

If an "abstract frame" (which has to be a "MOD_abs frame") within a configuration file of any type is marked as defining a tool, the following steps are performed:

1. create a '!TOOL' entity and open its scope;
2. map the transformation parameters to an 'AXIS2_PLACEMENT', which will represent the "unmounted_position" attribute of the '!TOOL'. This "unmounted position" will show the tool in the STEP model exactly at the same place as it has been stored in the KISMET model, i.e., it is related to the ground frame;
3. map all "work frames" defined within the MOD_abs (including the "tool base frame") to further 'AXIS2_PLACEMENT' entities, accompanied by 'INDEX_ENTRY' entities indicating their usage. These placements will be addressed by the "tool_centre_points" attribute and are related to the local co-ordinate system of the MOD_abs that is used as a tool. Therefore, they are mapped directly from the "trans_par" parameters of the "work frames", without any transformation concatenation;
4. all geometry references within the MOD_abs are mapped to 'SOLID_INSTANCE's whose 'TRANSFORMATION' entities correspond again directly to the transformations given in the references, without any concatenation. These 'SOLID_INSTANCES' correspond the "geometric_representation" attribute of the '!TOOL';
5. close the '!TOOL' scope, add the keyword and the list of attributes. For the last attribute "delay_parameters" a suitable number of more or less meaningful default values (selected by the program) is inserted immediately, since this information is not available in KISMET.

This mapping procedure assumes that the "tool base frame" is a unit transformation. In other words, it assumes that the local origin of the MOD_abs defining a tool position coincides with the tool attachment frame on which it is mounted. This convention (which is not implied in KISMET) has been introduced to facilitate the mapping.

In the STEP model, all '!TOOL' entities are "unmounted", no matter, whether their "unmounted position" corresponds to a mounting position or not. '!MOUNT_TOOL' entities are not created by the pre-processor.

Special mapping problems

Obviously, the mapping procedures described in the previous section cannot be performed sequentially. Rather, some data have to be stored temporarily until all information required for the proper mapping has been retrieved from the input files. Furthermore, the hierarchical structure of the input model requires some recursive actions, and some data have to be kept across several levels of recursion.

This applies in particular for the contents of the configuration files. As one may expect, mapping of KISMET geometry files is rather straight-forward, since all information needed is contained implicitly or explicitly in one single file. Nevertheless, some problems had to be solved.

First, because NIRO has restricted itself to the 'FACETED_BREP' representation of shape models, only 'POLYHEDRON' and 'EXTENDED_POLYHEDRON' models can be transferred to the STEP file. If another type of shape model supported by KISMET is input, the pre-processor ignores it.

Furthermore, KISMET does not require its polyhedral models to form two-manifolds or even to have closed shells (as its primary goal is the real-time visualisation). Thus, faces of a polyhedron that may never be seen in a scenario, are often omitted to yield better performance. In such a case, the polyhedron degenerates to a "faceted oriented surface", which cannot be converted to a valid STEP 'FACETED_BREP' entity. Therefore, each KISMET polyhedron model read by the pre-processor is (optionally) checked for the necessary (though not sufficient) condition whether all of its edges are referenced exactly twice with opposite directions. A violation to this rule causes an error message to be issued to the LOG-file (and to the screen) and the pre-processing to be stopped. This test does not include a check for manifolds, i.e. two disjoint polyhedra combined in one KISMET 'POLYHEDRON' model cannot be detected by this check.

Another test concerning polyhedral models concerns different vertices having (almost) identical co-ordinates. In most cases, such coincidences have been introduced unintentionally, but it is hard to detect them in the KISMET environment, especially in a huge vertex list. Therefore the pre-processor includes an (optional) check for such coincidences (where "coincidence" is defined by a reasonable small distance tolerance, which can be input in the pre-processor control file). If two vertices of a polyhedron coincide, the second one is removed from the vertex list, and all references to it are replaced by references to the first one. This correction, however, does not always result in valid polyhedra. Therefore, a warning message is issued to the LOG-file. In most cases, the user would be well advised to correct the polyhedron in the modelling environment.

Most mapping problems with the configuration files arose from the different topological and kinematical modelling approaches in KISMET and STEP. They caused many transformations to be concatenated before the mapping, and sometimes the proper transformation chaining cannot be determined before a huge amount of other information has been read and stored suitably in internal data structures.

A very special mapping problem arose with the planar kinematical meshes provided by KISMET. As mentioned above, the characteristics of the closing joint of such a mesh are defined only implicitly, and some of them cannot be found at all in the KISMET file, because KISMET does not need them for this passive joint. In this case, the implicit values have to be evaluated, and the missing values are substituted by some reasonable defaults. From the usage of such a mesh it is known that this joint will never experience a full rotation. It is sufficient therefore, to set the limits to two full rotations and the initial value to the middle of this interval.

Embedding the processor in the KISMET environment

The link of the KISMET pre-processor to the KISMET environment is given by the control file that is read a the beginning of a pre-processor run. In this file, the paths, where the various model file types are to be found, and where the STEP file is to be stored are specified. If these paths correspond to the ones used by KISMET, an immediate connection is established (see also Sect. 5.2.3, where the usual KISMET subdirectory structure is shown).

5.2 From STEP to Robot Programming and Simulation

5.2.1 GRASP

P. Sorenti

Overview

A general introduction to the GRASP system is given in chapter 5.1.1. There are two approaches developing a post-processor from a neutral file format to a proprietary application package, namely:
1. modifying the application so that it can read the new file format directly,
2. translating, by the use of a stand-alone program, the neutral file format into the input language and format recognised by the application.

BYG Systems elected to use the latter approach for the STEP to GRASP post-processor. The program reads ASCII STEP files as input, and converts the data into a GRASP source file that may subsequently be read into GRASP. Thus the processor is a stand-alone tool that may be used independently from the simulation system. The programming language used is 'C' and thus the software can easily be ported to a UNIX workstation. Current platforms tested and supported include the SUN SPARCstations, Hewlett Packard and IBM Risc System 6000 workstations.

Method of model output from a STEP data file to GRASP

The post-processor performs three main actions:

1. Stage 1 scanning - Syntax checking
2. Stage 2 parsing - Reference checking
3. Output of data to GRASP source file

(1) Stage 1 scanning - Syntax checking. The STEP input file is scanned through and the post-processor builds up an internal data structure for the model held in the STEP input file. The user is warned of any errors as they occur. If the errors are not fatal, an automatic recovery is attempted and the remaining section of the file will be processed as normal. For example, a syntax error in the specification of a 'SOLID_INSTANCE' in the STEP file is not considered fatal. However a syntax error in the data keyword section would cause processing to be aborted.

The primary aim of the STEP to GRASP post-processor is to maximise the amount of information retrieved from STEP data files even if errors are found. Therefore, wherever possible, syntax errors or referencing errors are not considered to be fatal errors.

(2) Stage 2 parsing - Reference checking. Since forward referencing is valid in a STEP file, it is impractical for stage 1 scanning to check for invalid referencing. An example of forward referencing is a 'SOLID_INSTANCE' occurrence in the STEP file which references a faceted boundary representation ('FACETED_BREP') that is defined later in the file.

During stage 2 parsing the internal data structure built up by the STEP to GRASP post-processor is traversed and checked for invalid referencing. Invalid referencing means that either the referenced entity does not exist at all in the STEP file, or it is not of the appropriate type.

Stage 2 parsing also checks for invalid scope referencing. For example, the referencing by an 'AXIS2_PLACEMENT' occurrence of a 'DIRECTION' occurrence that does not lie in the scope section of the 'AXIS2_PLACEMENT' entity is invalid. Invalid scope usage is not a fatal error. The user is warned and given details of the invalid referencing.

(3) Output of data to GRASP source file. Once the scanning and parsing stages have been completed by the STEP to GRASP post-processor, the internal data structure is output to a file in the GRASP source file format.

There is a simple and efficient mapping between the entities held in the STEP data files and the equivalent model entities held in the output GRASP source file. All STEP solids model occurrences are output as a GRASP-specific geometric entity known as a General Module. All STEP mechanisms in kinematic model occurrences are output as GRASP robots. 'INDEX_ENTRY' occurrences are accounted for when the subject occurrence is output. Therefore all entities are output to the GRASP source file under their 'INDEX_ ENTRY' name if applicable.

Most other entity types have a simple mapping to a GRASP geometric entity. However, some STEP user defined entities do not have any GRASP equivalent, for example, the 'RENDER_FACE' definition. The STEP to GRASP post-processor will parse such entities but they will not subsequently be used within GRASP.

Capabilities of the software

All entities defined in the NIRO-STEP specification are recognised by the STEP to GRASP post-processor. However, some entities have no equivalent in the GRASP source file and are ignored. This does not affect the geometric model read in and used by the simulation system.

The principal focus on the work of the NIRO project in the development of STEP processors has been the definition of the kinematic structure of the robot in the neutral file format and the post-processing of this information to the robot simulation system. Open loop mechanisms, mechanisms owning mechanisms, tool and tool centre point structures defined in the STEP file are also post-processed. The resulting kinematic structure has been shown to work correctly in the simulation system using joint level, tool level and object level control. Object level control involves defining the position of the robot-tool-centre-point relative to the reference frame of some other object, or the global reference frame. Individual joint positions are then calculated using inverse kinematics. The methodology adopted by the GRASP system is that of Denavit - Hartenberg [Denavit and Hartenberg 1955].

Known limitations and problems

The major outstanding problem that has not been addressed within the scope of this project is the handling of closed kinematic chains. Such mechanisms may be specified in the GRASP source file by defining the driving mechanism and its relationship to the controlling mechanism. Both driving and controlling mechanisms are thus simulated correctly using all levels of tool-centre-point (TCP) control. Closed kinematic chains may also be defined in STEP but there is no method of determining which joints from the controlling mechanism and which joints from the driving mechanism. This problem must be addressed in future revisions of the STEP neutral file format.

The STEP to GRASP post-processor is able to identify closed loop mechanisms during the second stage of parsing the input file. Since it is not possible to derive the kinematic structure, only the geometry of the robot is transferred with a warning issued to the user during the output stage. An alternative approach would be to sever the kinematic chain using some arbitrary rule but this is seen by BYG as a misleading approach since the final model imported to the receiving system will be incorrect whilst the user will assume it is correct. It is far better to detect conditions such as this that cannot be consistently or accurately resolved and warn the user that such a problem has been detected.

Demonstration systems

The STEP to GRASP processor features in two of the industrial demonstration systems set-up in NIRO (see chapter 8). The first of these demonstrations is based at the Odense Steel Shipyard, Denmark. STEP files describing the section of a ship to be welded were generated by the HICADEC-H system for ship hull design and translated by the GRASP post-processor. The model of the robot cell was taken from the CATIA system, again via STEP, and imported into GRASP. These two models originating from two completely different systems were then used to generate the robot programs required for post processing to ICR (Intermediate Code for Robotics, a low-level neutral interface for robot programs).

In the other demonstration, a Reis robot was modelled in the BRAVO3 CAD system. This was output to a STEP file along with the model of a "solitaire" game board and the playing pieces. In this case, the robot programs generated in GRASP were post processed to IRL (a higher level , human readable form of robot program definition). A detailed description of both of these demonstrations may be found in Chapter 8.

Library of test models

The STEP to GRASP processor has been tested using a library of test models set up within the project. The results on the processing of these test files by GRASP and other software systems in the project are described in detail in Chapter 5.3. A summary of the results however is that a high degree of success has been achieved in the transfer of data between systems via the STEP interface.

Summary

The STEP processor to the GRASP planning and simulation system has been successfully tested using the library of models set up within the project. Further, the industrial demonstrations have shown the practical need for, and success of the data transfer within the wider context of the complete production process from computer aided design to real robot control.

5.2.2 CATIA Robotics

R. Lutz

System description

Please refer to Sect. 5.1.4.

Description of the post processor

The CATIA post processor as one BACKEND module of a common post processor schema in the project is written in FORTRAN 77 and employs the

CATGEO/CATMSP FORTRAN subroutine library. It is operational as well under UNIX/AIX on an IBM RISC System/6000 as under VM/SP on an IBM 9375. Several files are used to manage the INPUT/OUPUT of the postprocessor (fig. 5.2.2-1) :

STTREE.DATA:	STEP parse TREE file, generated by the SCANNER/PARSER
CATPOST.ERR:	List of possible error messages
CATPOST.KEY:	List of all used NIRO/STEP keywords
CATPOST.PRM:	input parameter list for boundaries and extensions of the dynamic arrays
CATPOST.TMP:	temporary data file which stores the input parameters like the name of the neutral STEP file, the CATIA USER, GROUP, PASSWORD, MODEL name, ... It is used to combine the two parts of the common post processor schema (SCANNER/PARSER and CATIA backend) to have only one processor run.
CATPOST.MEM:	All input parameters of the interactive dialog are stored for the next processor run
CATPOST.LIST:	Protocol and Statistics file

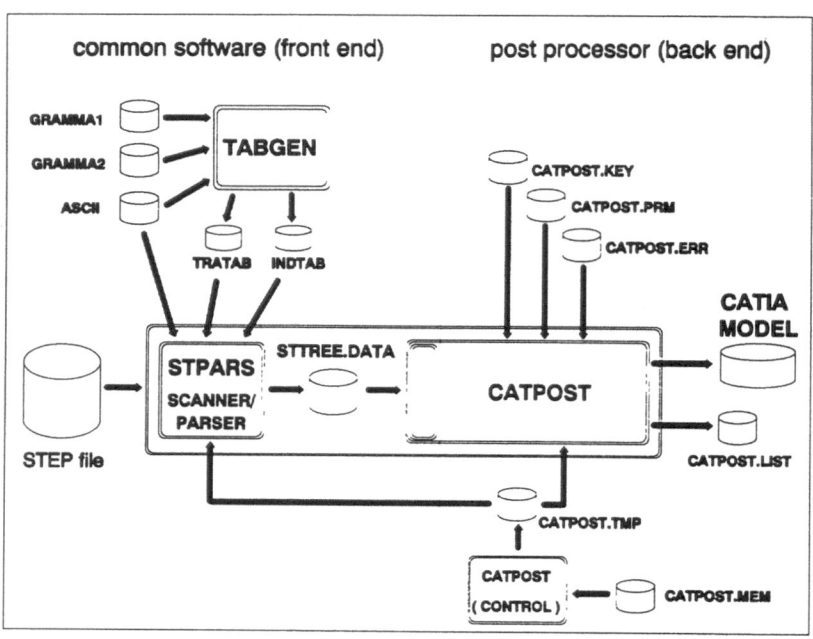

Figure 5.2.2-1: Environment of the CATIA NIRO-STEP post processor

The post processor is started using a commands file (AIX: shell script, VM/SP: REXX exec file). This commands file combines the two parts of the common post processor schema to have only one processor run (SCANNER/PARSER and CATIA backend, refer to fig. 5.2.2-1). First the SCANNER/PARSER module is invoked which produces the direct access tree file. This file carries the neutral STEP file contents in a structured and syntactic tested form (refer to chapter 5.2.4: SCANNER/PARSER for CATIA and KISMET). Then the CATIA backend module is started if no error messages occured while the SCANNER/PARSER run. First the post processor scans the tree file until a relevant NIRO/STEP entity is found (FCTBRP, SOLID_INSTANCE, KINEMATIC_MODEL,..). This entity is created while the entity specific structure is built using dynamical allocated reference and parameter arrays. A reference table carrying the already created entity names is used to refind the backpointers of following entities in the TREE file.

All referenced SOLIDs are organized in CATIA SETs according to their grouping in sets of SOLIDs in the STEP file (corresponding either to a KINEMATIC_LINK or to the GROUND representation). CATIA requires LINE entities to define a revolute or a prismatic KINEMATIC_JOINT. These lines are created within the PAIR_PLACEMENT frame which is related, in turn, to the respective LINK_FRAME of a KINEMATIC_LINK.

The post processor creates CATIA ROBOTs, because it is not possible to create any CATIA KINEMATICS entity using CATGEO/CATMSP subroutines. Hence it is necessary to work upon the created CATIA model to get a KINEMATIC model. Fortunately the creation of a KINEMATICS model is quite similar to the creation of a CATIA ROBOT. It is based on the linking of two CATIA SETs (being the KINEMATIC_LINKs) with a KINEMATIC_JOINT. The first SET is represented by a CATIA LINE being the KINEMATIC_PAIR of the KINEMATIC_JOINT (i.e. this LINE resides in this SET (LINK)). To ease the creation of KINEMATIC_SETs in CATIA all KINEMATIC_LINKs (CATIA SETs) and KINEMATIC_PAIRs (CATIA LINEs) of the STEP file get an own CATIA identifier when they are created:

CATIA identifier:	Meaning:
"KL_xx/Ryy"	means the "XX"th KINEMATIC_LINK of the ROBOT "yy" (Example: "KL_03/R01", the 3rd KINEMATIC_LINK of ROBOT 1)
"P_xxxxx" or "R_yyyyy"	PRISMATIC_PAIR with the NIRO/STEP name "xxxxx" or REVOLUTE_PAIR with the NIRO/STEP name "yyyyy" (Example: "P_00382" represents the NIRO/STEP entity: #382= PRISMATIC_PAIR(0.000E+00,-6.000E-01,6.000E-01);)

"GROUND"	identifier of the SET representing the GROUND. All SOLIDs and SOLID_INSTANCEs which are not referenced by a KINEMATIC_LINK reside within this CATIA SET.

Mapping the NIRO/STEP data into CATIA

The relevant NIRO/STEP entities are mapped as follows:

NIRO/STEP entity	CATIA entity	remarks
INDEX_ENTRY	CATIA identifier	used for SOLIDs, MECHs MECHANISMs, SOLID_INSTANCES
SOLID_INSTANCE	DETAIL and DITTO	a SOLID is transferred into a DETAIL and a DITTO is created
FACETED_BREP	SOLID	SOLID (POLYHEDRON) without history
KINEMATIC_MODEL	ROBOTIC SET	
GROUND	SET	with the SET-ID: "GROUND"
MECHANISM	CATIA ROBOT	i.e. a CATIA ROBOTICS SET
KINEMATIC_JOINT	JOINT	
KINEMATIC_LINK	SET of SOLID(s)	with the SET-ID: "K_L_..."
PAIR_PLACEMENT_ STRUCTURE	local AXIS system	
KINEMATIC_PAIR	JOINT	the kind of JOINT implies the kind of KIN_PAIR combined with the occurence of a CATIA line
COLOR_ATTRIBUTE	color attribute	the color attribute is assigned as a graphic attribute

All other NIRO/STEP entities are ignored (FILE_DESCR., FILE_NAME, . . .) or they are considered as auxiliary low-level entities to define another entity of a higher level (CARTESIAN_POINT, POLY_LOOP, ...).

To ease the association of the created CATIA entities with the NIRO/STEP entities of the neutral file most of the entities get an own identifier if there is no INDEX_ENTRY entity which assigns an identifier explicitly to them. This standard identifier is composed of a character string and a number which represents the STEP name or the type enumeration of the entity. In the following table "xxxxxx" means the STEP name of the entity and "yy" and "zz" means the type enumeration of the entity:

STEP entity	CATIA identifier
FACETED_BREP	"SOLID_xxxxxxx" or "WIREF_xxxxxxx" (if the SOLID creation failed, the identifier "WIREF_..." is assigned to the CATIA DETAIL and DITTO)
SOLID_INSTANCE	"SOL_xxxxxx" (or "WIREF_xxxxxx") is assigned to the created CATIA DETAIL and "SOL_I_xxxxxx" is assigned to the created CATIA DITTO
MECHANISM	"ROBOT_yy" assigned to the ROBOTICS SET
KINEMATIC_LINK	"K_L_yy/Ryy" (Example: "K_L_03/R01": 3rd K_LINK of the 1st robot)
KINEMATIC_PAIR	"P_xxxxxx" or "R_xxxxxx"
GROUND	"GROUND"

Mapping problems in the processor

Usually the post processor works successful. It turned out, however, that CATIA is extremely sensitive to polyhedron creation. CATIA expects an absolutely correct input table with the CATIA-specific polyhedron description (references, face normal vectors, coordinates ...). Containing one wrong normal vector (i.e. a wrong loop orientation) or one point out of the plane of the face to which it belongs, CATIA refuses this table whith no clear error message. In these cases the post processor replaces the corrupted polyhedron by a corresponding wireframe to recognize the shape of this polyhedron. This replacement has no influence on the ROBOTICS contents and structure of the CATIA model because CATIA only moves all the entities whithin a CATIA (geometric) SET.

Another problem is caused by the system restrictions of the CATIA ROBOTICS module. It does not allow the creation of kinematic loops. Although the post processor in principle is able to support closed kinematic loops when creating the corresponding KINEMATIC_LINKs it is not possible to create the resulting KINEMATIC_JOINT with the ROBOTICs-related subroutine.

Embedding the post processor in the CATIA environment

To get an executable post processor module it is necessary to link the CATGEO/CATMSP library subroutines of the CATIA modules BASE, 3D DESIGN and ROBOTICS with the backend source code. During a post processor run a CATIA database (model) is opened and all entities are created in this CATIA-specific database with an own address and entity name. This database is stored in a single data file using the CATIA binary form.

Several input files are used when the post processor is started (refer to Sect. 5.2.2). They contain the runtime variables and parameters for the allocation of the dynamic arrays and the CATIA-specific model parameters (index table and data table, model unit,...) which are necessary for the creation of a CATIA database (model).

5.2.3 KISMET

H.-P. Lorenz and S. Haas

System description

KISMET (Kinematic Simulation, Monitoring and Off-line Programming Environment for Telerobotics) is a software tool for effective planning, simulating, programming, and monitoring remote handling equipment, industrial robots, and various other types of mechanisms. KISMET was originally developed at the Nuclear Research Centre Karlsruhe (KfK) for nuclear applications to support remote maintenance in hazardous environment, in particular in fusion plants. The present KISMET version V4.0 requires a Silicon Graphics IRIS 4D Workstation running IRIS 3.3. KISMET allows for a real-time, synthetic rendering of any view of handling and manufacturing cells as shaded display (Gouraud shading), as wire frame or as transparent models. Furthermore a ray tracing module is implemented to produce photo realistic images. Through sensor signals from robot controllers it is possible to generate in real-time synthetic views from work cells with changing scenarios.

Mechanical structures are normally simulated with rigid links. The simulation module of KISMET offers a real-time simulation and the modelling of control characteristics of the robot system. Another feature of the most recent version of

KISMET is the modelling and real-time simulation of elastomechanic effects like robot link torsion and bending. The concurrent simulation of any number and types of robots and other mechanisms, including tree-like mechanical structures and planar closed loops is possible.

Description of the post-processor

The KISMET post-processor as one back-end of the common post-processor schema (detailed description in Sect. 5.2.4) is written in C. Only the routines to access the direct access input file generated by the front-end are written in FORTRAN 77. The post-processor was developed on a VAX 8350 using ANSI FORTRAN and ANSI C language and is operational under UNIX/IRIS on Silicon Graphic workstations as well as under VMS on DEC-VAX systems.

The post-processor translates geometric, topological and kinematic data from STEP format to the KISMET native file format. The geometry part of the specification is restricted to polyhedron models and the kinematics supports only "revolute pair" and "prismatic pair" types of joints. The post-processor can cope with polyhedron models and solid instances, linear and branched kinematic chains within a mechanism, comprising revolute and prismatic joints (including their relevant information about the actual positions and joint limits) as well as chains of mechanisms contained in a kinematic model. Closed mesh kinematics within a mechanism is translated as open branched chains into the native file format. Moreover, the transfer of colours and identifiers of the solids is implemented. Several files are used to manage the INPUT/OUTPUT of the post-processor (Fig. 5.2.3-1)

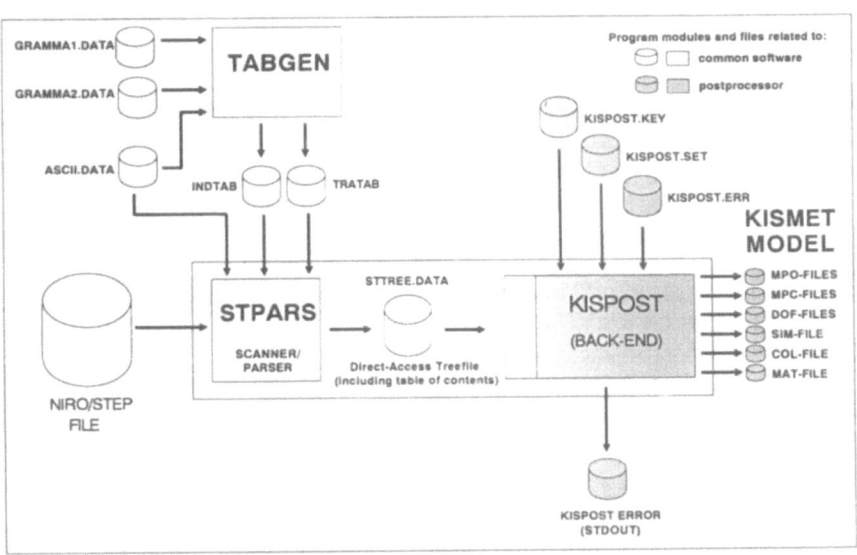

Source: KfK

Fig. 5.2.3-1. Environment of the KISMET STEP post-processor

INPUT-files:

STEP file	-	a direct access file generated by the post-processor front-end (SSTTREE.DATA, these names in parentheses are the names as shown in figure 5.2.3-1)
error table	-	containing the possible error messages (KISPOST.ERR)
set-up file	-	containing the input parameters (KISPOST.SET)
keyword file	-	containing the STEP keywords which are used in the NIRO-STEP specification (KISPOST.KEY)

OUTPUT-files:

control file	-	containing control information for the transferred model (simulation file)
colour file	-	containing the KISMET colour table of the transferred model (colour file)
material file	-	containing default surface properties (material file)
geometry file	-	containing the shape information (geometry files)
abstract file	-	containing topological and kinematics data (configuration files)
dof file	-	containing number and types of input degrees of freedom (degrees of freedom files)

Firstly, the post-processor reads the set-up file and the keyword file with processor relevant information. After that the STEP file is scanned until a relevant STEP entity is found. Then all its references are resolved and stored in STEP specific data structures. After reading the STEP file, the STEP data structure is mapped into the data structure of the KISMET native files and then the OUTPUT-files are written. If an error occurs during the processor run, the corresponding error message is searched in the error table and stored in memory until the program ends. Then all the stored error messages are sent to the standard output.

Mapping the STEP data into KISMET data

The following table shows the mapping of the relevant STEP entities of the NIRO specification to KISMET specific data as done by the post-processor [Kühnapfel 1991B]. The entities beginning with an exclamation mark are user-defined entities that have no counterpart in the original STEP specification.

STEP entity	KISMET data
INDEX_ENTRY	name of the GEOMETRY file
!RGBTBL	MATERIAL file/COLOUR file
!CLRATT	material_index
!S_MASS_PR	no mapping
!RGBCLR	COLOUR file/MATERIAL file
EXPORT_LIST	no mapping
SOLID_INSTANCE	GEOMETRY RECORDS
AXIS_PLACEMENT	allr_ABB
TRANSFORMATION	mod_trans
FACETED_BREP	GEOMETRY file
CLOSED_SHELL	no mapping
FACE	vertices_per_face_list
POLY_LOOP	vertex_index_list
CARTESIAN_POINT	3D_coord
!RENDER_FACE	facet_colour_list
!POINT_DIRECTION_PAIR	normal_index_list
KINEMATIC_MODEL	ABSTRACT file
GROUND	BASE file (BASE_file)
MECHANISM	ROBOT file (ROB_file)
KINEMATIC_JOINT	predecessor, successor
KINEMATIC_LINK	n_geos
PAIR_PLACEMENT_ STRUCTURE	DH_par, mod_trans
PRISMATIC_PAIR	kin_type, min_max, init_pos
REVOLUTE_PAIR	kin_type, min_max, init_pos
!TOOL	ABSTRACT records (abs_frame)
!MOUNT	abs_frame -> geo_spec
!ACTUATOR	DOF file (n_dof)
!L_MASS_PR	no mapping
!TAP_FRM	WORKFRAME records (wfr_spec)

Fig. 5.2.3-2: Mapping of STEP data to KISMET data. The keywords on the left-hand side are taken from the STEP specification; those on the right-hand side are specific to KISMET

The processor allows for the transfer of polyhedron models with any number of inner loops or "holes". Moreover, KISMET is very generous with inaccurate models. It is not necessary for usage in KISMET to define the polyhedron as a closed surface, unless the part is used for Boolean operations.

The use of the entity '!POINT_DIRECTION_PAIR' together with '!RENDER_FACE' enables the processor to generate an EXTENDED_POLYHEDRON in KISMET. This kind of geometrical shape is defined through a list of Cartesian points, a list of index pointers to describe the facet structure referring to the point list, a list of normal vectors that is used to define the surface normal at any defined surface point in the point list, and another index list for the surface normals. Such a representation can be used to define soft textures called GOURAUD-shading on the GEOMETRY surface.

Furthermore, the use of the '!RENDER_FACE' entity allows to define a different colour for any surface facet in KISMET.

The post-processor resolves any number of mechanisms inside a kinematics model and maps mechanism to mechanism references.

Mapping problems in the processor

As already mentioned the mapping of geometric data causes no problems. A mapping problem occurs if a colour attribute refers to a 'SOLID_INSTANCE' and not to a 'FACETED_BREP', because KISMET cannot assign a new colour to an instance of shape; the colour of the instance is the same as for the template.

A more complex problem is the mapping of a kinematic model. In STEP SU-parameters, proposed by Sheth and Uicker, are used to describe the kinematics (ISO Part 105 1990). This notation defines the relative position of adjacent joint co-ordinate systems for kinematic structures with prismatic and revolute joints by six parameters per joint co-ordinate system. KISMET describes the translational and rotational relationships between adjacent links based on the notation by Denavit and Hardenberg (the so-called DH-parameter) in a modified form by (Paul 1981). In this notation the transformation to the next joint frame is defined by four parameters, relative to the position of the z-axis of the predecessor frame.

Because the usage of a pure DH-notation is limited to simple kinematic chains, KISMET allows in open chains the definition of additional 6 degrees of freedom 'AXIS_PLACEMENT' transformations. Thus it is possible to connect any number of joints to a kinematic link. These transformations are used by the post-processor for mapping the SU-transformations. The DH-parameters created for KISMET is set to correspond to the unit matrix.

This approach enables to map any kinematics to KISMET, except for closed kinematic loops which require special treatment in KISMET. 3D-kinematic closed loops are converted to 3D kinematic open chains and have to be worked on interactively in KISMET to be properly integrated in the entire kinematic model.

Embedding the post-processor in the KISMET environment

Embedding the post-processor optimally in the KISMET environment requires the use of a few parameters that are defined in the set-up file. To achieve a smart model structure for KISMET it is useful to create for each KISMET model its own directory structure. The set-up file offers the possibility to set the home directory of a model and its subdirectories of the corresponding KISMET model files for the post-processor.

Another important parameter for KISMET is the orientation of the z-axis of the world co-ordinate system in KISMET. KISMET prefers the z-axis set perpendicular to the screen towards the user. The setting of this parameter called WORLD_FRAME depends on the sending system.

Other features of the post-processor are optional reading of unreferenced solids in the kinematic model and to read also CADEX-files, another ESPRIT project for the exchange of geometry data.

5.2.4 Scanner/Parser for CATIA and KISMET

H.-P. Lorenz and S. Haas

Introduction

A post-processor can normally be subdivided into two parts: The first one is the so-called front-end consisting of the scanner/parser for the physical STEP files. It reads the STEP data from the file and generates an intermediate data file which is syntactically correct. The second part is the so-called back-end (here the CATIA and KISMET post-processor), which reads the intermediate file, evaluates the data and writes it in CAD/CAE-system specific form. The interface, the intermediate file, between front-end and back-end is called treefile.

The common part of the software for the post-processors of CATIA and KISMET consists of a table generator program TABGEN, the scanner/parser STPARS (representing the front-end of the post-processors), and the treefile (a direct access file) access routines. This software package is based on the common software of the CAD*I project, a former project for the exchange of geometry data, and was written in FORTRAN 77.

The scanner/parser checks the grammar of STEP input files based on the tables generated by the table generator. Furthermore the front-end performs some useful tasks that normally are postponed to the back-end: The parser builds a table indicating the SCOPE structure of the STEP file. Another advantage is the enhancement for the back-end to deal with forward references.

Description of the table generator TABGEN

The table generator generates the parser input tables concerning the definition of the STEP specification.

The input to the table generator is:
- the alphabet definition
- the token definitions
- the statement structure on defined tokens

The input is realised by three files, ASCII.DATA, GRAMMA1.DATA and GRAMMA2.DATA.

The file ASCII.DATA defining the alphabet of the Front-end looks like this:

```
        C     B!S#$S&'()*+,-./DDDDDDDDDDS;S=SS
SUUUUEFUUUUUUUNUUUUUTUUUUUUS\SSU`LLLLLLLLLLLLLLLLLLLLLLLLLLLSSSSW
```

The position of the used letters corresponds to the position in the ASCII-code and the letters have predefined meanings (e.g., D = digit, U = uppercase, L = lowercase).

For environments that do not use the ASCII code (such as IBM mainframes), another alphabet definition has to be provided. A utility program called ASCEBC enables to generate a similar file with the EBCDIC alphabet definition, e.g., for computers using the VM or MVS operating system.

The token definitions in the file GRAMMA1.DATA look as follows:

```
INTEGER        = [ "+" ! "-" ] "D" { "D" } .

REAL           = [ "+" ! "-" ] "D" { "D" } "."
                 { "D" } [ "E" [ "+" ! "-" ] "D" { "D" } ] .

STRING         = "'" { "C" ! "B" ! "!" ! "S" ! "#" ! "$" ! "&" !
                 "L" ! "W" ! "(" ! ")" ! "*" ! "+" ! "," ! "-" !
                 "." ! "/" ! "D" ! ";" ! "=" ! "U" ! "E" ! "F" !
                 "N" ! "T" ! "\" } "'" .

NAME           = "#" "D" [ "D" [ "D" [ "D" [ "D"
                 [ "D" [ "D" [ "D" ] ] ] ] ] ] ] .

KEYWORD        = ( "U" ! "E" ! "F" ! "N" ! "T" ! "&" )
                 { "U" ! "E" ! "F" ! "N" ! "T" ! "D" } .

USER_DEF_ENT   = "!" ( "U" ! "E" ! "F" ! "N" ! "T" )
                 { "U" ! "E" ! "F" ! "N" ! "T" ! "D" } .

ENUMERATION    = "." ( "U" ! "E" ! "F" ! "N" ! "T" )
                 { "U" ! "E" ! "F" ! "N" ! "T" ! "D" } "." .

DEFAULT_TOKEN  = "$" .

DELIMITER      = ( "," ! ";" ! "=" ! "(" ! ")" ) .

COMMENT        = "/" "*" .

PRINT_CONTROL  = "C" ( "F" ! "N" ) "C" ..
```

Finally, the grammar rules defined in file GRAMMA2.DATA look as follows:

```
NIRO/STEP, Version L02V00"+CADEX"(ext.), S.Haas,H.-
P.Lorenz,KfK, 21.7.92


..
"STEP" ";" .

"HEADER" ";" .

"FILE_NAME" "(" "STRING" "," "STRING" "," STRING_LIST "," STRING_LIST
        "," "STRING" "," "STRING" "," "STRING" "," "STRING" ")" ";" .

"FILE_IDENTIFIER" "(" "STRING" "," "STRING" "," STRING_LIST ","
        STRING_LIST "," "STRING" "," "STRING" "," "STRING" ")" ";" .

"FILE_DESCRIPTION" "(" STRING_LIST "," "STRING" ")" ";" .
```

```
"FILE_SCHEMA" "(" STRING_LIST ")" ";" .

"CLASSIFICATION" "(" "STRING" ")" ";" .

"DATA" ";" .

"USE_AP" "(" "ENUMERATION" ")"  ";" .

"ENDSCOPE" ANY_ENTITY .

"NAME" "=" ANY_ENTITY .

"END_AP" "(" "ENUMERATION" ")" ";" .

"ENDSEC" ";" .

"ENDSTEP" ";" ..

ANY_ENTITY = "&SCOPE"                                        !
"!ACTUATOR" "(" "NAME" ","
    "DEFAULT_TOKEN" "," "DEFAULT_TOKEN" ")" ";"              !
"APPLICATION_PROTOCOL"
    "(" "STRING" "," "INTEGER"  "," "NAME" ")" ";"           !
"APPLICATION_INTERPRETED_MODEL"
                 "(" "STRING" "," "NAME" ")" ";"             !
"AXIS2_DIRECTION"
    "(" "NAME" "," "NAME" "," "NAME" ")" ";"                 !
"AXSPLZ" "(" "NAME" "," "NAME" "," "NAME" ")" ";"            !
"AXIS2_PLACEMENT" "(" "NAME" "," "NAME" "," "NAME" ")" ";" !

"CARTESIAN_POINT" "(" "REAL" "," "REAL" ","
    REAL_OR_DEFAULT ")" ";"                                  !
"CRTPNT" "(" "REAL" "," "REAL" "," "REAL" ")" ";"            !
"CLOSED_SHELL" "(" "(" REFERENCE_LIST ")" ";"                !
"!CLRATT" "(" "NAME" "," "INTEGER" ")" ";"                   !
"CLSSHL" "(" "(" "NAME" REFERENCE_LIST_REST2 ")" ";"         !

"DIRECTION" "(" "REAL" "," "REAL" "," "REAL" ")" ";"         !
"DIRCTN" "(" "REAL" "," "REAL" "," "REAL" ")" ";"            !

"EXPORT_LIST" "(" "(" REFERENCE_LIST ")" ";"                 !

"FACE" "(" DEFAULT_OR_OMIT "NAME" REFERENCE_LIST_REST2 ","
    NAME_OR_DEFAULT ")" ";"                                  !
"FACETED_BREP" "(" "NAME" "," "(" ")" ")" ";"               !
"FCTBRP" "(" "NAME" "," "DEFAULT_TOKEN" ")" ";"              !

"GROUND" "(" "NAME" "," REFERENCE_LIST_OR_DEFAULT ")" ";"  !

"INDEX_ENTRY" "(" "STRING" "," "NAME" ")" ";"               !

"KINEMATIC_JOINT" "(" "NAME" "," "NAME" "," "NAME" ")" ";" !
"KINEMATIC_LINK" "(" "(" "NAME" REFERENCE_LIST_REST2 ","
    REFERENCE_LIST_OR_DEFAULT ","
        REFERENCE_LIST_OR_DEFAULT ")" ";"                    !
"KINEMATIC_MODEL" "(" "NAME" "," "(" "NAME"
    REFERENCE_LIST_REST2 "," "DEFAULT_TOKEN" ","
                    "DEFAULT_TOKEN" ")" ";"                  !
"KINEMATIC_STRUCTURE" "(" "(" REFERENCE_LIST ","
                    "DEFAULT_TOKEN" ")" ";"                  !

"!L_MASS_PR" "(" "NAME" "," "REAL" "," "REAL" "," "REAL" ","
                 "REAL" "," "REAL" "," "REAL" "," "REAL" ","
                 "REAL" "," "REAL" "," "REAL" ")" ";"              !
```

```
"MECHANISM"  "("  "NAME"  ","  "NAME"  ","
                  "NAME"  ","  "NAME"  ")"  ";"                      !
"!MOUNT"  "("  "NAME"  ","  "NAME"  ","  "NAME"  ")"  ";"            !

"PAIR_PLACEMENT_STRUCTURE"  "("  "NAME"  ","  "NAME"  ")"  ";"   !
"PLANE"  "("  NAME_OR_DEFAULTNAME  ")"  ";"                         !
"!POINT_DIRECTION_PAIR"  "("  "NAME"  ","  "NAME"  ")"  ";"         !
"POLY_LOOP"  "("  "("  "NAME"  ","  "NAME"  ","
                  "NAME"  REFERENCE_LIST_REST2  ")"  ";"            !
"PLYLOP"  "("  "("  "NAME"  ","  "NAME"  ","
                  "NAME"  REFERENCE_LIST_REST2  ")"  ";"            !
"PRISMATIC_PAIR"  "("  "REAL"  ","  "REAL"  ","  "REAL"  ")"  ";"  !
"PRODUCT"  "("  "STRING"  ","  "STRING"  ","
                "DEFAULT_TOKEN"  ","  "DEFAULT_TOKEN"  ")"  ";"  !
"PRODUCT_VERSION"  "("  "STRING"  ","  "DEFAULT_TOKEN"  ","  "NAME"  ","
                        ENUMERATION_OR_DEFAULT  ")"  ";"             !
"PRODUCT_DEFINITION_CONTEXT"  "("  "STRING"  ","
                        "("  REFERENCE_LIST  ")"  ";"  !
"PRODUCT_DEFINITION"  "("  "DEFAULT_TOKEN"  ","
    "DEFAULT_TOKEN"  ","  "NAME"  ","  "("  REFERENCE_LIST  ")"  ";"  !
"PRODUCT_DEFINITION_SHAPE"  "("  "NAME"  ")"  ";"                    !

"RENDER_FACETED_FACE"  "("  "NAME"  ","  "INTEGER"  ","
                       "("  REFERENCE_LIST  ")"  ";"                 !
"!RENDER_FACE"  "("  "NAME"  ","  "INTEGER"  ","
                       "("  REFERENCE_LIST  ")"  ";"                 !
"!RGBCLR"  "("  "REAL"  ","  "REAL"  ","  "REAL"  ")"  ";"           !
"!RGBTBL"  "("  "("  REFERENCE_LIST  ")"  ";"                        !
"REPRESENTATION_CONTEXT"  "("  "INTEGER"  ")"  ";"                   !
"REVOLUTE_PAIR"  "("  "REAL"  ","  "REAL"  ","  "REAL"  ")"  ";"     !

"SHAPE_REPRESENTATION"  "("  "NAME"  ","  "NAME"  ")"  ";"           !
"SHAPE_DEFINITION_REPRESENTATION"  "("  "NAME"  ","
                                   "NAME"  ")"  ";"                  !
"!S_MASS_PR"  "("  REFERENCE_LIST_OR_DEFAULT  ","
              "REAL"  ","  "REAL"  ","  "REAL"  ","  "REAL"  ","
              "REAL"  ","  "REAL"  ","  "REAL"  ","  "REAL"  ","
              "REAL"  ","  "REAL"  ","  "REAL"  ")"  ";"             !
"SURFACE_LOGICAL_STRUCTURE"  "("  "NAME"  ","
                             "ENUMERATION"  ")"  ";"                 !
"SOLID_INSTANCE"  "("  "NAME"  ","  "NAME"  ")"  ";"                 !

"!TAP_FRM"  "("  "NAME"  ")"  ";"                                    !
"!TOOL"  "("  "NAME"  ","  "("  REFERENCE_LIST  ","
         REFERENCE_LIST_OR_DEFAULT  ","  "("  REAL_LIST  ")"  ";"    !
"TRANSFORM"  "("  "NAME"  ","  "NAME"  ")"  ";"                      !
"TRANSLATION"  "("  "REAL"  ","  "REAL"  ","  "REAL"  ")"  ";"       !
"TRNSFR"  "("  "NAME"  ","  "DEFAULT_TOKEN"  ","  "NAME"  ","
               "NAME"  ","  "DEFAULT_TOKEN"  ")"  ";"                !

"YEAR_OF_APPLICATION_PROTOCOL"  "("  "INTEGER"  ","
                                ENUMERATION_OR_DEFAULT  ")"  ";"  .

DEFAULT_OR_OMIT = "("  !  "DEFAULT_TOKEN"  ","  "("  .
STRING_OR_OMIT = ")"  !  ","  "STRING"  ")"  .

NAME_OR_DEFAULTNAME = "NAME"  !  "DEFAULT_TOKEN"  ","  "NAME"  .

ENUMERATION_OR_DEFAULT = "ENUMERATION"  !  "DEFAULT_TOKEN"  .
ENUMERATION_LIST_OR_DEFAULT =
                        "("  ENUMERATION_LIST  !  "DEFAULT_TOKEN"  .
INTEGER_OR_DEFAULT = "INTEGER"  !  "DEFAULT_TOKEN"  .
INTEGER_LIST_OR_DEFAULT = "("  INTEGER_LIST  !  "DEFAULT_TOKEN"  .

NAME_OR_DEFAULT = "NAME"  !  "DEFAULT_TOKEN"  .
```

```
REAL_OR_DEFAULT = "REAL" ! "DEFAULT_TOKEN" .
REAL_LIST_OR_DEFAULT = "(" REAL_LIST ! "DEFAULT_TOKEN" .
REAL_LIST_LIST_OR_DEFAULT = "(" REAL_LIST_LIST ! "DEFAULT_TOKEN" .

ENUMERATION_LIST = ")" ! "ENUMERATION" ENUMERATION_LIST_REST .
ENUMERATION_LIST_REST =
                   ")" ! "," "ENUMERATION" ENUMERATION_LIST_REST .

INTEGER_LIST = ")" ! "INTEGER" INTEGER_LIST_REST .
INTEGER_LIST_REST = ")" ! "," "INTEGER" INTEGER_LIST_REST .

REAL_LIST = ")" ! "REAL" REAL_LIST_REST .
REAL_LIST_REST = ")" ! "," "REAL" REAL_LIST_REST .
REAL_LIST_LIST = ")" ! "(" REAL_LIST REAL_LIST_LIST_REST .
REAL_LIST_LIST_REST = ")" ! "," "(" REAL_LIST REAL_LIST_LIST_REST .

REFERENCE_LIST_OR_DEFAULT = "(" REFERENCE_LIST ! "DEFAULT_TOKEN" .
REFERENCE_LIST = ")" ! "NAME" REFERENCE_LIST_REST2 .
REFERENCE_LIST_REST2 = ")" ! "," "NAME" REFERENCE_LIST_REST2 .
REFERENCE_LIST_LIST = ")" ! "(" REFERENCE_LIST
                               REFERENCE_LIST_LIST_REST .
REFERENCE_LIST_LIST_REST = ")" ! "," "(" REFERENCE_LIST
                               REFERENCE_LIST_LIST_REST .

STRING_LIST = "(" STRING_LIST_REST1 .
STRING_LIST_REST1 = ")" ! "STRING" STRING_LIST_REST2 .
STRING_LIST_REST2 = ")" ! "," "STRING" STRING_LIST_REST2 ..
```

Description of the scanner/parser

The scanner/parser, the so-called STPARS, generates the treefile which is a direct access file with fixed record lengths. To fill up the records, all parse-tree data are written to an internal buffer. As soon as the buffer is full, the buffer content is written to a treefile record beginning with the second treefile record. At last the first record is filled with administrative information about the treefile. Besides the STEP data STPARS produces a table of contents for the backend. The table of contents consists of the STEP name of each entity instance and the name of the enclosing instance if the instance is situated within the scope of another entity instance. Moreover it contains the storage address of the instance in the treefile. The table of contents is written onto the last records in the treefile. The treefile access routines of the backend read these records containing the table of contents automatically.

Backend interface routines

The interface routines to access the treefile are mandatory for the backend. They constitute a common part of the post-processor and consist of following subroutines:

NPSRBU	reads a single treefile record into an internal buffer
NPSRCT	reads the contents of the treefile
NPSRIA	reads an integer array from the internal buffer

NPSUGR	reads the record number and offset of an entity in the treefile
PARS_UTIL	routines to handle the tokens of a parse-tree including the routine GET
READ	reads a whole parse-tree from the treefile

Each parse-tree represents an entire entity and each entity is represented by a single parse-tree. In the post-processor only three routines are used. First of all the routine NPSRCT has to be invoked once in the post-processor. It establishes the first access to the treefile and reads the table of contents.

The other routines implemented to access the treefile are READ and GET. The READ routine is called for each entity or rather for each grammatical construct in the STEP file. The input argument is the STEP name of the entity asked for or zero if just the next entity should be read. The return values of the routine are the STEP name of the entity and the entity name of the scope.

The GET routine reads the tokens of a parse-tree in sequential order and the return values depending on the token type correspond to its position defined in GRAMMA1.DATA. It has to be called once for each token of an entity. The return values are a flag, showing the type of the token found, and four arguments (type integer, real, logical, or character) showing the value of the actual token.

5.3 Test of Processors and Consistency of the Specification

T. Sørensen and U. Kroszynski

This chapter presents the results of an activity designed primarily to show the feasibility of the exchange of robotic models between dissimilar CAD and robot simulation and programming systems.

At the same time, the results from exchange events have served to test and correct the pilot processor programs development NIRO partners' sites. The processor test also imply a consistency test of the specification (Kroszynski, Sorensen, and Schlechtendahl 1991) on which the processors are based (refer to Fig. 5.0-1 at the beginning of Chap. 5).

5.3.1 Test Suites, Library of Test Models

Exchange test suites

The exchange test suites consist on a series of steps to recover the models from the sending systems, described in the form of STEP files, into the receiving CAD and robot simulation and programming systems, by means of post-processor pilot programs developed within the project.

The measure of how successful an exchange event is, is appraised by whether the modelling activity can be resumed in the receiving system on a model imported via the STEP interface. There are several features that are checked on the receiving system. Among others, we can mention:

- Visual appearance of solids
- Parallelism and perpendicularity
- Correct location and orientation of objects in space
- Conservation of integral properties, such as volume, location of the centre of mass, and moments of inertia.

These visual, geometric, and integral properties can be directly compared on the model both at the sending and receiving systems. Concerning kinematics, the measurements are done on the receiving system by interactively verifying that links can move separately and in an articulated manner about the defined joints. Here, the checks concern mainly:

- Minimum and maximum displacements of links about their current position
- Connectivity and consistency of link movement within a mechanism
- Verification of correct position and orientation of tool-attachment-frame, tool definition, and the associated tool-centre-point frames.

Once recovered in the receiving system (normally a robot simulation and programming system) the robot task can be modelled, simulated, and output on a file. The task so described can the be converted into ICR language (refer to Chap. 7), and transferred to the production equipment for calibration, fine tuning, and execution.

The exchange suites also include "cycle tests", in which models generated and pre-processed by a system, are post-processed and recovered into the same system. Another kind of "cycle tests" concerns the KISMET system. Although no models originated directly from KISMET, BRAVO3/robot models were imported into KISMET via a dedicated (non STEP) interface. This is the only instance in this project where exchange events of the type *SYSTEM A to SYSTEM B to SYSTEM C* have been undertaken, though with the remark that the A to B transfer was not via the STEP interface. In the exchange suites, the reported model transfer from BRAVO to KISMET should be interpreted essentially as a "cycle test" within KISMET on models that were originally created on BRAVO3/robot.

Library of test models

The components of the test library, were generated in the different CAD and simulation systems at the partners' sites, as indicated in Figure 5.3.1-1, namely CATIA (DTH, KfK), BRAVO3/robot (KfK), CADDS-5 (FIAT/SESAM), and GRASP (BYG). Also, a model of a ship-section, used in one of the demonstrations (refer to chapter 8.4), was generated in the HICADEC-H CAD system for ship design at Odense Steel shipyard in Lindoe Denmark. The models were pre-processed, into STEP files, by pilot programs developed within NIRO. The STEP files comprise the test model library.

A set of STEP files was kindly provided by ESPRIT Project 2195, CADEX, (CAD Geometry Data EXchange), featuring polyhedral solid geometry only.

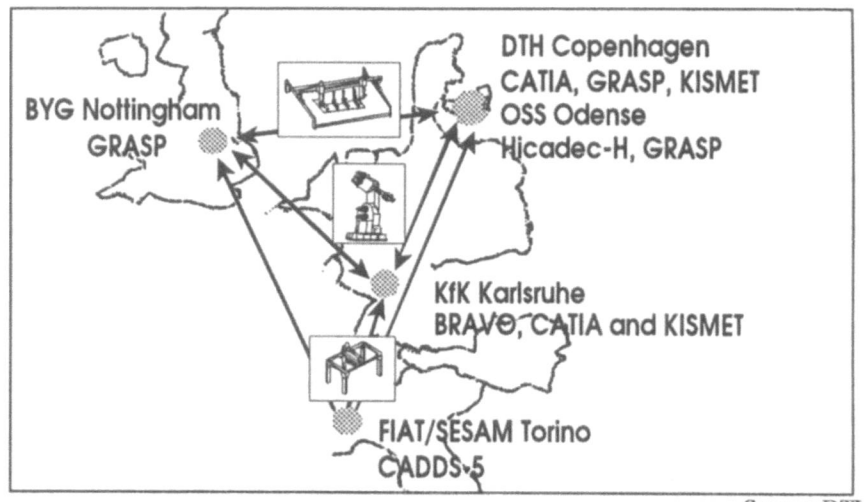

Source: DTH

Fig. 5.3.1-1: Transfer of test models between systems

These files correspond to models generated in CATIA and SICAD (a CAD system by SIEMENS) They were included in the test model library for the sake of testing compatibility with other STEP dialects.

Finally, STEP models of the library were post-processed into the diverse receiving systems available to the NIRO partners.

The exchange tests were documented in (Kroszynski and Sorensen 1992).

As part of this document, diskettes containing the STEP files produced by the sending systems' pre-processors (in the dialect corresponding to the NIRO specifications L02V00) were enclosed. This was in fact a selection of representative models from the library of test models.

As these STEP files, stored in DOS format, would be of little use for the readers of (Kroszynski and Sorensen 1992), (unless post-processors and the corresponding receiving systems hardware and software platform are available to them), it was considered useful to include in the diskettes a *STEP file viewer* program. This program, developed for the benefit of the readers of (Kroszynski and Sorensen 1992), can be invoked under DOS on a Personal Computer, and shows the header and contents of the STEP files in a convenient manner. A STEP file viewer program with enhanced capabilities is currently under development at Control Engineering Institute, DTH, Denmark, featuring scanning/parsing of the STEP file, visual appearance of the model on a graphical screen, and facilities for driving mechanisms that are red in from a STEP file.

The models of the library, are categorised according to the complexity of features included, and each model has one or more distinct features that was tested during processor program development.

The *categorisation of models* is as follows (refer to Fig. 5.3.1-2a - 5.3.1-2g):

1.a Simple geometry models, without kinematics, featuring only objects modelled as faceted-Breps (polyhedral representation of solids). The range of STEP geometry and topology entities covered is: 'CARTESIAN_POINT', 'POLY_LOOP', 'FACE', 'SHELL' and 'FACETED_BREP' (Fig. 5.3.1-2a).

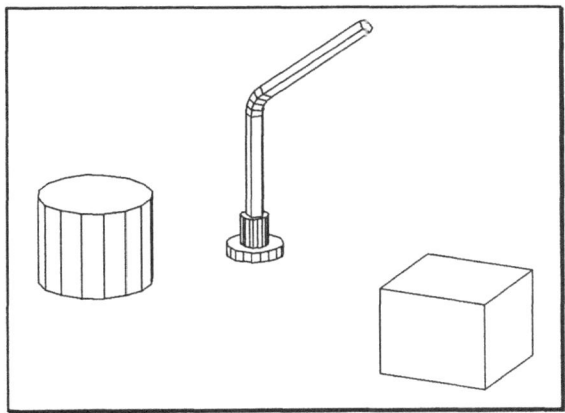

Source: DTH

Fig. 5.3.1-2a: Models from category 1.a of the library.

Besides, the entire skeleton of STEP file structure is present even for such simple models. This includes the 'HEADER' and 'DATA' sections of the STEP file, according to the agreed syntax in the NIRO specification L02V00. General entities such as 'INDEX_ENTRY', as well as user-defined-entities considered within NIRO are also covered under this category, namely: 'POINT_DIRECTION_PAIR', 'RENDER_FACE', 'COLOUR_TABLE', 'COLOUR_ATTRIBUTE', and 'SOLID_MASS_PROPERTIES'.

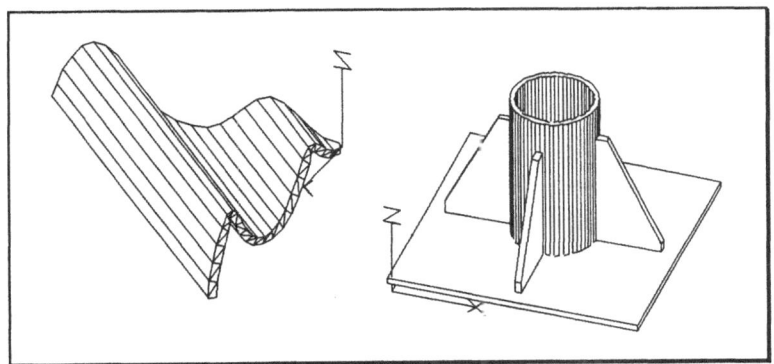

Source: DTH

Fig. 5.3.1-2b. Models from category 1.b of the library

1.b More complex solids, featuring objects with through holes, as well as solids with "thin" portions, curved surfaces approximated by many planar facets, tangent surfaces, etc. (Figure 5.3.1-2b).

1.c The inclusion of the 'SOLID_INSTANCE', entity representing an object that refers to a 'FACETED_BREP' solid which is placed and oriented in space according to a given transformation matrix. (Fig. 5.3.1-2c)

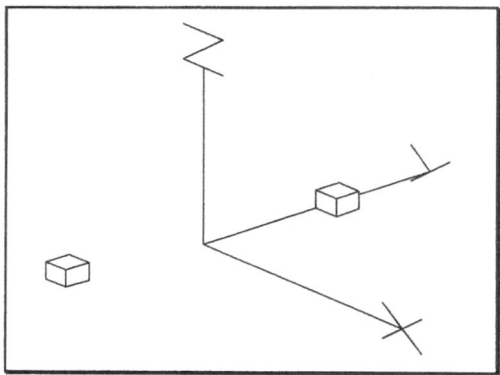

Source: DTH

Fig. 5.3.1-2c. A model from category 1.c of the library (two instances of a simple box defined at the origo of the absolute co-ordinate system shown)

Models within the above mentioned categories do not include any kinematics description but are employed to transfer the geometric shape of objects in the scene of the workplace, e.g. objects to be assembled, welded, etc., as well as the work cell geometry. Moreover, these solid models are used to represent the flesh of robot arms in models of robotic mechanisms.

More advanced categories, which are the primary interest within NIRO are:

2.a Simple kinematic models featuring mechanisms with one or two joints represented by prismatic or revolute kinematic pairs. In this category, all the skeleton entities conforming a kinematic model are present, namely: 'AXIS2_PLACEMENT', 'KINEMATIC_PAIR', 'PAIR_PLACEMENT_STRUCTURE', 'KINEMATIC_LINK', 'MECHANISM', and 'KINEMATIC_ MODEL'. Besides, the 'GROUND' entity is also present (Fig. 5.3.1-2d).

2.b More complex mechanisms, featuring models of real robots with up to six Degrees of Freedom (DOF) in up to nine or more joints in open as well as closed kinematic chains, as encountered in some robotic models. Here, the user defined entity 'ACTUATOR' is included (Fig. 5.3.1-2e).

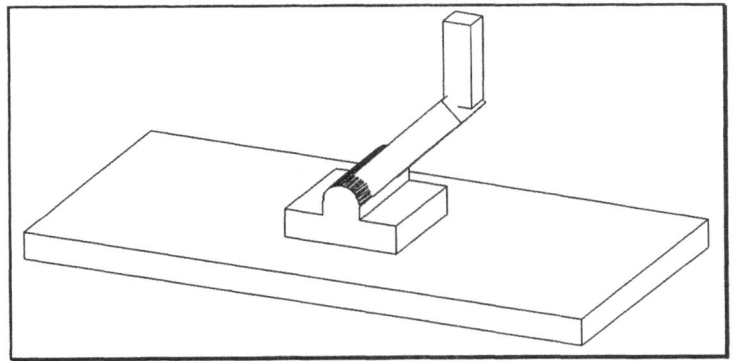

Source: DTH

Fig. 5.3.1-2d. A model from category 2.a of the library

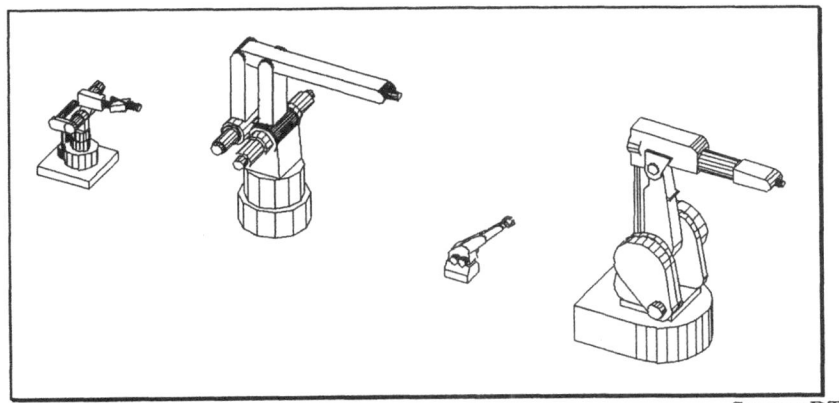

Source: DTH

Fig. 5.3.1-2e. Models from category 2.b of the library

3. Mechanisms featuring models of the former category, but in which also the user defined entities 'TOOL_ ATTACHMENT_FRAME', as well as 'ADDITIONAL_ FRAMES' in links are included in the kinematic model. Also models of robotic tools, with one or more 'TOOL_CENTRE_ POINT' frames, and the 'MOUNT' entity, indicating which tool is mounted on which robot, belong in this category (Figure 5.3.1-2f).

4. Models of entire scenes, featuring several robotic mechanisms, tools, and workplace objects are included in this category. Also mechanisms hanging on other mechanisms (rather than being attached to the ground) belong in this category (e.g. gantries with robots, etc.) (Figure 5.3..1-2g).

80

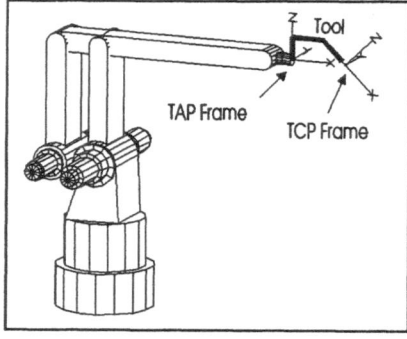

Source: DTH

Fig. 5.3.1-2f: A model from category 3 of the library

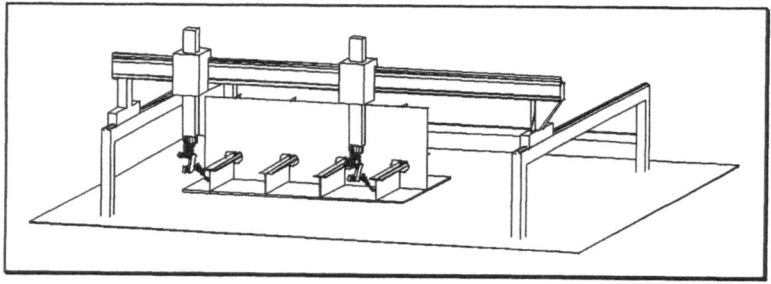

Source: DTH

Fig. 5.3.1-2g: A model from category 4 of the library

The library of test models contains some "50 selected entries", representative of the categories mentioned above. An evaluation of the processor test based on this exchange of the model library, and some experiences gained are outlined in the next chapter.

5.3.2 Evaluation of Test Results. Experiences

Presentation and evaluation of results

A full documentation of the test suites, the model library, and the exchange results is given in (Kroszynski and Sørensen 1992) it occupies 126 pages. Thus, it is much too comprehensive to include in this context. Instead, some representative test models are chosen here and listed in Table 5.3.2-1 with one entry per model.

The entry consists of the model name associated with a STEP file, its complexity category, the originating system, and the system or systems where the model was recovered. The grouping within the table is according to which organisation contributed the entries. The models are shown in Figure 5.3.2 (-1a to -1e) as indicated in the table:

Table 5.3.2-1: Overview of the library of test models and exchange suites.
**) See Sect. 5.3.1 for an explanation of the categorisation.

Model (*.stp)	Category **)	Sending System	Recovered in		
			CATIA (KfK)	GRASP (BYG)	KISMET (KfK)
		CATIA (DTH)			
zippo	2 a)	Fig. 5.3.2-1a	no	yes	yes
abb2000	2 b)	Fig. 5.3.2-1b	yes	yes	yes
		GRASP (BYG)			
pera	1 b), 1 c)	Fig. 5.3.2-1c		cycle test	
		CADDS 5 (FIAT)			
pmast-25	1 b)	Fig. 5.3.2-1d	yes	yes	yes
		BRAVO3 (KfK)			
solitaire	4	Fig. 5.3.2-1e	yes	yes	yes

Remarks on the transfer of the models shown in Fig. 5.3.2-1:

- The *geometry* of all the models, a) to e), were transferred successfully into the receiving systems. This success reflects very well the success of the majority of the geometric transfer of the model library.

With respect to the transfer of *kinematic* information:

- the 'Zippo' model, a), was transferred successfully into GRASP and KISMET (no attempt was made to transfer it into CATIA). The transfer was in fact successful for most of the 7-8 different models of category 2 a) from the library.

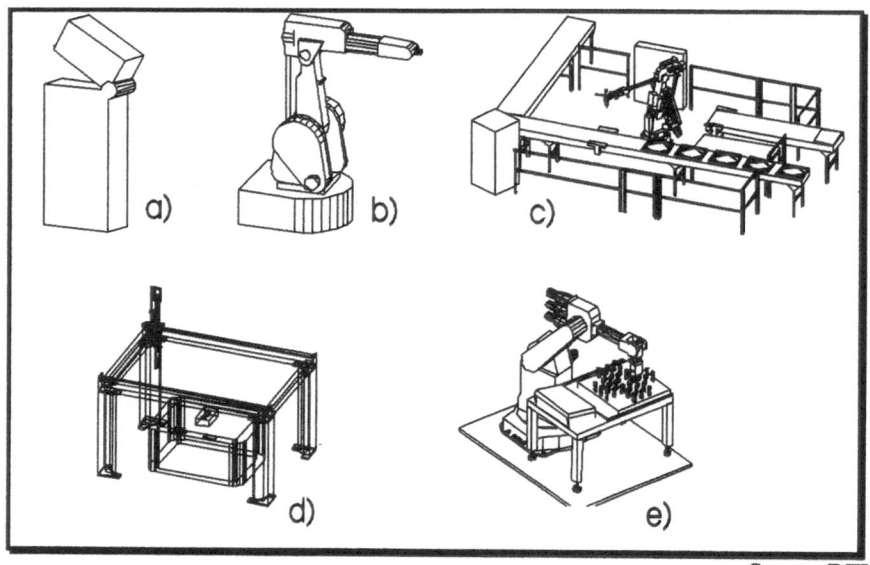

Fig. 5.3.2-1: Models from the test library

- the 'ABB IRB-2000' model, b), has a so-called 'closed kinematic chain', this chain was not recovered in any receiving system from the error free STEP file. Non systems were able to recover 'closed chained' models from the library.
- the cell 'pera', c), originally defined in the GRASP system, was post-processed into a STEP file and pre-processed into the GRASP system again (i.e. a 'cycle-test'). Only the geometry of the cell was recovered since the post-processor was not able to read and write kinematics information in the test period of the project. This is now solved and the post-processor is fully tested and operational (the 'solitaire' model, e), was successfully transferred after the test period).
- the 'PMAST-25' model, d) (geometry only), was converted from a solid model to a wire-frame when written into CATIA's database, due to inaccuracies in the point co-ordinates. This implies that four or more points did not define a planar 'facet' of the model as they did in the sending system. CATIA is very sensitive to the accuracy of the solid model that is written into its database (the point co-ordinates must in fact often have higher accuracy that the interface to the CATIA database supports!). A solution to this problem has not been found within the NIRO project. The model was successfully transferred into GRASP and KISMET.
- the 'solitaire' model, e), from BRAVO3, was successfully transferred into CATIA, GRASP, and KISMET after the test period of the project. The

model was used in the Reis demonstration described in chapter 8.5. It consists of a kinematic model and a large number of solid instances.

Experiences

Once the processor programs for conversion of models from a sending system to the neutral file and from the neutral file to a receiving system are coded based upon the specification of the neutral reference scheme, model transfer between the systems can be attempted. Quite frequently, a degradation of the model data occurs, even with an error free translation, due to:

1) Limitations in the neutral description reference scheme; Upon mapping features from the sending system that are not covered in the neutral description, pre-processors must either approximate these features in terms of functionalities that are supported in the neutral description reference scheme, or just not map them.

2) Limitations in the sending system as compared with the reference scheme of the neutral description; In this case, some conventions might be applied concerning modelling with alternative features which are otherwise not used (e.g. an auxiliary axes set, which has no counterpart in the information reference scheme, can be pre-processed to mean a Tool-Attachment-Point frame).

3) Limitations in the receiving system with respect to the neutral description; Here, the post-processor can either ignore the entities that are not supported, or recover them by means of other entities as an approximation. An example of this is a system that has no solid modelling capabilities. The solid models can be recovered as groups of collected wire-frame entities like points and curves).

4) Inaccuracies in numeric data; The decimal coding of binary data, in particular on textual ASCII format with a fixed number of significant digits, gives rise to truncation and/or round-off errors. The redundant information of 3D co-ordinates of points that have to lie on planes, as employed in faceted-boundary-representations, can give rise to inaccuracies that are sometimes larger than the tolerance allowed for in the receiving system. Thus, a fourth point may be slightly off the plane determined by the first three points. Similarly, shapes with parallel and perpendicular lines or surfaces might appear slightly deformed. A sensitive system (like Dassault's CATIA, employed by several project partners as their main CAD platform), requiring tight tolerances, sometimes necessitates data with a precision that the sending system is unable to deliver. We have experienced that solid models in polyhedral representation are sometimes recovered as wire-frames in CATIA due to this problem.

All these issues of mapping approximations and inaccuracies are unavoidable and irreversibly change the original model information. Nevertheless, the main functionalities are still successfully transferred, so modelling can be resumed in the receiving system.

A further discussion on the causes for the loss of information in model transfer events, and issues in modelling techniques in general, is found in [Kroszynski, Sørensen and Trostmann 1993].

6. Results related to IRL and ICR

Chapter Editor: *W. Jakob*

6.1 Industrial Robot Language (IRL)

U. Schmiedecke

Whereas in general programming, machine independence and structured programming have been state of the art for several decades, robot programming is done (in most cases) on a robot specific machine language-like level.

The idea of IRL is to introduce the advantages of high-level general purpose programming languages to robot controlling. The advantages to win are robot-independent programs that are easier and quicker to write, read, understand and restructure. The price to be paid is still execution time, but compared to the time taken for robot arm movement, it is hardly relevant.

6.1.1 Present Status

The language IRL was first introduced on the 21st International Symposium for Industrial Robots (ISIR) in Copenhagen on Oct. 23-25, 1990 [1].

IRL is now standardised in Germany, and is described in "DIN 66312 Teil 1" [2]. "Teil 1" (part 1) means that it is planned to create a second part, adding some important features, which we note in Sect. 6.1.3 below. As far as the ISO is concerned the work on PLR (Programming Language for Robots), the ISO counterpart to IRL, is paused due to the overall political situation within ISO as described in Sect. 6.2.5. PLR was strongly influenced by IRL. The American position is to standardize a language based on C or nothing. This meets the Japanese opposition against PLR; Japan prefers a Japanese language based on BASIC called SLIM.

At least three IRL-Compilers already exist, which translate IRL to ICR and IRDATA respectively (see Sect. 6.2):

- one made by "PSI Berlin" (Gesellschaft für Prozessteuerungs- und Informationssysteme mbH, Germany),
- one by "AEDVS der RWTH Aachen" (Allgemeine Elektrotechnik und Datenverarbeitungssysteme der Rheinisch Westfälischen Technischen Hochschule Aachen, Germany).
- one by KfK (Kernforschungszentrum Karlsruhe, Germany).

6.1.2 The Language Features

IRL has all major features of a high-level general purpose programming language, like PASCAL, including modular programming. In addition, it contains a set of robot specific constructs that mainly deal with the fact that a robot arm is moving at a certain speed in a three-dimensional space, and that sensors of different kind may be used to control it. Special attention was given to enable "teaching" of robots, t#o control several robots from one program, and to provide structured means for system adaptation.

Readers familiar with PASCAL or similar languages may skip the following Sect. (or just scan the bold face keywords) and start reading at Sect. 6.1.2, where the robot specific features are explained.

6.1.3 General Purpose Language Features

The syntax of IRL is similar to that of PASCAL, many constructs even being identical.

Constants, Variables, and Data Types

Variables: Variables of different types can be defined to hold values of the respectively data types. The assignation of a value of a different type is an error - a great help in spotting programming mistakes.

Constant declarations of any data type are used to define names for constants. These values cannot be changed by the program. So the programmer can be sure that this values will never (incidentally) get wrong.

IRL has the following **pre-defined data types**: BOOL (values True and False), CHAR (any ASCII character), INT (from -2147483647 to +2147483647), REAL (from -1.7E+38 to -1.7E-38, 0.0 and from +1.7E-38 to +1.7E+38), and STRING (any sequence of ASCII characters enclosed in single quotes).

Type declarations are the means to define further (problem specific) data types. They may be used just for "renaming" other data types or for creating new ones.

An **Array** is a group of elements of the same data type. The elements may be grouped in more than one dimension. The number of dimensions and the index range of each dimension may freely be specified in the declaration of the array. The elements of that array are accessed using an integral index expression for each of the dimensions: "arrayname [indexexpression1, indexexpression2, ...]".

A **record** is an aggregate of any number of fields of other data type in any order. Even an array or another record may be used as a record field. Each field has a unique name and is accessed by "recordvariablename.fieldname".

A **file** is a sequence of any number of elements of the same data type. They are stored on a mass storage device. To access its elements, a **sequential** file must be opened for reading (or for overwriting or appending). After that, the elements of the opened file may be read (or written) sequentially one after another. After opening a **random access** file, its elements may be read or written in any order by specifying an element number. Files should be closed, when no access is wanted any more. The predefined type TEXT is used for a sequential file of characters. Here complete lines may be read or written with special predefined functions. There are also some special predefined sequential character files called **channels**.

One main advantage of user defined data types is that the compiler will do type checking, i.e. it will check whether both sides of an operation match in data type, and whether the parameter types of procedure and function calls match the corresponding procedure (or function) declaration.

Furthermore, type declarations allow the construction of so-called **abstract data types**, i.e. a data type plus a set of type specific operations, usually "hidden" in a module. This is of great importance for constructing (application oriented) libraries.

Procedures and functions

Procedures and Functions are named groups of statements which perform some specific action. A **procedure** serves as an abbreviation of this group of statements, whereas a **function** computes a function result, to be used within expressions. A procedure or function declaration defines its name, its parameters and the set of statements it stands for. A function declaration additionally defines its result data type. For each parameter, its data type must be specified and whether it is an input, output or transient parameter. Recursion is allowed in IRL, but static nesting of procedures (or functions) is not permitted. When calling a procedure (or function):

- an **Input parameter** is used to pass a value to the procedure. This value can be an expression and its data type must be compatible as that of the formal parameter, i.e. the value will not be converted to the required data type.
- the name of a variable (or an array element, or a record field) must be used to specify the destination of an **output parameter**,
- the same holds for a **transient parameter** but additionally the passed variable must have a defined value before the procedure is called, since it is used by the called procedure before assigning a new value to it. So it is an input parameter and output parameter as well.

Array parameters may either be passed **fixed** or **conformant**. A fixed array parameter is an array of fixed size, i.e. all bounds and the number of dimensions are explicitly specified. A conformant array parameter has "variable" bounds that are passed as implicit parameters and are accessible within the procedure (or function).

Modules

An IRL Program can be split up into separate modules which may be compiled separately. This way, the programmer can implement abstract data types or modules that may be used in different programs. An **export** specification allows to specify what is visible outside. An **import** specification specifies, what is required from outside.

Control structure

IRL offers the standard set of high level control structures, as required for "structured programming":

- IF ... THEN ... ELSE ... ENDIF
- CASE ... WHEN ... WHEN ... ... DEFAULT ... ENDCASE
- FOR ... TO ... STEP ... ... ENDFOR
- REPEAT ... UNTIL
- WHILE ... ENDWHILE
- GOTO label

File I/O

The **OPEN** statement is used to open a file or a channel. Files may be opened for reading, writing, appending or for random access. It is possible to open text files (ASCII) or binary files. The path and name of the file is specified as STRING.

The **WRITE** statement is used to write data to an opened file or channel. The first parameter of the WRITE statement may be a FILE OF ... identifier, which then is used as file to be written to. If it is missing, the predefined FILE OF CHAR variable STANDARD_OUT is used. All other parameters must be of some simple data type (BOOL, CHAR, INT, REAL or STRING) or a geometric data type.

For INT and REAL values in text files it is possible to specify the number of characters to be used to represent the printed value and the precision of REAL values. To represent an orientation value (see Sect. 6.1.4.1 below) in a text file it is necessary to specify its underlying orientation mode using the predefined identifiers RS, XYZ, YXZ, ZYX, ZYZ2, VANG and QUAT. So orientation values are always written in the same form as they are generated within the IRL program text (e.g. "ORIRS(30.0, 45.0, -15.1)", see Sect. 6.1.4).

Geometric data types in text files are written by writing the type identifier followed by its contained data in parentheses (e.g. "POSE(POSITION(123.456, 789.012, 345.678), ORIXYZ(901.234, 567.890, 123.456))".

For binary files, the values are written in their internal representation, so that writing and reading is very efficient, but "unreadable" by human beings.

WRITELN is only allowed for TEXT files (i.e. FILE OF CHAR) and does the same as WRITE, except that a new line character is appended to the written line.

For random access files **WRITERAC** must be used. A file variable must be specified as first parameter to specify the file that should be written into. The element number (i.e. the record number of the file contents) must be specified as second parameter. All other parameters must have the same data type as in the definition of the associated file variable.

The **READ** statement is used to read data from an opened file or channel. The first parameter of the READ statement may be a FILE OF ... identifier, which then is used as file to be read from. If it is missing, the predefined FILE OF CHAR variable STANDARD_IN is used. All other parameters must be variables of some simple data type (BOOL, CHAR, INT, REAL or STRING) or of geometric data type. The data is read from the file and assigned to the objects. Orientation values must appear in the input in the form "ORI...(val1, val2, ...)" as they would be written with the WRITE statement (see Sect. 6.1.4). The number of digits for INT and REAL values and the precision for real values is unimportant for READ.

READLN is only allowed for TEXT files and does the same as READ, except that the rest of the line is ignored after assigning the read values to all parameters.

READRAC is used for reading random access files. The first parameter must specify a FILE OF ..., the second parameter the element number to be read. All other parameters must be of the same data type as the FILE OF ... file variable and are assigned the data read from the binary file as it comes.

The **CLOSE** statement is used to close an opened file or channel.

String handling

A comfortable set of functions to manipulate strings is supplied:

- concatenation,
- searching of one string in another,
- comparing two strings,
- inserting one string into another,
- deleting characters within a string,
- extracting a substring from another string,
- getting the actual and the maximal length of a string.

Expressions on simple data types

The standard set of dyadic operators to be used in expressions is supplied:

arithmetical operators	+ - * / DIV MOD
logical operators	AND OR EXOR
relational operators	= < > <> <= >=

as well as monadic operators +, - (arithmetical) and NOT (logical).

Arithmetical and relational operators operate on CHAR, INT, and REAL values. They will be converted automatically to the next data type of the following "order": CHAR -> INT -> REAL until both operands have the same data type using a minimum of conversions. There are only two exceptions:

- both "/"-operands are always converted into a REAL value,
- if the operands of MOD and DIV are both CHAR values, the result will also be a CHAR value, otherwise they are converted to an INT value.

Relational operators always return a BOOL result.

Logical operators operate on BOOL or INT values with no automatic conversion, i. e. both operands must be of the same data type and the result value of the operation is of the same data type.

The priority order of the operators is as expected (highest to lowest):

monadic,   * / MOD DIV,   + -,   < > <= >=,   = <>,   AND,   EXOR OR.

The INPUT and TEACH attributes (see below Sect. 6.1.4) are irrelevant for operating on values. In assignations, the value on its right hand side will also converted automatically to the data type of the left hand side in the following cases: INT := BOOL,   CHAR/BOOL/REAL := INT,   INT/REAL := CHAR. Variables with type attribute OUTPUT are allowed on the left hand side of an assignation.

6.1.4 Robot Specific Features

The special means for robot control are embedded in IRL as a set of predefined data types and operations which describe the geometry of the working space, and a set of special instructions with optional keywords allows comfortable programming.

Predefined geometric data types

In IRL the following predefined geometric data types are implicitly needed by several language constructs:

A POSITION is a record with three real fields (x, y, z) that serve as carthesian coordinates in the space.

An ORIENTATION is a record that describes the direction of a coordinate system in space in relation to a reference system. The zero orientation is identical with the orientation of the reference system. The internal representation of that data type is left open to enable the compiler to use optimal representation. There are seven predefined functions to generate a result of type ORIENTATION which represents a coordinate system that is derived from the reference system by rotating the reference system around three axes by the given angle parameters (as degree angles):

ORIRS(a,b,c) a, b and c are the rotation angles around the x-, y- and z-axis of the reference system.

ORIXYZ(a,b,c) first rotate the reference system a degrees around its x-axis, then b degrees around the new y-axis, then c degrees around the new z-axis.

ORIYXZ(a,b,c) first rotate the reference system a degrees around its y-axis, then b degrees around the new x-axis, then c degrees around the new z-axis.

ORIZYX(a,b,c) first rotate the reference system a degrees around its z-axis, then b degrees around the new y-axis, then c degrees around the new x-axis. This is equal to ORIRS(c,b,a).

ORIZYZ2(a,b,c) first rotate the reference system a degrees around its z-axis, then b degrees around the new y-axis, then c degrees around the new z-axis.

ORIVANG(v,a) rotate the reference system a degrees around the vector v, which is represented by a POSITION value, where its components describe the vector v within the old reference system.

ORIQUAT(a,b,c,d) a, b, c, and d are the parameters of a quaternian specification.

A POSE is a record with a POSITION and an ORIENTATION field to describe the position and orientation of the tool center point (TCP) within space.

A JOINT is a record that describes the position of the TCP in the robot coordinate system. It contains a number of REAL values, that depend on the number of axis of the robot, where each REAL value describes the position of an axis. The REAL value for a rotation axis is an angle, for a translation axis it is a distance. The REAL values of a JOINT are grouped together in two groups (i.e. two internal record types): the values for all main axes are put into a record of type MAIN_JOINT, those for all additional axes are put into a record typed ADD_JOINT.

ROBTARGET is a record with a POSE, some configuration information and an ADDJOINT. The internal structure of the configuration info is left open to the target robot controller and is probably a record containing some INT values. The precise structure is specified in the system specification file (see Sect. 6.1.4).

Orientation specific predefined functions

There are special predefined functions to retrieve the angle values of rotations in ORIENTATION type expressions:

ANGLEX(RS,ori_val) returns the rotation angle around the x-axis of the orientation value ori_val under ORIRS definitions. The functions ANGLEY and ANGLEZ have the same parameters and return the rotation angle around the y- and z-axis respectively.

When the first parameter of these three functions is XYZ, YXZ, or ZYX, the angle x, y, or z is returned when the underlying orientation definition is ORIXYZ, ORIYXZ, and ORIZYX respectively.

To retrieve the angles of ori_val=ORIZYZ2(a,b,c) the functions ANGLEX(ZYZ2,ori_val) can be used to retrieve a, ANGLEY(ZYZ2, ori_val) can be used to retrieve b and ANGLEZ2(ZYZ2, ori_val) can be used to retrieve c. When ori=ORIVANG(v,a), ROTAXIS(ori) returns v and ROTANGLE(ori) returns a.

When ori=ORIQUAT(a,b,c,d), QUATA(ori) returns a, QUATB(ori) returns b, QUATC(ori) returns c and QUATD(ori) returns d.

In short terms the following relations hold:

Let o=ORIRS(a,b,c) then the following three expressions are TRUE:
ANGLEX(RS,o)=a and ANGLEY(RS,o)=b and ANGLEZ(RS,o)=c.
Let o=ORIXYZ(a,b,c) then the following three expressions are TRUE:
ANGLEX(XYZ,o)=a and ANGLEY(XYZ,o)=b and ANGLEZ(XYZ,o)=c.
Let o=ORIYXZ(a,b,c) then the following three expressions are TRUE:
ANGLEY(YXZ,o)=a and ANGLEX(YXZ,o)=b and ANGLEZ(YXZ,o)=c.
Let o=ORIZYX(a,b,c) then the following three expressions are TRUE:
ANGLEZ(ZYX,o)=a and ANGLEY(ZYX,o)=b and ANGLEX(ZYX,o)=c.
Let o=ORIZYZ2(a,b,c) then the following three expressions are TRUE:
ANGLEZ(ZYZ2,o)=a and ANGLEY(ZYZ2,o)=b and ANGLEZ2(ZYZ2,o)=c.
Let o=ORIVANG(v,a) then the following two expressions are TRUE:
ROTAXIS(o)=v and ROTANGLE(o)=a.
Let o=ORIQUAT(a,b,c,d) then the following four expressions are TRUE:
QUATA(o)=a and QUATB(o)=b and (QUATC(o)=c and QUATD(o)=d.

Robot specific geometric expressions

Geometric expressions allow comfortable computation of geometric data types. The predefined data type names POSITION, ORIENTATION, POSE and ROBTARGET are used like functions to express an explicit conversion of an expression into a value of the desired data type (also known as "casting"). The first two may be applied to POSE and ROBTARGET expressions only and they just select the desired part of it.

POSE(robtargetvalue) just selects the POSE-part. POSE(positionvalue) adds a zero-orientation to form a POSE-value. POSE(orientationvalue) adds a zero position to create a POSE-value.

ROBTARGET applied to a position or orientation value first does the same as POSE but then adds zeroes to create a robtarget value. This casting is often performed automatically, if the geometric data type supplied differs from the one required.

Some dyadic operators are available to handle geometric expressions. The following expressions yield a POSITION type ("pos" is a POSITION and "r" a REAL variable):

pos + pos	is vector addition, i.e. tho components are added,
pos * r, r * pos	are vector stretching,
pos / r	is vector shortening.

In the following expressions, ori is an ORIENTATION, pse is a POSE and robt is a ROBTARGET:

robt @ robt   yields a ROBTARGET,

but there are a lot of meanings, when applied to other geometric data types that are casted to ROBTARGET:

ori @ ori	is chaining of two rotations,
pos @ pos	is vector translation (same as addition),
pse @ pos, robt @ pos	is POSE translation,
pos @ ori	is vector rotation,
pse @ ori, robt @ ori	is POSE rotation,
pos @ pse	is vector tranlation and rotation,
pse @ pse, robt @ pse	is POSE tranlation and rotation,
pos @ robt	is vector tranlation and rotation,
pse @ robt, robt @ robt	is POSE tranlation and rotation.

Other robot specific features

A **system specification** file is used to specify all robot dependent items as: predefined data types, variables, constants, functions, and procedures.

In the beginning of his IRL program, the programmer may specify the name of the system specification file he wants the compiler to use. If the file name is not specified, the compiler will read a default one.

A **Signal** is a type attribute, that allows the mapping of variables to digital and analog input and output. The following variable types may be mapped:

    BOOL: to binary I/O,   INT: to digital I/O,   REAL: to analog I/O.

Channels can be used to read from and to write to serial or parallel interfaces of the robot controller at runtime. There are a lot of predefined channels, e.g. SER_1, SER_2, ... PAR_1, PAR_2, ... PRINTER, SCREEN, KEYPOARD, PANEL, TTY, HPU, MCP. The complete list of supported channels has to be specified in the system specification file.

A **datalist concept** allows global variables to "survive" the execution of a program. This way, teach programs may be written that fill variables with taught positions or other data and to reuse the values of these variables in other application programs. Global variables with the attribute TEACH or PERMANENT will be used for that purpose (permanent and teach are synonyms).

The surviving variables may be grouped together in different data lists. A data list has a name, given by the programmer. Before the first TEACH variable may be declared, a datalist specification mentioning the data list name must be specified textually by the programmer. All subsequent TEACH variables will then be taken from that data list. The next data list specification "opens" a new data list.

Before the first statement of a program is executed, all TEACH variables are read from the data list. After execution of the (dynamically) last statement of the program all TEACH variables are written back into the appropriate data lists.

A **WAIT** statement allows to wait for a specified time (e.g. 5 seconds) or to wait for a specified input signal to become a specified value. In the latter case a timeout value might be specified together with a sequence of statements that will be executed, when the timeout becomes true.

A **PAUSE** statement is used to write a message to the operator. The IRL program waits until the operator gives some signal (e.g. a button or some special key on the panel, depending on the system environment) to continue the IRL program.

A **STOP** statement stops program execution.

The RETURN statement terminates the execution of a function or procedure. Functions require a RETURN statement with a parameter, that is used as a function result. This parameter must be of the required function result type. In a procedure, the **RETURN** statement has no parameter and may be omitted at the end of the procedure.

A set of MOVE statements is provided to describe robot arm movements with several keywords describing variants. The target position of a **MOVE** statement is always specified with a ROBTARGET parameter, the value of which is absolute, where the parameter value of a **MOVE_INC** is relative to the TCP just before the move.

Furthermore, it has to be specified, whether a move should be **PTP** ("point-to-point" move), **LIN** (linear move), or **CIRCLE** (move along a part of a circle). LIN is slower in computing and moving, since a lot of intermediate points must be computed and moved to, so that the TCP is moving along the shortest (linear) path between the current point and the destination point. A PTP-move is faster, but normally moves along a "strange" curve that is the result of several independent moves of all joints.

In the CIRCLE-move, the TCP moves along a circle that is specified by two robtarget parameters instead of one. The current TCP position and position of these two parameters are used to compute a circle in space. The TCP will move along that circle beginning at the current position. The position of the first robtarget parameter is only used to specify the coordinates of the intermediate point on the circle to move along. The second robtarget parameter is the destination point (including orientation) to move to.

The keyword **PATH** allows to specify a list of ROBTARGET parameters instead of just one. The TCP will be moved to all points, where the last point in the specified list is the final destination point. All other points specify intermediate points, where smoothing is possible, i.e. as soon as the TCP is near enough at one of the intermediate points, it begins to move to the next point in the list. The distance "near enough" is specified by the smooth parameter explained below.

Instead of specifying a list of distinct points, IRL allows to specify an array (array_name[]) or only a part of it (array_name[from .. to]).

For a MOVE CIRCLE, the PATH parameter is a list of pairs, where the position of the first point of each pair is used to specify the intermediate point of the circle to move along. If an array was specified, its number of elements must be even so that each two specify such a pair.

The parameter **TIME** allows to specify a time (in seconds) for the whole movement. If this param is specified, the param(s) SPEED (and SPEED_PTP) may not be specified.

The parameter **UNTIL** allows to specify a logical expression. This expression will be computed permanently while moving. As soon as this expression becomes TRUE, the movement will stop. Only signals and constants are allowed to be compared in this expression. This feature allows to search a special position depending on sensors. Additionally a sequence of statements may be specified after the parameter **ERROR**. They are executed, if the movement has stopped because the expression became TRUE.

A multi robot feature allows to handle a number of different robots with one program. The system specification defines a special record variable for each robot, where its fields contain all robot dependent information necessary. So the additional param **ACT_ROB** allows to specify the desired robot by specifying the name of that robot variable that is predefined in the system specification. If this parameter is missing the default robot variable used will be the predefined variable R_ACT_ROB.

For a PTP move, the following additional parameters are possible:

SPEED_PTP	This parameter allows to specify a factor of the wanted speed of an axis in relation to its maximal movement speed
ACC_PTP	This parameter allows to specify the factor for the acceleration of an axis in relation to its maximal acceleration of the movement
C_PTP	This parameter specifies the factor of the axis distance, where the smoothing starts

For a LIN or CIRCLE move, the following additional parameters are possible:

SPEED	This parameter allows to specify an absolute movement speed for the TCP
ACC	This parameter allows to specify the TCP acceleration of the movement
C_CP	This parameter specifies the distance of the moving TCP to the position of an intermediate point where smooting should start

C_SPEED	This parameter specifies the smoothing speed value. If C_SPEED is specified, C_CP must not be specified
SPEED_ORI	This parameter specifies the speed at which the orientation may be changed, if the path is empty

If one of the above parameters is missing, the default value (if needed) is taken from a predefined variable with the name "R_" followed by the name of the missing parameter. These predefined variables are defined and initialized in the system specification and may be modified by the program as needed.

The parameter WOBBLE allows a superimposed movement. This is needed in welding processes. The welding parameters must be specified in additional special data structures specified in the system specification.

6.1.5 Recommendations for Further Enhancements

The following features are to be treated in "part 2" of the IRL standard:

- syncron action during a movement
- more general smoothing specifications
- specification of interpolation of an orientation movement
- multitasking, parallel processing
- more different sensor instructions
- interrupt handling
- exception handling
- analog output as a function of the actual velocity
- nesting of comments or conditional compiling
- specification of jerk (acceleration as a function of time)
- waiting for end of move
- specification of oscillation of TCP along a move path
- a message statement for error messages with priority that must be answered

All these features are planned to be included into "Teil 2" of DIN 66312. This second part is expected to be finished in 1995.

6.1.6 NIRO Contributions to IRL

The following concepts are contributed from the NIRO project: data list handling, string handling, multi robot handling, and a proposal for exception handling.

Data list handling

Here the following items were influenced by the NIRO project:

TEACH CONSTants: it is possible to specify some distinct elements of a data list as constants rather than as variables. This enables the possibility to write to these variables by one program but only to read them in other programs. The IRL compiler will check, that writing to TEACH constants is not permitted.

TEACH (or PERMANENT) variables of other types than geometric data types (e.g. REAL, INT). This way, other simple values may be transported from one program to another.

Sequential reading and writing of variables in a data list instead of accessing them by its name. This allows very efficient reading of a large number of variables, where the name of them might be unimportant. An example might be: a long array of POSEs which forms a path that the TCP has to move along.

String handling

It is obvious, that a comfortable string handling will ease the construction of better messages where some variable information is included.

Multirobot handling

It will happen more and more, that several robots will work on one workpiece. This can be organized far easier in one program controlling them all together than in several programs which need to synchronize themselves via digital I/O ports.

Exception handling

A proposal for exception handling was worked out. This might be included in the language in a later state.

6.1.7 Contextfree Grammar of IRL

This section describes the grammar of IRL in terms of syntax rules. Each rule consists of: an ordinal rule number, the rule name, a colon, a possibly empty sequence of tokens, and a dot terminating the rule. Each rule has an ordinal number that is referred to in an alphabetic sorted list of rule names at the end of this chapter. In the right hand side of a rule the following notations are used:

"BEGIN"	text within double quotes is a terminal symbol. Letters are case insensitive.
(...)	grouping of a sequence of tokens to limit the range of an alternative
[abc]	optional tokens: abc may occur zero or one time
{ abc }	repeated tokens: abc may occur zero or more times
{ abc }+	repeated tokens: abc may occur one or more times
abc | def	Alternative (either abc or def) of a rule. To express the alternative of tokens within a sequence of other tokens, Parentheses, brackets, and braces must be used.

Example: a (b c | d) e
possible alternatives in the example: a b c e or a d e.

Any number of comments, blanks, tabs and new line characters are allowed between any tokens except within a terminal symbol, within an identifier, and within a constant. All characters within a string constant are significant, letters are case sensitive. A quote within a string constant must be specified twice to distinguish it from the string terminating quote.

Syntax rules

1   Program: ProgramHead SystemSpecification DeclarationPart
                        RoutineList ["BEGIN" StatementBlock]
                        "ENDPROGRAM" [";"].

2   ProgramHead: "PROGRAM" ProgramIdentifier ";" CompilerDirective.

3   CompilerDirective:
                        ["DECLON" ";" | "DECLOFF" ";"].

4   SystemSpecification:
                        ["SYSTEM_SPECIFICATION" Expression ";"].

5   DeclarationPart: {   "CONST" { ConstantDefinition } |
                              "TYPE" { TypeDefinition } |
                              "VAR" { VariableDeclaration } |
                              ImportDeclaration | ExportDeclaration |
                              DataListSpecification | RobotSpecification |
                              "STANDARD_IN" ":=" Identifier ";" |
                              "STANDARD_OUT" ":=" Identifier ";"
                        }.

6   ConstantDefinition;:
                        Type ":" ConstantIdentifier [":=" Expression] ";".

7   TypeDefinition:
                        StructuredTypeDefinition "=" TypeIdentifier ";".

8   VariableDeclaration:
                        Declaration | SignalDeclaration.

9   Declaration: DeclarationWithoutInit | DeclarationWithInit.

10 DeclarationWithoutInit:
                        Type ":" VariableIdList ";".

11 DeclarationWithInit:
                        Type ":" VariableIdList ":=" Expression ";".

12 VariableIdList:
                        Identifier { "," Identifier }.

13 SignalDeclaration:
                        Type ":" AddressList ";".

14 AddressList: Identifier "AT" Address
                        { "," Identifier "AT" Address }.

15 Address: SingleAddress |
                    ContinuousAddress |
                    NonContinuousAddresses.

16 SingleAddress:
                    DecimalConstant.

17 ContinuousAddress:
                    "(" DecimalConstant ".." DecimalConstant ")".

18 NonContinuousAddresses:
                    "(" DecimalConstant { "," DecimalConstant } ")".

19 ImportDeclaration:
                    "FROM" ProgramIdentifier "IMPORT"
                    ("ALL" | ImportIdList) ";".

20 ImportIdList: Identifier { "," Identifier }.

21 ExportDeclaration:
                    "EXPORT" ("ALL" | ExportIdList)
                    ["TO" ProgramIdList] ";".

22 ExportIdList: Identifier { "," Identifier }.

23 ProgramIdList: ProgramIdentifier { "," ProgramIdentifier }.

24 DataListSpecification:
                    ["IMPORT" "DATALIST" Expression ";"].

25 RobotSpecification:
                    "ROBOT" Expression ";".

26 RoutineList: { (ProcedureDeclaration | FunctionDeclaration)
                       [";"]
                    }

27 ProcedureDeclaration:
                    "PROCEDURE" ProcedureIdentifier
                    "(" [FormalParameter] ")" ";"
                    DeclarationPart "BEGIN" StatementBlock "ENDPROC".

28 FunctionDeclaration:
                    "FUNCTION" FunctionIdentifier
                    "(" [FormalParameter] ")" ":" Type ";"
                    DeclarationPart "BEGIN" StatementBlock "ENDFCT".

29 FormalParameter:
                    ["IN" | "OUT" | "INOUT"] ParameterSpecification
                    { ";" ["IN" | "OUT" | "INOUT"] ParameterSpecification }.

30 ParameterSpecification:
                    Type ":" ParameterIdList |
                    ConformantScheme ":" Identifier.

31 ParameterIdList:
                    Identifier { "," Identifier }.

32 ConformantScheme:
  "ARRAY" "[" IndexTypeSpecList "]" "OF" Type.

33 IndexTypeSpecList:
  IndexTypeSpecification { "," IndexTypeSpecification }.

34 IndexTypeSpecification:
  TypeIdentifier ":"
  ConformantParameter ".." ConformantParameter.

35 ConformantParameter: Identifier.

36 StatementBlock:
  { LabelIdentifier ":" | Statement ";" }.

37 Statement: [Assignment |
                       ProcedureCall |
                       ProgramflowStatement |
                       MovementStatement |
                       I/O-Statement
                  ].

38 Assignment: Object ":=" Expression.

39 ProcedureCall:
  ProcedureIdentifier "(" ValueList ")".

40 ProgramflowStatement:
  GotoStatement | ConditionalBranchStatement |
  CaseStatement | LoopStatement |
  WaitStatement | PauseStatement |
  ReturnStatement | ProgramStopStatement.

41 GotoStatement:
  "GOTO" LabelIdentifier.

42 ConditionalBranchStatement:
  "IF" Expression "THEN" StatementBlock
  ["ELSE" StatementBlock]
  "ENDIF".

43 CaseStatement:
  "CASE" Expression "OF"
  CaseList ["DEFAULT" [":"] StatementBlock]
  "ENDCASE".

44 CaseList: Case { ("|" | "WHEN") Case }.

45 Case: [CaseConstantList ":" StatementBlock].

46 CaseConstantList:
  CaseConstant { "," CaseConstant }.

47 CaseConstant: Expression.

48 LoopStatement:
  ForStatement | RepeatStatement | WhileStatement.

49 ForStatement: "FOR" Object ":=" Expression "TO" Expression
            ["STEP" Expression]
            StatementBlock "ENDFOR".

50 RepeatStatement:
            "REPEAT" StatementBlock "UNTIL" Expression.

51 WhileStatement:
            "WHILE" Expression StatementBlock "ENDWHILE".

52 PauseStatement:
            "PAUSE" [Expression].

53 WaitStatement:
            "WAIT" Expression "SEC" |
            "WAIT" "FOR" Expression
            ["TIMEOUT" Expression "SEC"
                 ["ERROR" StatementBlock "ENDERROR"]
            ].

54 ReturnStatement:
            "RETURN" [Expression].

55 ProgramStopStatement:
            "HALT".

56 MovementStatement:
            LinearMovement | PointToPointMovement |
            CircleMovement.

57 LinearMovement:
            ("MOVE" | "MOVE_INC") "LIN" ToPoint
            ContinuePathParameter.

58 PointToPointMovement:
            ("MOVE" | "MOVE_INC") "PTP" ToPoint
            PointToPointParameter.

59 CircleMovement:
            ("MOVE" | "MOVE_INC") "CIRCLE" Circle
            ContinuePathParameter.

60 Circle: Expression "," Expression | Path.

61 ToPoint: Expression | Path.

62 Path: "PATH" [Expression ":"]
            (Objekt "[" [IndexType] "]" |
            "(" GeometricExprList ")").

63 GeometricExprList:
            Expression { "," Expression }.

64 ContinuePathParameter:
            ["ACT_ROB" ":=" Expression]
            {   "C_CP" [":=" Expression] |
               "C_SPEED" [":=" Expression] |
               "SPEED_ORI" ":=" Expression |

```
                    "SPEED" ":=" Expression |
                    "ACC" ":=" Expression |
                    "TIME" ":=" Expression |
                    "UNTIL" Expression
                            [ "ERROR" StatementBlock "ENDERROR" ]
              |
                    "WOBBLE"
            }.
```

65 PointToPointParameter:

```
            [ "ACT_ROB" ":=" Expression ]
            {      "C_PTP" [":=" Expression] |
                   "SPEED_PTP" ":=" Expression |
                   "ACC_PTP" ":=" Expression |
                   "TIME" ":=" Expression |
                   "UNTIL" Expression
                            [ "ERROR" StatementBlock "ENDERROR" ]
            }.
```

66 I/O-Statement:

```
            OpenStatement | CloseStatement |
            WriteStatement | ReadStatement.
```

67 OpenStatement:

```
            "OPEN" "(" Object "," Expression ","  Expression "," Object
            ")".
```

68 CloseStatement:

```
            "CLOSE" [ "(" Expression ")" ].
```

69 WriteStatement:

```
            ( "WRITE" | "WRITELN" ) "(" FormatList ")" |
            "WRITERAC" "(" Expression "," ElementNumber ","
            ValueList ")".
```

70 FormatList:

```
            [ OutputValue
              [ ":" Expression [ ":" Expression [ ":" Expression
              ]]]
              { "," OutputValue
                  [ ":" Expression [ ":" Expression [ ":" Expression
                  ]]]
              }
            ]
```

71 OutputValue: Expression.

72 ReadStatement:

```
            ( "READ" | "READLN" ) "(" ObjectList ")" |
            "READRAC" "(" Expression "," Elementnumber ","
            ObjectList ")".
```

73 ObjectList: Object { "," Object }.

74 ElementNumber:
    Expression.

75 StructuredTypeDefinition:
    RecordTypeDefinition | ArrayTypeDefinition |
    FileType | SignalType | TeachType.

76 .i.RecordTypeDefinition;:
    "RECORD" ComponentList "ENDRECORD".

77 ComponentList:
    { Type ":" ComponentIdList ";" }.

78 ComponentIdList:
    ComponentIdentifier { "," ComponentIdentifier }.

79 ArrayTypeDefinition:
    "ARRAY" "[" IndexTypeList "]" "OF" Type.

80 IndexTypeList:
    IndexType { "," IndexType }.

81 IndexType: Expression ".." Expression.

82 FileType: "FILE" "OF" Type.

83 SignalType: ("INPUT" | "OUTPUT") [Type].

84 TeachType: ("TEACH" | "PERMANENT") Type.

85 Type: TypeIdentifier |
    StructuredTypeDefinition.

86 Expression: "+" Expression | "-" Expression |
    "NOT" Expression | Expression Operator Expression |
    Object | IntegerConstant | RealConstant | CharacterString |
    CallOrRecordOrPrecedence | ArrayInit.

87 Object: Identifier | ArrayElement | RecordComponent.

88 ArrayElement: Object "[" ArrayExpressionList "]".

89 ArrayExpressionList:
    Expression { "," Expression }.

90 RecordComponent:
    Object "." ComponentIdentifier.

91 CharacterString:
    SingleQuote
    { Any-Character-Except-Single-Quote |
     SingleQuote SingleQuote }
    SingleQuote.

92 CallOrRecordOrPrecedence:
    [Object] "(" ValueList ")".

93 ValueList: [Expression { "," Expression }].

94 ArrayInit: "[" InitValueList "]".

95 InitValueList: InitValue { "," InitValue }.

96 InitValue: [[Expression "OF"] Expression].

97 TypeIdentifier: Identifier.

98 ComponentIdentifier: Identifier.

99 LabelIdentifier: Identifier.

100 ProgramIdentifier: Identifier.

101 ProcedureIdentifier: Identifier.

102 FunctionIdentifier: Identifier.

103 ConstantIdentifier: Identifier.

104 Identifier: Letter { Letter | Digit }.

105 IntegerConstant:
          DecimalConstant | BinaryConstant |
          OctalConstant | HexadecimalConstant.

106 RealConstant:
          DecimalConstant Exponent |
          DecimalConstant "." { Digit } [Exponent] |
          "." DecimalConstant [Exponent].

107 Exponent:
          ("E" | "e") ["+"|"-"] DecimalConstant.

108 DecimalConstant:
          Digit { Digit }.

109 BinaryConstant:
          "2#" { "0" | "1" | "_" }+.

110 OctalConstant:
          "8#" { "0" | "1" | "2" | "3" | "4" | "5" | "6" | "7" | "_" }+.

111 HexadecimalConstant:
          "16#" { Digit | "A" | "B" | "C" | "D" | "E" | "F" | "_"}+.

112 Operator: "NOT" |
          "*" | "/" | "DIV" | "MOD" | "@" |
          "+" | "-" |
          "<" | ">" | "<=" | ">=" |
          "=" | "<>" |
          "AND" |
          "OR" | "EXOR".

113 Letter: "a" | "b" | "c" | "d" | "e" | "f" | "g" | "h" | "i" |
          "j" | "k" | "l" | "m" | "n" | "o" | "p" | "q" | "r" |
          "s" | "t" | "u" | "v" | "w" | "x" | "y" | "z" |
          "A" | "B" | "C" | "D" | "E" | "F" | "G" | "H" | "I" |
          "J" | "K" | "L" | "M" | "N" | "O" | "P" | "Q" | "R" |
          "S" | "T" | "U" | "V" | "W" | "X" | "Y" | "Z" | "_".

114 Digit: "0" | "1" | "2" | "3" | "4" | "5" | "6" | "7" | "8" | "9".

115 Comment: "{" { Any-Character-Except-"}" } "}".

116   SingleQuote: "'".

Alphabetical index of syntax rules

Address	15	FormalParameter	29
AddressList	14	FormatList	70
ArrayElement	88	ForStatement	49
ArrayExpressionList	89	FunctionDeclaration	28
ArrayInit	94	FunctionIdentifier	102
ArrayTypeDefinition	79	GeometricExprList	63
Assignment	38	GotoStatement	41
BinaryConstant	109	HexadecimalConstant	111
CallOrRecordOrPrecedence	92	I/O-Statement	66
Case	45	Identifier	104
CaseConstant	47	ImportDeclaration	19
CaseConstantList	46	ImportIdList	20
CaseList	44	IndexType	81
CaseStatement	43	IndexTypeList	80
CharacterString	91	IndexTypeSpecification	34
Circle	60	IndexTypeSpecList	33
CircleMovement	59	InitValue	96
CloseStatement	68	InitValueList	95
Comment	115	IntegerConstant	105
CompilerDirective	3	LabelIdentifier	99
ComponentIdentifier	98	Letter	113
ComponentIdList	78	LinearMovement	57
ComponentList	77	LoopStatement	48
ConditionalBranchStatement	42	MovementStatement	56
ConformantParameter	35	NonContinuousAddresses	18
ConformantScheme	32	Object	87
ConstantDefinition	6	ObjectList	73
ConstantIdentifier	103	OctalConstant	110
ContinuePathParameter	65	OpenStatement	67
ContinuousAddress	17	Operator	112
DataListSpecification	24	OutputValue	71
DecimalConstant	108	ParameterIdList	31
Declaration	9	ParameterSpecification	30
DeclarationPart	5	Path	62
DeclarationWithInit	11	PauseStatement	52
DeclarationWithoutInit	10	PointToPointMovement	58
Digit	114	PointToPointParameter	65
ElementNumber	74	ProcedureCall	39
Exponent	107	ProcedureDeclaration	27
ExportDeclaration	21	ProcedureIdentifier	101
ExportIdList	22	Program	1
Expression	86	ProgramflowStatement	40
FileType	82	ProgramHead	2

ProgramIdentifier	100	StatementBlock	36
ProgramIdList	23	StructuredTypeDefinition	75
ProgramStopStatement	55	SystemSpecification	4
ReadStatement	72	TeachType	84
RealConstant	106	ToPoint	61
RecordComponent	90	Type	85
RecordTypeDefinition	76	TypeDefinition	7
RepeatStatement	50	TypeIdentifier	97
ReturnStatement	54	ValueList	93
RobotSpecification	25	VariableDeclaration	8
RoutineList	26	VariableIdList	12
SignalDeclaration	13	WaitStatement	53
SignalType	83	WhileStatement	51
SingleAddress	16	WriteStatement	69
SingleQuote	116		
Statement	37		

6.2 Intermediate Code for Robots (ICR)

T. Clausen.

This chapter introduces the robot code "Intermediate Code for Robots", or in short ICR [1]. ICR is an ISO draft proposal to a standardized code designed for robot off-line programming. The described code is called "code" because it is a low level language with similarities to assembler code.

The ICR proposal has roots back to the CNC programming code CLDATA (Cutter Location DATA). CLDATA was one of the first codes for programming CNC machines, and it is basically used to define the path which the tool shall follow. ICR's major advantage is machine independency, i.e. it is not designed for any specific machine. Instead it reflects the basic functionality of a typical CNC machine. The German VDI adopted this idea in the design of the robot programming code IRDATA (Industrial Robot DATA) but extended IRDATA to include a large instruction set for arithmetic operations, program flow control, and I/O operations. In the late eighties ISO recognized the need for a standardized robot code between an off-line programming system and the robot and the development of ICR was initiated. Since the goal was much the same as with IRDATA, IRDATA was used as an important baseline for the ICR work. ISO published the first ICR documents in 1989.

This section addresses various aspects related to the ICR code. Section 6.2.1 gives the background, philosophy, and purpose of ICR. Section 6.2.2 deepens the technical aspects of ICR. Section 6.2.3 presents NIRO contributions to ICR. Section 6.2.4 evaluates ICR on basis of the NIRO experiences. Section 6.2.5

explains the present status of ICR. Finally, sect. 6.2.6 states the NIRO projects recommendations to future enhancements of ICR.

6.2.1 Introduction to ICR

The trend in today's robot programming is aimed at off-line programming because only limited time can be taken from the production time to program the robot. It is therefore important to focus on the functionality of the interface between the robot programming system and the robot itself. This interface is based on the robot off-line programming language.

Most robot languages are build on the same philosophy: They shall be easy to use by humans. However, off-line programming by means of computerized tools demands for other qualities of a robot language. It is no longer a need to make it easy for humans to understand since programs are to be exchanged between computers. Thus, the language has to be optimized with respect to other demands. It must be capable of expressing general robot tasks, be easy to generate, and easy to read and execute by machines.

ISO's draft proposal for a robot code interface meets these demands. The draft proposal is called "Intermediate Code for Robotics", ICR. The goal of ICR is to have a standardized way to transfer programs from off-line programming systems to robot controllers, which supports all the features that a robot controller usually performs and all the elements to describe a task.

Figure 6.2.1-1 illustrates the way ICR is intended to be used. ICR is strictly designed as an interface between off-line programming systems and robots. ICR is not intended to be the language which the operator uses in the programming of the robot. The code is simply not suitable for this purpose since it is optimized with respect to fast machine execution. ICR is machine oriented instead of user oriented, and it is intended to be the output of a software tool rather than the language which the programmer use.

The programmer uses a high level interface to the robot programming system in the creation of programs. It may, for example, be a "point and click" facility in a graphical environment or a high level programming language like IRL. The result of the programming is an internal description of the robot task. This internal description is compiled to an ICR program and transferred to the robot for real-time execution. Figure 6.2.1-1 includes other elements which will be addressed later in this chapter.

Besides this ICR is intended to:

- Be a *neutral interface* between off-line programming systems and robots. By making ICR system independent it is possible to combine the most suitable programming system and robot for a given task.
- Reduce the implementation effort in constructing robot controllers. The ICR processor on the robot consists of two parts: A front end that manages the program read-in and robot independent instructions, and a robot specific

part which takes care of robot dependent instructions. It is only necessary to implement the front end once. It can then be reused in the next installations.
- Reduce the programming by using redefined program modules. Some types of tasks include standard sub-tasks which can be programmed once for all.

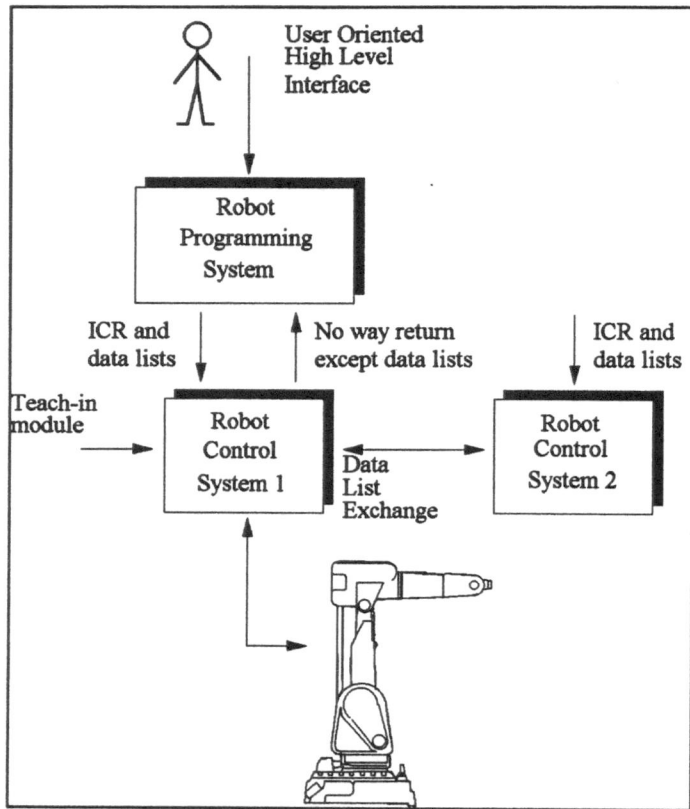

Fig. 6.2.1-1: The exchange of ICR programs and data lists between subsystems in a robot environment [1]

The basic philosophy of ICR is to have a simple, yet powerful code for expressing the task for real-time robot execution. At the same time the code has to be system and machine independent. This goal is fulfilled by designing ICR as an intermediate code.

An intermediate code in this context is a universal, system independent, assembler code at the level between machine code and high level languages like PASCAL and C. As a consequence, intermediate codes do not include high level instructions such as C=A+3*B. Instead, it is possible to develop these instructions out of a combination of more basic instructions. The post fix notation is used for this purpose.

The most important quality of the intermediate level for off-line programming is that all higher level languages can be mapped onto the same intermediate code. The intermediate code is, therefore, suitable for a neutral interface.

The robot independence is ensured by designing ICR for a reference robot model which reflects the general functionality of an industrial robot. Functions common to most industrial robots are included directly in the instruction set of ICR. Examples are motion, digital I/O control, and gripper control. More product specific functions are not supported directly, such functions are constructed by combining the basic instructions instead.

It is important to realize that a robot language consist of two different types of commands: A robot dependent part and a robot independent part.

The instructions belonging to the robot dependent part are used to control the motion of the robot, the I/O ports, the operation mode et cetera. That is, this part of the commands are used to control the physical behaviour of the robot.

The robot independent part is used for program flow control and general data manipulation just as it is done in any ordinary computer language as for example FORTRAN or C. Typical elements in this part are arithmetic operations, memory management, data types, program flow control and so on.

The robot independent part of the robot language is normally quite simple. This is more because of historical reasons rather than practical limitations. Robot languages are from the early days used to control the movements of the robot, the robot related part is therefore the most developed part. Besides, it is often possible to carry out the necessary calculations in the off-line programming system before the robot program is generated. These pre-calculations reduce the complexity of the motion program. However, instructions for data manipulation and program flow control are becoming more important since these instructions give a more flexible robot system. This flexibility can be used to create user specific solutions instead of being dependent on the solutions supported by the vendor.

The ICR code itself is line oriented and every line holds one instruction. All code and data elements are represented by ASCII-characters. The block structure of a high level programming language is supported for structured programming. Variables and other programming objects are identified by symbols with respect to their scope in the block structure. For some special use, absolute addresses may be used, e.g. to handle memory mapping I/O directly.

ICR consist of a set of basic instructions. The following list gives a brief summary of the ICR instruction set. The list is ordered in groups reflecting the different classes of instructions:

- *Memory and data management.*
  Instruction for loading and storing of variables and memory allocation. ICR has the following basic types: Integer, real, boolean, character, string, position, orientation, pose, joint, main_joint, add_joint, and robtarget. Besides this it is possible to construct structured variables like arrays and records.

- *Program flow control.*
  These functions are used to control the program flow. Examples are function definition, loops, branching, jump, and wait.
- *Boolean and arithmetic operations.*
  These instructions include the boolean and arithmetic operations as they are known by high level languages. Examples are: Add, subtract, divide, trigonometric functions, logarithmic functions, string operations, type conversions, and boolean operations. Robot related instructions do also exist. Examples are: Distance between two points, conversion between cartesian position and joint coordinat, geometrical transformations, and construction of a position from reals.
- *Technical specifications.*
  This group of instructions is used to control the condition for the motion. Examples are: Acceleration control, velocity control, motion accuracy, selection between robots, and reading of current position.
- *Move and end effector control.*
  These instructions allow the description of robot moves and end effector operations. Examples are: Joint motion, linear motion, circular motion, calibration, selection of tool, activation of tool, TCP definition, and TCP selection.
- *Interchange of data.*
  Instruction statements for communication with peripheral devices, operator control, or process control. Examples are: Reading from the console, writing to the console, control of communication channels, reading/writing communication channels, digital I/O, and analog I/O.
- *Data list management.*
  These instructions are used to manage data lists. Examples are: Create data lists, read from a data lists, writing to a data lists, erasing data lists.
- *Description of robot facilities.*
  These are instructions for limitation of work space, robot kinematics, tool description, units of measure, length and angles, scaling factors, and others.

It is important to recognize that programming languages belong to different abstraction levels. Native robot languages are normally on a high abstraction level compared to ICR because they normally contain commands that are stronger than the basic movement and data manipulation commands found in ICR.

ICR is designed for the intermediate level, the level between machine code and high level languages like IRL. As a consequence of the intermediate level ICR does not include high level commands but it is instead possible to construct these out of combination of more basic orders. This situation can be illustrated by the following example:

Example:

An object shall be found. It is done by moving a contact sensor slowly towards the object. When the sensor touches the object the robot stops and the object is found. Such a search command is often programmed by a special move command in a high level robot language. It could for example look like the following command. The example is taken from ABB's robot language ARLA:

```
POS V=100% SEARCH S6
```

Where POS is the move command. SEARCH tells that it is a search command and the "object is found" signal level is stored in S6. The same command could be written in ICR by combine a set of more fundamental commands. This is illustrated by the following example. It is written in a high level language instead of ICR for the readability:

```
REPEAT
  POS := POS + (1,0,0);
  MOVE(POS);
UNTIL SENSOR=HIGH;
```

The idea is to move the robot in small steps and test for the stop condition between each step. It is here important to recognize that the ARLA command is on a higher abstraction level than the "ICR" example.

The key intention of ICR is to interface off-line programming systems with robots, i.e. to interface machines. ICR is therefore designed for the intermediate abstraction level which is a prerequisite for fast machine execution.

Nevertheless it should be possible to display the "low-level-instructions" and their parameters on a panel or a display device on the robot controller. The ICR code is therefore designed with mnemonic abbreviations for the various instructions. This makes it possible follow the execution of the ICR program on the robot controller and debug it.

Another important quality of ICR is it's flexible data lists. The basic idea of using data lists is to separate the data associated with the program from the program itself. It is then possible to access the data in the data lists from the ICR program. Various types of data can be stored in data lists. Examples are positions, technological data, numbers of work pieces, and quality statistics from the robot. By using data lists it is for example possible to:

- Create the program in one system and the data lists in another system. The data list could for example hold the positions and be created on-line on the robot, while the logical part of the program is created off-line. This way it is possible to enter the exact location into the program. It is also possible to bring the positions (the data list) back to the off-line programming level and thereby update the geometrical model of the work scene. See figure 6.2.1-1.
- Change data. This is for example relevant if the positions in the task have to be changed frequently. Another example is welding parameters which may be optimized under the running-in.

- Exchange data between two robot systems. The first robot system is for example used to find the seam with a laser scanner and to write the seam positions in a data list. The second robot is then used to read the positions of the seam in the data list and to execute the weld operation on basis of the data from the first robot. See Fig. 6.2.1-1.
- Write quality control lists. The robot controller may carry out quality control tasks and the results can be written in data lists.

ICR file example

The following examples illustrate what an ICR program and a data list look like. The ICR program does two tasks. First it calculates the speed on basis of input from an analog port. Then it commands the robot to move to a new position:

Example:

```
1,PBEG,2,"A program example";
2,PUSHI,#5
3,AIN;
4,PUSHR,#31;
5,MULI;
6,FLOAT;
7,PUSHI,#0;
8,W_VEL,T,2;
9,PUSHS,#"START_POSITION";
10,PUSHS,#"EXAMPLE";
11,DLEIN,11;
12,LMOVE,1,J,W;
13,PEND;
```

Line 1 and 13 are the start and end of the program. A value is read in from an analog port in line 2-3. Line 2 is used to indicate the port number and the AIN command returns the value on the stack. In line 4 the integer 31 is pushed on the stack and the value from the analog port is multiplied by this value in line 5. The result is converted to a real in line 6. Line 7-8 set the velocity on basis of the calculated velocity. Line 9-11 is related to the move instruction in line 12. First two strings are pushed on the stack. They give the name of the data list element holding the position to move to and the name of the data list. The data list itself is shown in the next example. The DLEIN instruction reads in the data list element (the position) which is used by the LMOVE command in line 12.

Data lists have a structure much like an ICR program. A data list consists of lines and every line holds one "data" entity. A data can be any of the basic types defined in ICR. A name is associated with every data entity by which it is addressed. The following example shows a data list:

Example:

```
1,DLHEAD,"EXAMPLE",3;
2,DLDAT,"WORK_PIECE_NAME",5,#"Muck up";
3,DLDAT,"CYCLES",1,#42;
4,DLDAT,"CURRENT",2,#100.0;
5,DLDAT,"START_POSITION",14,#(14,923.0,373.9,....);
6,DLEND,"EXAMPLE";
```

Line 1 defines the beginning of the data list. The data lists name is EXAMPLE and it contains four elements. Line two to five contain the four data entities. After the DLDAT keyword follows the name of the data list element, a number indicating the type of the data, and the data itself. Line six concludes the data list.

6.2.2 Technical Aspects of ICR

This chapter deepens central technical aspects of the philosophy behind ICR. Four topics are addressed as follows:

- Two approaches to execute ICR on a robot
- Consequences of choosing the intermediate level
- An argument for that most other robot languages can be mapped onto ICR
- Two central reasons for having computational power in ICR.

Two approaches to execute ICR

There are fundamentally two ways for running an ICR program on a robot:

- The ICR program can be translated into the native robot language and then be down loaded to the robot controller.
- The ICR program can be executed on an interpreter which passes the movement commands to the path planner of the robot controller.

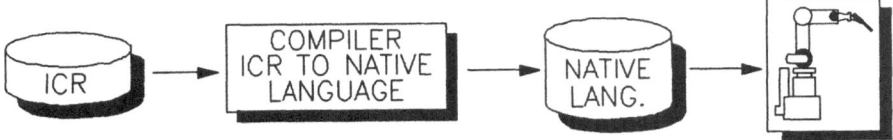

Fig. 6.2.2-1: The compiler solution

A compiler is a program that is able to translate from one language to another, see Fig. 6.2.2-1. The compiler solution is often preferred in a CIM environment because it offers a straightforward solution for the system designer. It gives a good answer to the problem: "I have one language from the off-line programming system and I want to use this language on a robot that understands another language". The major drawback of compiling the ICR program to the native format is that the native format may not be able to express all that ICR can express.

ICR is more suitable for interpretation because of it's low level nature. An interpreter is a program that is able to read another program line by line and execute it's contents stepwise. A robot controller does always include at least one interpreter: The one that reads the native robot language.

ICR interpreters within the NIRO project consist of two main parts: A general purpose ICR interpreter and a robot specific module. This structure makes it possible to reuse the general purpose part of the ICR interpreter in other installations. Figure 6.2.2-2 shows a graphic overview of an interpreter.

The general part reads the ICR program line by line and handles the robot independent instructions. Only robot related instructions are sent to the robot specific module where they are executed by the robot manipulator and the additional equipment. Examples of robot related instructions for this interpreter are: Movement, velocity, digital I/O (incl. gripper), and TCP selection control.

A shortcoming of the interpreter solution is that an existing robot controller must be equipped with an extra controller card for ICR. Alternatively an interpreter may be implemented on an external computer.

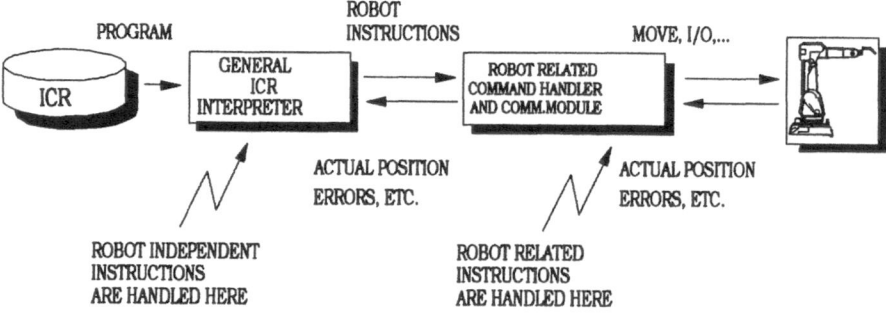

Fig. 6.2.2-2: The interpreter solution

To summarize: The strength of a compiler solution compared to an interpreter solution is that compilers do not require real-time access to the robot controller. An interpreter solution is capable of supporting a larger part of the ICR instruction set since it actually executes the instructions. It is only limited to the physically limitations of the robot (move types, I/O ports, etc.), but it requires direct access to the robot controller. On the long view the interpreter solution is the most advantageous approach since is not necessary to translate the neutral robot program before it can be executed. However, today's industrial robots are not designed to be used with interpreters (since real-time access to the controller is not available) and therefore the compiler approach is still relevant.

Consequences of choosing the intermediate level

ICR is designed for the intermediate level which is a quite unusual approach in robot languages. This choice affects ICR in so many ways that it perhaps is the one most significant characteristic of the code. The NIRO project has worked intensively with ICR and the following text introduces important consequences of choosing the intermediate level, summarized by the following statements:

- It is difficult to read an ICR program for a human.
- ICR is more voluminous than an ordinary high level language.
- It is relatively simple to make an interpreter for ICR.
- It is relatively complicated to make a compiler for ICR.
- It is, in the general case, impossible to translate ICR to a high level language.

Some of these statements are deepened in the following.
ICR is not designed for human interaction. Anyway, it is important to notice that ICR is more difficult to use for a human than a high level language like IRL because of the low abstraction level. In ICR each instruction has a very basic function. Therefore, it takes several ICR lines to express what may take up only one line in a high level language. ICR is consequently more difficult to overview, read, and write for humans, because the abstraction level does not fit our preferred abstraction level. Furthermore, ICR programs are also more voluminous. Imagine for example the operation to set the variable R2 equal to five times R1 minus one. It will look like the following in ICR:

Example:

```
4,PUSHI,R1;
5,PUSHI,#-1;
6,SUBI;
7,PUSHI,#5;
8,MULI;
9,POPI,R2;
```

High level languages focus on a more operator friendly abstraction level where the general intention is to have the complete instruction on one line. This makes it easier to use the language for humans. The same program fragment will look like the following in a high level language:

Example:

```
R2 = 5 * (R1 - 1);
```

On the other hand, the low abstraction level is very suitable for fast machine processing because each command has a simple, very precise meaning. This is not the case with high level languages. A high level languages statement is much more complicated seen from a machine's point of view. For example, the operator hierarchy is most important. The multiplication operator has higher precedence than the subtraction operator, but the parenthesis forces the subtraction to take place first. A high level language processor must therefore include routines that analyses the statement and evaluates the operator hierarchy and uses the operators on the operands in the correct order. This is not a problem in ICR where the operator hierarchy is given indirect by the sequence of the instructions. ICR is therefore executed faster because it is not necessary to make complicated analyses of the operator hierarchy.

Low level languages as ICR are very suitable for interpretation because every instruction has a clear and simple meaning. It is therefore relatively easy to construct an interpreter for such a language. An interpreter for a low level language is, in principle, one big case statement in a loop as illustrated in Fig. 6.2.2-3. The interpreter repeats the loop until it reads an "END OF PROGRAM" instruction. In every loop a new ICR instruction is read and the CASE structure is used to find the corresponding code. The code implements the actions that are necessary to execute the instruction.

```
REPEAT
    READ A NEW ICR INSTRUCTION
    CASE ICR INSTRUCTION OF
        PUSHI:   Code to execute command
        PUSHR:   Code to execute command
        ADDI:    Code to execute command
         ...
         ...
        W_VEL:   Code to execute command
        LMOVE:   Code to execute command
    END CASE
UNTIL END OF PROGRAM
```

Fig. 6.2.2-3: The structure of an ICR interpreter

It is more complicated to construct a translator from ICR to a native robot language. The complexity of the problem does, of course, depend on the complexity of the ICR program. In case of the ICR program only consist of move instructions it is quite straightforward to translate since a "one to one" mapping exists. Such a compiler simply translates the same instruction to other words of the same meaning. The problem becomes much more complicated when the goal is to compile more complex program structures to a high level language.

The problem is, probably, best introduced by using an example from the CAD world. The off-line programming system is represented by CAD system A, ICR is represented by the "drawing exchange format", the robot is represented by the plotter, and the native robot language is represented by CAD system B. See Fig. 6.2.2-4.

The CAD system A is able to draw various objects such as triangles, squares, and straight lines. The "drawing exchange format" represents any kind of drawing by combining lines and is thereby a kind of low level format. By combining lines it is possible to build up other objects as triangles and squares. The "drawing exchange format" is therefore able to carry any kind of drawing and keep it's appearance. But the description of the drawing is on a lower abstraction level than it was in the sending CAD system A since all information about the connection between the lines are lost now. This is a fact, even though the drawing still has the same appearance.

This is not a problem when the drawing is send to a plotter. In fact, it is quite an advantage because this format is able to represent practically any kind of drawing in a simple way. It is also an advantage for the plotter because it gets a simpler job and it is possible to optimize the plotter routines for drawing lines. Hence, the loss of information is unimportant in this case since the "drawing exchange format" holds information enough to make a correct plot.

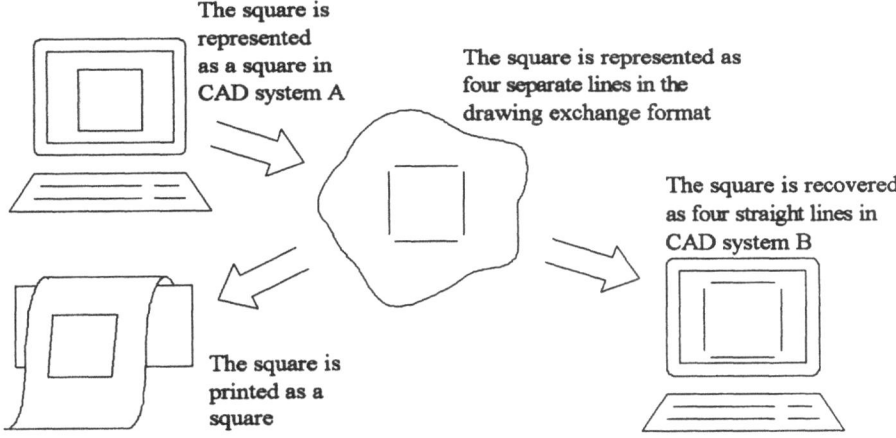

The square is represented as a square in CAD system A

The square is represented as four separate lines in the drawing exchange format

The square is recovered as four straight lines in CAD system B

The square is printed as a square

Fig. 6.2.2-4: Information is lost under the transfer

The drawing can also be read into CAD system B and edited there but it will not be a drawing made of squares and triangles. Instead it will be a drawing made of straight lines. It is, for example, not possible to re-create a square because it is impossible to determine if the four lines should form a square or are mend to be four separate lines. The "drawing exchange format" holds no information about this matter and the information is therefore lost.

This is a central problem in compiling a low level language to a high level language. The two programs may do the same job when they are executed, but the high level program holds more information about the internal structure of the program than the low level language.

Turning back to ICR, Fig. 6.2.1-1 includes an arrow from the robot to the off-line level that says "No way return except data lists". This arrow indicates that it is problematic to bring an ICR program up on the off-line level and translate it back to a high level representation.

A compiler tool may be able to recognize fragments of the ICR program like move instructions, loops, or calculation of variables, but a deeper meaning of the total code would be very difficult to recognize. The problem is again that it is impossible to determine from which set of high level orders a given low level program is made. (It could for example have been written directly in the low level language).

That is, it is not possible to raise the abstraction level in the general case. The conclusion of this language level discussion is therefore that it is very difficult to *translate* ICR to a native robot language on a higher level (and use the higher abstraction level) in the general case. ICR is basically designed to be interpreted and it is well fitted for this purpose.

ICR is usable to represent most known robot languages

It is of some importance that ICR can represent other robot programming languages. "Represented" does here mean that it is possible to translate programs written in other languages to ICR without loosing the *functionality* of the program.

Programming languages can be compared by their *strength* [2]. By comparing the strength of different programming languages is it possible to determine if the one language can represent the other. That is, language A can represent language B if the strength of language A is higher or equal to language B:

$$Strength(A) \wedge Strength(B)$$

The strength of ICR is as high as the strength of other ordinary robot languages because ICR includes all common program structures, data types, and robot control instructions, i.e. there is a very fine fit between the instruction set of ICR and other languages. ICR is therefore able to, and well fitted to, represent most other robot programming languages.

Reasons for having computation power in ICR

ICR differs from many conventional robot languages by having comprehensive computing facilities. This is very useful because it causes a more flexible interface. Two key reasons for having calculation capabilities in the robot language are:

a) It is possible to make calculations on sensor measurements
b) Complicated operations can be stored as formulas instead of a high number of steps.

These two key reasons are addressed in the following.

As the demands for flexibility become more present the need for sensor information from the environment of the robot increases. It is therefore useful to be able to make calculations on the sensor measurements. A vision system can for example count the number of work pieces on a pallet and calculate their positions. This information can be used in a repeat loop until there is no more work pieces.

Another example is corner finding. The problem is to find a corner for a welding operation. The corner can be found by touching the three plates forming the corner. The search command in example 6.2.1-1 can be used for this purpose. With the knowledge of the three wall positions it is possible to *calculate* the location of the corner. This will result in a program with the following structure:

```
1 Move to the start point for the first stop-move
2 Do the first stop move
3 Get the x position of the robot
  ...
  The same for y and z
  ...
10 Calculate the corner position on basis of the
   measured x, y, and z values
```

The second key reason for having computational capabilities in a robot language is the ability to *calculate* the needed operation instead of having a long program with the specified operations.

Consider the following task. A robot has to move along the surface of a three dimensional object carrying out a process (welding, gluing, etc.). The path to be followed is described as a Bezier curve [3] which is generated on basis of a CAD model of the object. The Bezier spatial curve is described by the following equation:

$$P(t) = (1-t)^3 P_1 + 3t(t-1)^2 P_2 + 3t^2(1-t)P_3 + t^3 P_4 ,$$

where the P_1, P_2, P_3, and P_4 are four known points in space which define the curve. See Fig. 6.2.2-5.

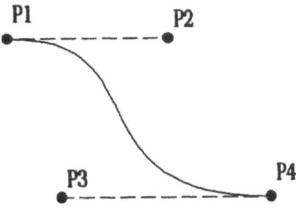

Fig. 6.2.2-5: A Bezier curve and its four control points

There are two ways to program the motion along the curve. The first possibility is to calculate a high number of intermediate points on the curve and write as many move statements in the robot program. This will result in a large program.

The second possibility is to include the formula defining the Bezier curve in the ICR program. This leads to a more compact program since the many intermediate points are calculated on the run as illustrated in Fig. 6.2.2-6. Besides this, it is very simple to modify the path by modifying the control points.

```
t = 0
REPEAT
  P(t) = (1-t)³P₁ + ......
  MOVE_TO(P)
  t = t + 0.1
UNTIL t=1
```

Fig. 6.2.2-6: A program for controlling Bezier motion

6.2.3 NIRO Contributions to ICR

The NIRO project has contributed in various ways to the standardization of ICR. In the beginning much effort were put onto an intensive review of the ICR draft proposal. This resulted in a number of enhancement ideas:

- The ICR document included a number of formal errors. Examples are missing brackets, incomplete instruction specifications, and the like. These were identified and registered.
- Proposals for making the draft proposal easier to read. An introduction and an implementation specification were for example proposed.
- New instructions within the scope of ICR. An example is the instructions to convert between integers and strings. These instructions are necessary to print an integer on the console, but they did not exist in the proposal.

The NIRO project has also contributed to ICR beyond the present scope of standardization. The practical ICR implementations with specific tasks in mind (see chapter 8) showed the need for more application specific instructions. For example, one of the tasks is to weld large ship sections. The existing instructions could be used for the basic robot operations as motion and I/O activities. However, the task did also require sensor operations and welding. Thus, it was necessary to develop new ICR commands for these operations. The NIRO project has especially contributed to enhance ICR within the following areas:

- General instructions for welding operations. Much effort have been put in defining general usable instructions for controlling welding operations. The developed instructions can be used to control: The path, start welding data, main welding data, end welding data, weave data, and sensor operations related to welding.

- Definition of PLC instructions. Robots are normally integrated with a number of other machines. The overall control is typically handled by a Programmable Logic Controller (PLC). In smaller working cells it may be beneficial to use the robot for this synchronization to achieve a simpler system. As a NIRO result, PLC functionality is included in ICR.
- Definition of general sensor instructions. External sensors enable robots to react on changes in the environment in an intelligent way. The developed instructions can be used to specify the way a sensor shall control the robot motion.

6.2.4 Experiences with ICR

This chapter summarises central experiences with ICR achieved in the NIRO project. The preceding two chapters have also included fragments of experience, but they have been more background related. The experiences presented in this chapter are related to the neutral robot language concept, it's use and it's implementation.

The most important experience with ICR in the NIRO projects is: ICR does work!

The NIRO project has tested ICR through several implementations (called demonstrations) which are described in detail in Chap. 8. These implementations have shown that it is possible to use ICR as a neutral interface between off-line programming systems and robots. Our experience build on one off-line programming system, one IRL to ICR compiler, two robot simulation systems, and six different robots.

The first official NIRO demonstration in 1991 at Odense Steel Shipyard in Denmark demonstrated how it is possible to execute the same neutral program on three different robots. The task were a fictive welding operation where the robots moved along the correct paths but no welding was actually performed. The program included motion and geometrical transformations. The task and the robots were selected carefully to make sure that the same program could be executed on all three robots.

This leads to another experience. ICR is *not totally neutral* when it comes to the practical execution of a program on a robot. In theory, it should be possible to generate a program off-line and then transfer the program to a arbitrary robot supporting ICR and execute it there. However, this is difficult to do in practice because robots differs very much in their physical capabilities. Examples are:

- The work space of different robots are not alike. A program including two meter long paths can not be executed by a miniature robot.
- A five axis robot can not solve a problem that requires six axes.
- Port characteristics. The amplification by A/D and D/A converters differs.

- Process equipment.

It is necessary to have these general restrictions in mind. ICR is not totally neutral in the sense that the same program can be executed on any robot that supports ICR. Neutral means that ICR is a system and machine independent *media*. That is, ICR guarantees that the program from the sender system represent meaningful information to the receiver.

Another facet of this problem is the need for *application specific* instructions in ICR. Sect. 6.2.4 presented NIRO contributions to ICR beyond the present standardization scope. These extensions indicate that robot control is more than simple motion control. Robot control does also include areas such as process control and control over external equipment. Instructions to control a given application is therefore most relevant. However it is a field that is difficult to standardize because of high complexity and many different solutions. Each robot vendor has his own ideas about process control etc. and it is consequently a "never ending story" to standardize the applications related to robotics. In fact, vendors are often using their special process control facilities as a key sales argument. Nevertheless, it is our experience that it is an absolute must to find a way to include instructions for application control, because motion is not to much use by itself.

Within the NIRO project we have used both the interpretation approach and the compiler approach, see Chap. 7. It is our experience that it has been relatively straight forward to construct the interpreters because of the low level nature of ICR. However, the interpreters support nearly the full ICR instruction set and the interpreter software is consequently relatively large (app. 1 MB source code). Our ICR interpreters are divided into a robot independent part and a robot dependent part as described in Sect. 7.2.2. It is our clear experience that it is a useful strategy because it makes it fairly easy to port the interpreter to new robots.

It has been more difficult to construct the compilers because of the reasons addressed in Sect. 6.2.2. The compilers do not support the full ICR instruction set. Instead the compilers support the instructions necessary for the given application. This way it was possible to limit the problems related to compiling and achieve an operational compiler

It have also been the experience of NIRO project that ICR is designed strictly for off-line programming and computer manipulation. It has been rather difficult to correct errors on the robot level. The hindrances have been of various kinds. For example, a small but very annoying point is the line numbers. If lines are added or removed it is necessary to renumber the program. This makes it cumbersome to correct even small errors. Another annoying problem is the lack of readability of ICR programs. ICR programs tends to become complicated to survey because the structure of the program is not visible as in most high level languages (empty lines, indentations, etc.).

As we worked with ICR we learned to use the data lists. They have offered a convenient way to handle data that are being changed frequently and they saved us from editing ICR programs to some extend. The data lists have been a pleasant experience.

123

6.2.5 Present Status

Various inputs from the NIRO project have been officially dealt within the ISO and have been integrated in the last ICR document (version 2.0). The discussions within the ISO was strongly influenced by these contributions. Especially the practical implementations gave valuable hints for improvements. The results of these inputs were improved technical documents for ICR.

ICR was in a state ready for becoming an international standard. In this situation the USA and Italy started to participate in the ISO working group (ISO TC184/SC2/WG4). The overall discussion was restarted by the new member countries and they formed together with Japan a strong opposition against the existing proposals and more or less any standardization which differs from their own view point. In this situation the Swedish national committee decided to reject the proposal although the Swedish delegation had supported and agreed the common decisions up to that moment. Thus balloting on ICR failed due to USA, Japan, Sweden, Italy, and Switzerland although more than two third of the countries voted positive. As the complete work of the recent years was questioned by several delegations the convenor decided to pause work within WG4 for one year until summer 1993.

Regardless of the ISO situation standardization work continues. In Germany the DIN/NAM decided to take the already published VDI recommendation IRDATA (Industrial Robot Data), which influenced ICR very much, as a German standard for the code level. IRDATA will become a German standard in 1992 and will be input to the European standardization bodies in 1993.

Japan was also active. Japan Industrial Standards (JIS) did in February 1992 start work on another proposal called STROLIC (Standard Robot Language in Intermediate Code).

This is only two of several national standardization proposals that are being prepared. Hence, there is a great deal of competition between the various proposals and a significant effort is required to achieve an international standard.

6.2.6 Recommendations for Further Enhancements

This chapter lists recommendations for further enhancements of ICR based on the experiences achieved in the NIRO project.

A better introduction material is needed

ICR is rather unapproachable for the novice because the ICR documents do not include a comprehensive introduction. Furthermore, the specification part of the ICR draft proposal is also quite difficult to use because of inexact definitions. We do believe that it is very important to bring the ICR documents up on a level where it is more "novice friendly" and clear in it's specifications, because unclear documents will prevent ICR becoming a standard.

Reload of ICR code

ICR is not designed to be reloaded back in the off-line programming system as described previously. However it is a general experience in the NIRO project that such a facility is very valuable, especially because some types of programming are made best on the robot itself. For archiving and programming purposes is it preferable that it is possible to have this reload facility. The NIRO project has no proposals to how such a facility can be implemented, but we will simply state that this field need more investigation.

Administration of user defined instructions

The NIRO work has shown a clear need for extending ICR beyond it's present scope by defining new instructions. The question is then whether an ICR program which uses many user-defined instructions is neutral. On the other hand, if the ICR code is expanded with new instructions whenever robots are used for a new application, the number of instructions will increase heavily. A way to administrate user defined instructions is therefore needed. One solution may eventually be to use libraries.

Development of more compact commands

ICR programs have a tendency to be quite voluminous compared to high level languages because of it's very explicit nature. This is a disadvantage in several ways:

- The ICR code takes more program space up in the robot controller.
- ICR files are slower transferred to the robot, which results in a higher standby time for the robot.
- ICR programs are more difficult to overview

Therefore, ICR statements should be as short as possible and as powerful as possible to shorten the length of the ICR programs.

A library concept

It would be of great importance to be able to merge a number of sub-programs into one "ready to run" program. The sub-programs can come from various sources and may be produced at different times. Sub-programs can also be reused in the production of new programs. Sub-tasks can be generated by different programming systems and languages and finally linked together to a complete robot program on the intermediate level. It is thereby possible to choose the most suitable programming tools for different sub-tasks.

The current state of ICR gives a suitable language kernel, but when one want to implement ICR in a production system the need for more application specific facilities raises (e.g. process control). One solution might be to develop new ICR instructions. Another solution might be to use libraries instead. The basic idea is to implement the needed instructions in subroutines kept in libraries. The subroutines can then be linked to the programs from libraries when needed.

A library concept might also be useful in the combination of off-line programmed and on-line programmed ICR fragments.

Multitasking

ICR includes a simple type of multitasking where it is possible to give move instructions a "no wait" parameter. This parameter controls the ICR interpreter to execute the lines following the move instruction immediately, i.e. to *not* wait for the motion to end before the program execution is continued. This facility can for example be used to prepare the next moves while the robot is moving.

However, true multitasking facilities as it is known in UNIX would be preferable since it makes it possible to run several processes in parallel. One process could control the motion while other processes monitors the process, takes care of communication with external equipment, quality control, etc.

Sensors and robot calibration

Deviations between programmed and needed motion are often the direct cause to inadequate robot performance. There exist two fundemental strategies to improve the accuracy: The robot system and environment can be calibrated to achieve a correct model or the robot can be equiped with sensors by which it is capable of correcting it's motion to the actual conditions. A robot language should include facilities for such operations.

7. Development of Processors for IRL and ICR

Chapter Editor: *D. Ball*

7.1 Processors for IRL

7.1.1 IRL Preprocessor for GRASP

P. Sorenti

GRASP for offline programming - GRDATA

GRASP is a general simulation tool from BYG Systems Ltd using full three dimensional graphical modelling and simulation techniques to address a wide variety of application areas. In addition to 3D graphics, its use as a tool for offline programming is already well established. Its ability to offer real time path generation for kinematic structures and on-line collision detection make it an obvious candidate for the generation and validation of robot programs remote from the robot system itself. Robot programs in GRASP are referred to as 'tracks' which contain individual instructions known as 'steps'.

Once a robot track has been generated and verified to the satisfaction of the operator there needs to be a mechanism by which the programs can be transferred to the actual robot controller for subsequent execution. GRASP can create a file describing the robot program in its own native language format called GRDATA.

The GRDATA language is regarded as a 'neutral file', albeit an in-house format. It is independent of the robot type at this stage. It is only when the robot type is specified that the GRDATA file passes through a robot-dependent postprocessor which produces robot-specific code. Each of these robot specific postprocessors has a common task at the start of the conversion: scanning and parsing the GRDATA file and creating a representative data structure. This forms the so-called 'front-end' of each postprocessor which is common amongst all postprocessors. It is only the 'back-end', or robot code generation and output part that needs to be different. This greatly reduces the amount of work required when developing a postprocessor for a new robot controller with yet another language type specific to it.

The generation of an external file to GRASP rather than creating the robot specific code directly, greatly improves the maintainability of each postprocessor since upgrades to each robot language processor do not require the release and supply of a new version of the entire GRASP software.

GRASP preprocessor to IRL implementation

In complete contrast to the existing method described above by which robot programs are created via the GRDATA 'neutral' interface, the generation of IRL code takes place entirely within the GRASP software. The reason for this is that it is intended that the future development of the IRL neutral interface will provide a stable platform to which the use of GRASP offline programming may migrate in order to complement the existing (GRDATA) protocol.

With a view to more long term developments of the GRASP software it seems prudent to incorporate the results of the IRL standard within it so that when the next generation of the GRASP software code is implemented the actual definition of the GRASP track and step database can directly reflect the IRL program code. After all, this is one of the major aspirations of work of this type: ie. that robot vendors and software suppliers adhere fully to a well-maintained and clearly defined standard for robot programming (and CAD data storage/transfer).

Further, since IRL is a high level language it is hoped that it will compliment the existing interfaces to robot controllers used by GRASP for interactive program generation. In this way users may be offered the choice to program directly in IRL at the graphical offline programming stage. This would ensure a consistent scheme by which information is generated and transferred.

This approach, i.e. that of devising output of program code directly from the GRASP simulation software, follows the same lines as that of the STEP preprocessor from GRASP described in Chapter 5.

GRASP and IRL data types

In considering the generation of IRL code from GRASP there is a need to review the basic ways in which information related to robot motion and program logic etc are defined. In this way the method by which data is mapped from the preprocessing system i.e. GRASP to the target programming language i.e. IRL can be appreciated more clearly. Here are the comparisons between them:

GRASP	IRL	
Integer	INT	
real	REAL	
signal	BOOL	
refobject		POSE

Those IRL types not included here do not have a direct equivalent in GRASP. There is however one important type which needs to be mentioned, the ROBTARGET data type. This incorporates information about the robot's configuration (the STATUS field of the ROBTARGET record) together with additional joint values (the ADD_JOINT field) where appropriate. Configuration information in GRASP is defined in a particular way by the use of a specific program instruction (step) called a configuration step. Thus, where a movement to a given pose is required the generation of the robtarget for IRL is made up of the contribution of the currently active (last defined) configuration for the robot together with the additional joint values which are defined explicitly with the GRASP movement step. (Note that at the time of writing the actual format of the configuration part of the IRL robtarget data type is not part of the proposed IRL specification since it has yet to be defined).

The orientation type defined in the pose data type is, by default taken to be the one with IRL orientation defined by the function ORIZYZ2 (rotate z, rotate around new y, rotate around new z). This may be changed at the time of creation of the IRL file from GRASP to one of any of the 7 types defined in the IRL specification. This is not strictly necessary since the receiving system (not necessarily a robot controller during these early days of IRL development, it is more likely to be a compiler, interpreter or translator) should be able to identify the orientation type by the function name and process the data accordingly either by converting to its own internal format or by using the data as it stands.

The IRL specification supports the use of array variables. This has been incorporated into the GRASP IRL preprocessor to the level available in GRASP, ie up to three dimensional arrays of any variable type can be used. As with IRL, the indices to each of the dimensions may also be variable expressions. This comes through to the IRL output program.

User-defined record types built up from the basic data types or other pre-defined types can be created in the IRL schema. Such user-defined records are not available with the GRASP programming system so these are not exploited in the preprocessor. It could be useful in the future however to take advantage of these user-defined types to allow the manipulation of GRASP specific variable data types by simply declaring them as user-defined in the IRL file and referencing them accordingly.

GRASP robot program definition and mapping to IRL

As described above the definition of robot programs in GRASP is achieved by the use of individual 'step' instructions forming program 'tracks'. The instruction set of the track programs is very powerful and extends beyond the scope of purely robot programming - GRASP is also a general discrete event simulation tool used for 3D graphical planning and simulation in application areas such as warehousing, conveyors and flexible manufacturing systems. The step types that can be converted from GRASP are shown below together with their corresponding IRL instruction:

GRASP	IRL
park	move
position	move
joint_position	move
set	push
message	writeln
pause	wait ... sec
wait	wait for ...
goto	goto
call	procedure name

return   return
if   if

IRL code generation

The mechanism by which GRASP tracks are output to IRL code is relatively straightforward. It basically requires an initial traversal of the internal data structure defining the program to validate that only those constructs that can be converted to IRL are present. The current implementation at the time of writing has a few limitations which must recognised. Such limitations include the omission of the high level 'weld' step which is a process specific instruction having no bearing on the demonstration environment (see below on the demonstration implementation). Other limitations are generally related to the incomplete implementation of all available step types in GRASP. This is not regarded as a handicap since the proving ground for the IRL concept is very much satisfied by the level of implementation that has actually been achieved. Further extensions to existing functionality are readily applicable to the existing software.

The first stage (of restriction checking) is very rapid since the step types and data types are readily available in the data structure. The next stage requires a step by step analysis, conversion and output to the IRL file. With regard to all but the movement specification this data conversion is simple and self explanatory.

In general GRASP specifies the target position for a robot TCP (tool centre point) in terms of a reference object and an additional transform. The product of the object's current position with the transform gives the final pose for the robot. In considering the output of the poses to the IRL file there were two ways in which this may be done. The concept adopted in GRASP could be preserved by using the reference object and modifying transform in combination in the movement statements, or the position could be resolved relative to the robot origin and used directly rather than resolving the position at run-time. The latter option would obviously result in much faster execution on a robot with a IRL interpreter, however, it is more likely that the IRL code will be compiled into ICR before actual execution. It was decided to adopt the former approach in order to reflect the flexibility inherent in the GRASP system whereby modifying a single reference object pose (ie. the position of an object relative to which positions are stored) can still produce the same desired relative positioning of the robot TCP. This system lends itself more readily to calibration procedures if the program is defined by the user in the simulation system in a rational way that allows for the repositioning of workpieces without disrupting the desired relative positioning. For this reason the GRASP IRL preprocessor generates a file where each movement instruction is built up from the product of the two poses defining the refobject position and the modifying transform.

Demonstration implementation of GRASP IRL preprocessor

In order to validate the generated IRL code from GRASP and to focus its development the preprocessor was used as part of a demonstration facility based around the use of a REIS robot playing the solitaire game (essentially a pick and place operation with fairly advanced logic). This demonstration is described in great detail in Sect. 8.5. The demonstration of the results for part of the NIRO project have been presented at the CIM Europe '92 Conference at Odense Steel Shipyard, Denmark on 29-30th September 1992 where the use of IRL output from GRASP to the receiving system was clearly validated.

The IRL functionality used within the demonstration was essentially one which did not rely greatly, if at all upon specific process related commands. This is in contrast to the demonstration for the ICR neutral interface where a strong emphasis was placed on a specific, complex process application, i.e. welding. Sects. 7.2.1 and 8.4 describe this in more depth.

The IRL demonstration placed greater emphasis on those features that are generic across all robot tasks such as variable declaration and use within complex expressions appearing in various complex contexts. This included program looping constructs and high level abstracted pose definition.

Example GRASP tracks and preprocessed IRL code

An example of a GRASP track with a representative selection of step types is shown in Appendix 2 together with the IRL code produced by the preprocessor.

Future developments

It is envisaged that the development of IRL into a full ISO standard will prompt both robot vendors and software suppliers to adopt it. Certainly within BYG Systems Limited and its GRASP product it is intended to follow this closely, adapting code accordingly. BYG will be actively promoting the IRL code generation output from GRASP as an option for the offline programming process.

Conclusions

The development of an interface from the GRASP 3D offline programming and simulation system to robot control via the IRL neutral language has been undertaken. It is clear that the demonstration of the results has validated the concept in general use.

Whilst the IRL language has been proven to be a working solution to a high-level neutral interface for robot offline programming, and that it will be actively promoted by members of the NIRO project it must still undergo extensive review and development into process specific application areas such as arc welding, palletising and paint spraying. It must also sustain close examination on a broad platform from robot vendors , software suppliers and end-users.

7.1.2 IRL to ICR Compiler

U. Schmiedecke

The IRL compiler translates programs written in IRL (Industrial Robot Language) into ICR (Intermediate Code for Robots). An IRL program is far easier to be read and understood by human beings than an ICR program with the same semantics. Some highlights of IRL are given below.

The IRL compiler was developed by PSI in 1990 to 1992 and is now available under SUN-OS 4.0.3 on SUN workstations 3/60 and on IBM-PC and compatibles under MS-DOS.

Anyone who would like to use this compiler, should contact Uwe Schmiedecke (PSI, Dept.PI-AT, Heilbronner Str. 10, 10711 Berlin) for a diskette with the newest version of the IRL compiler. We are also pleased to receive any feedback from users of the IRL compiler, or hints on compiler errors, or proposals for further enhancements.

Source and target languages

The IRL compiler reads (at least) 2 input files:

- the system specification file, which contains specifications of the underlying robot controller and
- the IRL user program, which may be spread over several files.

The compiler produces one output file:

- the ICR file which is a translation of the input file.

The system specification is described in the Sect. "The System Specification File" below.

The source language IRL is defined in "DIN 66312 Part 1", dated July 27th, 1992. Some highlights of IRL are:

- PASCAL-like syntax,
- strong type checking on all expressions, but an automatic conversion between closely related data types (e.g. INT to REAL),
- type definitions to introduce new data types (e.g. multidimensional arrays or structures) or to define new names for existing data types,
- procedure and function definitions with input, output and transient parameters and local variables,
- passing arrays with fixed bounds as parameters, as well as conformant arrays (with variable bounds and with access to the actual bounds from within the called procedure),
- constant definitions (including arithmetics at compile time),
- variable definitions with or without initialisation,

- digital and analog I/O, thus allowing additional external devices and sensors,
- input and output on channels (screen, keyboard) and files,
- special operators for translation and rotation of robot positions,
- program flow control like IF THEN ELSE, CASE, FOR, REPEAT, WHILE, GOTO, RETURN and procedure calls,
- special statements like WAIT some time, WAIT for some digital or analog input, MOVE (LIN, PTP or CIRCLE) with many additional optional parameters (e.g. UNTIL) to a specified point or along a PATH (i.e. a list of points),
- multi robot handling,
- separate compilation of program modules,
- highly sophisticated data list handling, to allow global variables to "survive" program execution by keeping their values in files, thus allowing separate teach-in programs,
- a separate system specification file to allow special controller features to be passed through to the IRL programmer.

The target language ICR is defined in the technical report "ISO/TR 10562", dated April 1992. Some important characteristics of ICR are:

- ASCII-file in assembler-like form,
- symbolic variables,
- user data type definitions,
- block nesting (dynamic variable allocation for procedures),
- block relative addressing for variables and structure fields,
- a large number of predefined functions to do arithmetics,
- many different move functions,
- analog, digital and file I/O,
- highly sophisticated data list handling (see IRL above).

Installation of the compiler

The distribution diskette contains a file named "README.TXT". This is an ASCII text file which describes the simple installation procedure consisting of 3 steps:

- Copy the file IRL.EXE into a directory that is specified in the current PATH-definition, so that it will be found by MS-DOS in any directory.
- Copy the file IRL.SYS into some directory.
- Assign the full path of that file to the environment variable "IRLSYS" (as a command in "AUTOEXEC.BAT").

Now you can work with the IRL compiler.

Any modifications to the file "IRL.SYS" in order to change compiler functions, e.g. for introducing predefined data types, constants and functions, should be made with great care and by trained personnel only.

Compiler invocation

An IRL user program must have the file extension ".IRL". The compiler can be invoked in the following way:

IRL [options] filename

where "filename" is the name of the IRL user program without the extension ".IRL" and "options" is a list of compiler options, as described below.

Compiler options

-h	will show a short help text about the compiler options,
-v	prints version and date of compiler, IRL and ICR definition,
-m	generates ICR operand numbers instead of symbols,
-n	generates ICR numeric opcodes and operand numbers numbers instead of symbolic opcodes and operand symbols,
-e\<fname\>	error messages and warnings are written into an ASCII file named "fname",
-s\<fname\>	path and filename of system specification file,
-t\<nn\>	tab length \<nn\>. Used for computing the source character position in error messages (default is 8).

Options without a parameter may be specified with one token, e.g. "-vhn".
For those with a parameter (e.g. -e\<fname\>) the next option must be specified as a new token, e.g. "-vex.x -t4".

Error messages

The error messages produced by the IRL compiler have the following format:

Error: fname.ext(linenr/column): error text.

Example:

```
Error:      prg2.irl(281/3): function 'abcde'
            has parameters
```

where "prg2.irl" is the file name of the IRL user program. The error occurred in line 281 character column 3 of that file (the first column in a line is column 1).

Warning messages are similar, except that the lines begin with the word "Warning:" instead of "Error:":

```
Warning: prg2.irl(89/17): variable xyz never used
```

The compilation process

The compiler can logically be split up into several parts:

1. *The scanner* splits up the source language file into a sequence of tokens (e.g. keywords, constants, identifiers), skipping all white space and comments.
2. *The parser* performs syntax analysis. During parsing, a stack of handles for partially parsed rules and tokens is organised.
3. *The semantic routines* are used for semantic checking. They also construct a program tree using a set of about 35 different nodes.
4. *The code generation* uses the program tree to generate code. This is done in two passes in order to handle forward jumps and forward calls.

The first three parts form the "front end" of the IRL compiler. Its task is to construct the program tree and to fill a number of global variables.

The fourth part is called the "back end". During this code generation, a number of optimisations are performed. For example: For an access to the first element of an array, the multiplication of the length of one array element with zero and addition of this product to the address of the first array element is suppressed.

Known bugs

1. It might happen in some cases that error messages show the wrong file name and line number information, especially if INCLUDE is used and the error is within an included file. Here the file name of the "main" file, i.e. the name specified in the compiler invocation command, is given but together with the line number and character position of the error position within the included file.
2. If the user wants to specify the file name of the system specification within the IRL source file, it must be specified *before* and not *after* the program heading "PROGRAM prog_name_idf;"
3. If a program contains more than about 2500 forward references (e.g. IF, CASE, WHILE, forward calls), the first pass of the code generation might produce an internal error because the storage for forward references in the output file is not dynamically allocated.

Extensions and enhancements

The compiler accepts the following language extensions:

- "from..to" is allowed in a MOVE statement to specify a sub array instead of the whole array. Example:

    ```
    MOVE LIN PATH array_name[nn+1 .. nn+mm]
    ```

- "from..to" is allowed in a CASE statement to specify a range of constants as a case element. Example:

```
CASE expr OF
10..15: <stmtlist>
...
ENDCASE;
```

The following features are not yet implemented:

- true separate compilation (not only INCLUDE 'file' feature) together with a library concept and linking,
- ACT_ROBOT handling,
- user record constants,
- address range and non continuous addresses,
- code generation of the address of x.y (for an OUT or INOUT parameter), when x itself is an address (an OUT or INOUT parameter). This applies to conformant array parameters only,
- CKON/CKOFF directive to control generation of code for boundcheck,
- the MOVE parameters C_SPEED, SPEED_ORI, ACT_ROBOT, WOBBLE,
- READ statement,
- READLN of geometric data types,
- Random access file I/O.

whereof the last four points cannot be implemented without extensions to the ICR language.

It would be very useful to add some general compiler features, such as

- production of a source listing file,
- production of a cross reference listing of objects,
- code production for other output languages.

The system specification file

The system specification contains (besides other information) the representation and structure of all primitive data types. Therefore the system specification file must be read by the IRL compiler before the user program can be parsed. Since the system specification is read by the parser and scanner, comments and layout are allowed between all tokens. In fact, the syntax of the system specification is treated by the parser as extension to the language IRL. The system specification file consists of the following parts:

1. First all system data types must be specified, all simple data types followed by all structured data types:
   a) simple data types, (BOOL, CHAR, INT, REAL, STRING). Each data type is specified in the following way:
      - data type name (e.g. REAL)
      - number of bytes occupied on stack (e.g. 8 for REAL)
      - number of variables occupied (1 for all simple types)

-     semi-colon as delimiter to the next data type
b) structured data types. Each of them is specified as follows:
-     data type name (e.g. POSITION)
-     number of bytes occupied on stack (e.g. 24 for POSITION)
-     number of variables occupied (e.g. 3 for POSITION)
-     a list of all components of that type. For each component the following informations must be specified:
    -     data type name (e.g. REAL)
    -     component name (e.g. y)
    -     byte offset of component (e.g. 8 for y)
    -     relative variable number (e.g. 1 for y)
-     semi-colon as delimiter to the next data type
c) number of bytes of a BRADDR variable on stack. This number terminates the list of predefined system data types.

2. Next all system functions must be specified. The order of them should be alpha sorted for efficiency reasons. Generic functions are allowed: when two (or more) functions have the same name, the correct function will later be identified using the data types of the actual parameters. In order to identify the correct one, the function with the more complicated data type(s) must be specified before the one with the simpler one e.g: ABS(REAL) must be specified before ABS(INT). Each system function is specified in the following way:
   - name
   - two integral numbers "code1" and "code2" which are used for code generation. See Sect. "Meaning of code numbers of system functions" below for explanation.
   - result data type for functions. For procedures this data type must be omitted.
   - all parameters of the function or procedure. Each parameter must be specified in the following way:
     - parameter direction (IN, OUT, INOUT)
     - data type name
     - parameter name
   - optional sequence of strings used for code production. These strings must be present, if "code1" is 6 or 7, and may not be present otherwise. If present, and "code1" is 7, these strings are generated after pushing all actual parameters of the function call.
   - semi-colon, to terminate the current function specification.

3. Normal declaration Sect. for all predefined data types, constants and variables. Note that some special variables and constants with a special name must be defined here. They are needed by the IRL compiler for code generation. These special definitions are the following:

name	what	data type
errmov	variable	record (int, string)
errspe	variable	record (int, string)
std_in	variable	file of char
std_out	variable	file of char
std_err	variable	file of char
ori_mode	const	array[1..7] of string
default_ori_id	const	int
w_vel_kind	const	int
w_acc_kind	const	int
w_res_kind	const	int
quat	const	string
vang	const	string

4. Semi-colon, followed twice by the keyword CONST followed by a sequence of identifiers, which are read/only for the user, like all other constants. These constants are coded into a special ICR-function call when applied by the user in any expression. These functions, when applied in ICR, leave a value on top of the stack, as a PUSH-instruction of a constant value would do. Each such special constant is specified as follows:
   - data type name
   - colon
   - constant name
   - integer1
   - integer2
   - string constant
   - semi-colon

where "integer1" is a combined value:
   - the right four bits are the data type code according to file DCL.H of a value, that must be pushed onto the stack before generating the special ICR-function. Data type 0 means: no value must be pushed,
   - the left four bits are the value that must be pushed first,

and "integer2" is the ICR-code of the special function which must be generated to implement the application of the constant. If the string constant is not empty, a comma is appended to the ICR-code generated, followed by that string.

Meaning of code numbers of system functions

The first integer ("code1") behind the name of a system function specification has the following meaning:

0: Normal simple function. First all parameters are pushed, then the code for the corresponding ICR operation ("code2") is generated.

1: Some "ORIxyz" function with 3 real parameters. "code2" is the corresponding ori_id. The code generated is:
   PUSHI ori_id;
   push all parameters;
   PUSHR 0.0;
   gen code "code1" (which should be "GENO").
2: The ORIVANG function. "code2" is the corresponding ori_id. The code generated is:
   PUSHI ori_id;
   push 1st parameter (a POSITION) as 3 separate REALs;
   push 2nd parameter (the rotation angle);
   gen code "code1" (which should be "GENO").
3: The ORIQUAT function. "code2" is the corresponding ori_id. The code generated is:
   PUSHI ori_id;
   push all 4 REAL parameters;
   gen code "code1" (which should be "GENO").
4: Some QUATx function. The parameter is pushed, converted to an orientation of type "QUAT", popped into a temporary ori-variable and the desired real component (according to "code2") is pushed again.
5: Function ROTANGLE or ROTAXIS. The ori-parameter is pushed, converted to an orientation of type "VANG", popped into a temporary orivariable and the desired result is pushed again: for ROTANGLE: the 4th REAL, for ROTAXIS: the 1st 3 REALs which are converted into a POSITION value.
6: Some ANGLExyz function. The 2nd parameter is pushed, converted to an orientation of type according to the 1st parameter, popped into a temporary ori-variable and the desired real component (according to "code2") is pushed again. Here a special way of checking the validity of the 1st parameter is used: The 1st parameter must be one of the constant strings of the array "ori_mode". The index of the found string is the corresponding ori-id, which is used to index the string constant that was specified in the system specification after the parameter list for the ANGLExyz function. If the indexed character is a digit (between 1 and 4), it denotes which REAL component of the ori-variable is wanted as result. If it is not a digit, an error message will be given, since the 1st parameter specified by the user does not fit the ANGLExyz function used.
7: These are special functions not predefined in the language IRL. They are implemented within the underlying ICR-interpreter and passed through to the IRL programmer. Here "code2" must be 0 and at least one string constant must be specified after the parameter list. After pushing all actual function parameters, these string constants are generated as specified, each string on a separate line.
other:
   The following error message is given to the user:

'system function "..." not implemented'

Conclusion

The IRL compiler has been successfully used in a demonstration at the "CIM Europe Conference" in Odense (DK) in September 1992, controlling a Reis robot.

Without separate compilation, IRL programs larger than about 7000 lines might exceed the storage available under MS-DOS, but this is probably beyond the scope of normal applications.

The compilation is moderately fast. A program of 500 lines requires about 5 seconds of compiling, a 3000 line program 35 seconds (which is a quarter of the speed of the Borland Turbo C-compiler). Due to some sophisticated optimisations in the back end of the compiler, the ICR code produced is concise and efficient.

However, the main advantage of using a high-level language for robot programming is twofold: The readability (and thus changeability) of the code, and the testing support due to error messages and warnings, which result from the semantic checks performed by the compiler.

7.2 Preprocessors for ICR

P. Sorenti

7.2.1 ICR Preprocessor for GRASP

GRASP for offline programming - GRDATA

GRASP is a general simulation tool from BYG Systems Ltd using full three dimensional graphical modelling and simulation techniques to address a wide variety of application areas. Its use as a tool for offline programming is already well established. In addition to 3D graphics, its ability to offer real time path generation for kinematic structures and on-line collision detection make it an obvious candidate for the generation and validation of robot programs remote from the robot system itself. Robot programs in GRASP are referred to as 'tracks' which contain individual instructions known as 'steps'.

Once a robot track has been generated and verified to the satisfaction of the operator there needs to be a mechanism by which the programs can be transferred to the actual robot controller for subsequent execution. GRASP can create a file describing the robot program in its own native language format called GRDATA.

The GRDATA language is regarded as a 'neutral file', albeit an in-house format. It is independent of the robot type at this stage. It is only when the robot type is specified that the GRDATA file passes through a robot-dependent postprocessor which produces robot-specific code. Each of these robot specific postprocessors has a common task at the start of the conversion: scanning and parsing the GRDATA file and creating a representative data structure. This forms the so-called 'front-end' of each postprocessor which is common amongst all postprocessors. It is only the 'back-end' or robot code generation and output part that needs to be different. This greatly reduces the amount of work required when developing a postprocessor for a new robot controller with yet another language type specific to it.

The generation of an external file to GRASP rather than creating the robot specific code directly greatly improves the maintainability of each postprocessor since upgrades to each robot language processor do not require the release and supply of a new version of the entire GRASP software.

GRASP preprocessor to ICR implementation

In complete contrast to the existing method described above by which robot programs are created via the GRDATA 'neutral' interface, the generation of ICR code takes place entirely within the GRASP software. The reasons for this are two-fold:

- It is intended that the future development of the ICR neutral interface will provide a stable platform to which the use of GRASP offline programming may migrate in order to compliment the existing (GRDATA) protocols,
- ICR is by its very nature a very low level language. The amount of work required to develop a back-end processor for manipulating the GRDATA data structure becomes so great that the advantage of the technique over developing the code within GRASP becomes minimal.

With a view to more long term developments of the GRASP software it seems prudent to incorporate the results of the ICR standard within it so that when the next generation of the code is implemented the actual definition of the GRASP track and step database can directly reflect the ICR program code. After all, this is one of the major aspirations of work of this type: ie. that robot vendors and software suppliers adhere fully to a well-maintained and clearly defined standard for robot programming (and CAD data storage/transfer).

This approach, that of devising output of program code directly from the GRASP simulation software, follows the same lines as that of the STEP preprocessor from GRASP described in Chapter 5.

GRASP and ICR data types

In considering the generation of ICR code from GRASP there is a need to review the basic ways in which information related to robot motion and program logic etc are defined. In this way the method by which data is mapped from the preprocessing system i.e. GRASP to the programming language i.e. ICR can be appreciated more clearly. Here are the comparisons between them:

GRASP	ICR
Integer	integer
real	real
signal	Boolean

refobject   pose

Those ICR types not included here do not have a direct equivalent in GRASP. There is however one important one which needs to be mentioned, the ICR robtarget data type. This incorporates information about the robot's configuration together with additional joint values where appropriate. Configuration information in GRASP is defined in a particular way by the use of a specific program instruction (step) called a configuration step. Thus, where a movement to a given pose is required the generation of the robtarget for ICR is made up of the contribution of the currently active (last defined) configuration for the robot together with the additional joint values which are defined explicitly with the GRASP movement step. (Note that at the time of writing the actual format of the configuration part of the ICR robtarget data type is not part of the proposed ISO standard since it has yet to be defined.)

The orientation type defined in the pose data type is, by default taken to be the one with ICR orientation id = 2 (rotate z, rotate around new y, rotate around new z). This may be changed at the time of creation of the ICR file from GRASP to one of any of the 6 types defined in the ICR specification. This is not strictly necessary since the receiving system (not necessarily a robot controller during these early days of ICR development, it is more likely to be an interpreter or translator) should be able to identify the orientation type by the id number and process the data accordingly either by converting to its own internal format or by using the data as it stands.

GRASP robot program definition and mapping to ICR

As described above the definition of robot programs in GRASP is achieved by the use of individual 'step' instructions forming program 'tracks'. The instruction set of the track programs is very powerful and extends beyond the scope of purely robot programming - GRASP is also a general discrete event simulation tool used for 3D graphical planning and simulation in application areas such as warehousing, conveyors and flexible manufacturing systems. The step types that can be converted from GRASP are shown below together with their corresponding ICR instruction:

GRASP	ICR
park	lmove (cmove, jmove)
position	lmove (cmove, jmove)
joint_position	lmove (cmove, jmove)
weld	lweld (cweld)
set	push
message	remark
pause	delay
wait	wait_b / if
goto	jump

call	call_s
return	subpend
if	if

(A discussion on the implementation of the weld step to ICR will be found below in the Sect. describing the demonstration of the processor results).

ICR code generation

The mechanism by which GRASP tracks are output to ICR code is relatively straightforward. It basically requires an initial traversal of the internal data structure defining the program to validate that only those constructs that can be converted to ICR are present. This first stage is very rapid since the step types and data types are readily available in the data structure.

The next stage requires a step by step analysis, conversion and output to the ICR file. With regard to all but the movement specification this data conversion is simple and self explanatory.

In general GRASP specifies the target position for a robot TCP (tool centre point) in terms of a reference object and an additional transform. The product of the object's current position with the transform gives the final pose for the robot. When considering the output of the poses to the ICR file there were two ways in which this may be done. The concept adopted in GRASP could be preserved by using datalists to store the reference object and modifying transform poses or the position could be resolved and used directly in the ICR code. It was decided to adopt the former approach in order to reflect the flexibility inherent in the GRASP system whereby modifying a single reference object pose (ie. the position of an object relative to which positions are stored) can still produce the same desired relative positioning. This system lends itself more readily to calibration procedures if the program is defined by the user in the simulation system in a rational way that allows for the repositioning of workpieces without disrupting the desired relative positioning. For this reason the GRASP ICR preprocessor generates two files, one containing the ICR program instructions, the other the datalist elements referenced in the previous file.

The output of ICR instructions is based on the instruction symbol as opposed to its code. The main reason for this is that, even though ICR is a low-level language as much readability as possible should be maintained. Should it prove necessary the inclusion of an option to output the instructions on a coded integer basis can be made available with little effort.

Demonstration implementation of GRASP ICR preprocessor

In order to validate the generated ICR code from GRASP and to focus its development the preprocessor was used as a key part of a demonstration facility located at Odense Steel Shipyard (OSS), Denmark, one of the demonstrator partner sites in the NIRO project. This demonstration is described in detail in Sect. 8.4. Essentially the development of the ICR code output has required the inclusion of an extended ICR instruction set. This is because the demonstration is based around the application of robots to welding of a ship Sect.. The ICR instruction set as it stood (the version submitted as a Committee Draft to the ISO, document number ISO/CD 10562.2, February 1991) made no provision for application specific processes such as welding. Thus, the NIRO project was compelled to define a proposal for the use of welding related instructions in ICR. These new instructions were developed internally to the project and implemented in order to provide a working solution for the demonstration site. Based on the successful implementation of these commands a proposal will be made to the ISO committee in order to enhance the ICR specification.

The new instructions incorporate a definition of the welding motion, welding process data (arc current, voltage, weaving etc) and seam searching routines. The experience gained has also resulted in the development of additional core functionality in the GRASP simulation software. In this way GRASP has benefited directly from its inclusion in the NIRO project beyond that of providing an interface to the ICR neutral format. A specific example of this is the GRASP 'sense' step. This feature allows the simulation of the search algorithm used commonly in the arc welding application area. The requirements of the demonstration site and the extensions to ICR prompted further positive development of this GRASP command, benefiting both the product and the project.

The demonstration of the results for part of the NIRO project have been presented at the CIM Europe '92 Conference at OSS on 29-30th September 1992 where the use of ICR output from GRASP to the receiving system was clearly validated, together with the inclusion of the proposed extensions to the instruction set.

Example GRASP tracks and preprocessed ICR code

An example of a GRASP track with a representative selection of step types is shown in Appendix 3 together with the ICR code produced by the preprocessor.

Future developments

It is envisaged that the development of ICR into a full ISO standard will prompt both robot vendors and software suppliers to adopt it. Certainly within BYG and its GRASP product it is intended to follow this closely, adapting code accordingly. BYG will be actively promoting the ICR as an option with GRASP.

Conclusions

The development of an interface from the GRASP 3D offline programming and simulation system to robot control via the ICR neutral language has been undertaken. It is clear that the demonstration of the results has validated the concept in general use and in the specific application of robotic welding. The success of ICR is further illustrated by the fact that it has been extended from its original draft proposal to incorporate nonexistent functionality viz the welding process.

Whilst the ICR language has been proven to be a working solution to a low-level neutral interface for robot offline programming, and that it will be actively promoted by members of the NIRO project it must still undergo extensive review and development into other application areas such as spot welding, paint spraying and palletising. It must also sustain close examination on a broad platform from robot vendors, software suppliers and end-users.

7.3 ICR Interpreters

7.3.1 Overview

W. Jakob and J. Reim

The ICR neutral file format is designed for interpretation in the robot controller. However, ICR contains a lot of functions that are independent from the robot to be controlled. To these belong the variable and data list administration, program flow control, arithmetics and parts of the I/O functions. The move statements and all related commands together with digital and analog I/O are dependent on the robot controller. The latter can be assigned to the independent part of the interpreter if the controller is designed for ICR.

Thus it is possible to divide an ICR interpreter into two parts. The robot independent part covers both the already mentioned ICR functionality, and a loader for ICR code and data lists (cf. Fig. 7.3.4-1). The loader performs a syntax check on the code that is read in and converts it from the ASCII representation to an internal binary one. The interpreter executes the ICR statements stepwise and communicates via a well defined procedural interface with the robot dependent part.

ICR is tested and evaluated through implementations on several different robot controllers, and also through the implementation of an ICR interpreter within the KISMET simulation system. These are described fully in the following sections. A summary of the level of functionality achieved is given here in Table 7.3.1-1.

ICR Command	KfK KISMET	ABB IRB 2000	ABB IRB 60	Reis RSIV
Checks Axes check (CHECK_AXES)	no	yes	yes	no
Orientation check (CHECK_ORI)	no	yes		no

Move parameters				
Velocity (R_VEL, W_VEL)	yes	yes	yes	yes
Duration (R_MOVETIM)	no	no		no
(W_MOVETIM)	no	no		no
Acceleration (R_ACCEL)	yes	no	no	yes
(W_ACCEL)	yes	no	no	yes
Accuracy on intermediate pose(R-RESIP, W_RESIP)	no	no	restr.	no
Accuracy on destination pose (R_RESDP, W_RESDP)	no	(1)	yes	no
Movements				
Joint interpolated movement (JMOVE)				
absolute movement	yes	yes	yes	yes
relative movement	no	yes	yes	no
Linear interpolated movement (LMOVE)				
absolute movement	yes	yes	yes	yes
relative movement	no	yes	yes	no
Circular interpolated movement (CMOVE)				
absolute movement	yes	no	no	yes
relative movement	no	no	no	no
Polynominal interpolated movement (PMOVE)				
absolute movement	no	(2)	no	no
relative movement	no	(2)	no	no
Movement on teached trajectory (MOVTRJ)	yes	no	restr.	no
Calibration (CALIB)	no	yes	restr.	no
Home position (GOHOME)	yes	yes	yes	no
Movement stop (MOVSTP)	no	no	yes	no
Movement continuation (MOVCONT)	no	no	yes	no
Movement cancellation (MOVCANCEL)	no	no		no
Path structure block (MOVBEG, MOVEND)	no	no		yes
Miscellaneous robot related commands				
Robot selection (SELECTIR)	no	yes	yes	no
Identification of actual robot (IDENTIR)	no	yes	yes	no
Limitation of axis range (LIMIT)	no	no	yes	no
Working space limitation				

Prohibited area	(WORKSP)	no	no	yes	no
	(PROSP)	no	no	yes	no
Tool and gripper commands					
End effector center point definition	(TCP_DEF)	yes	no		no
General call of end effector	(EFNA)	yes	no	no	yes
Gripper open and close (not defined in ICR, part 1)		yes		no	yes
Input/Output					
Chain administration (CNFGCHAN, OPENCHAN, CLOSECHAN, RESETCHAN, STATCHAN)		restr.	no	yes	no
Digital I/O (DIN, DOUT)		yes	yes	yes	yes
Analog I/O (AIN, AOUT)		yes	no	yes	yes
Character I/O (GETCH,		yes	yes	no	no
PUTCH)		yes	yes	no	no
Block I/O (GETBLK, PUTBLK)		yes	yes	no	no
Wait for binary input	(WAIT_B)	no	yes	yes	yes
Wait for analog input	(WAIT_LE, WAIT_GE)	no	no	yes	no
Wait for start signal	(PAUSE)	yes	yes	no	no
Miscellaneous					
Enter teach-in mode	(TEACHON)	no	no	no	no
Conversion of cartesian to joint coordinates	(CTTJ)	no	no	restr.	no
Conversion of joint to cartesian coordinates	(CTJT)	no	no	restr.	no

restr.: implemented with restrictions.
(1): only 2 levels of accuracy exist: precise and rough.
(2): The PMOVE is interpreted as if was the LMOVE command.

Table 7.3.1-1: Four implemented robot dependent interpreter parts and their interfaces to the robot independent part (KISMET is a simulation system)

7.3.2 Robot Independent Part

J. Reim

The implemented robot independent part of the ICR interpreter is based on ISO DP/10562.2. Besides the subroutine call interface to the robot dependent part the interpreter is integrated into a larger software environment using two function calls.

Call interface to the ICR interpreter

- The routine loader() initialises the interpreter and loads the ICR program.
- y calling the routine *icr_step()* a single statement is executed. The return value indicates finished execution.

Source file name and debug switches are transferred via global variables.

Debugging options of the robot independent part

For debugging routine *icr_step()* prompts on ANSI C device *stder*r for user input acquired from device *stdin*. Debug output such as stack or variable examination is printed to *stdout*. Therefore a terminal, console or alpha window is necessary to get the benefits of the debugging feature. The following functionality is implemented:

- I	run	run to PEND, break or interrupt
- Bn	break	set breakpoint at line n
- F		flush code to robot controller (reis robot only)
- S	step	interpret one line
- D	debug	toggle debug mode
- X	x-amin	examine stack from top to bottom
- T	TOS	examine top of stack
- Y	tYpes	print defined types
- M	Memory	print free memory size (on PC systems only)
- V	var	examine variable stacks
- @...	macro	take debug commands from file
- Q	quit	quit abort icr-program and interpreter
- H	HELP	these short hints

To enter debug mode while running a program an interrupt handler is integrated. On PC systems debug mode is activated by typing any key. The Unix version utilizes the SIG_QUIT which is activated on silicon graphics systems typing *ctrl backslash*.

The current version of the ICR interpreter is implemented in portable ANSI C and tested on PC (TURBO C), silicon graphics and VAX/VMS systems. It has been used in conjunction with two calling programs.

Stand alone ICR interpreter

The stand alone main program just interprets the command line for the debug with switch "-d" and the source filename, calls the *loader* and runs *icr_step* until program end is encountered. The name of the executable depends on the linked robot dependent part for example *icrnorob* or *icrreis*.

7.3.3 Robot Dependent Part for the KISMET Simulation System

J. Reim

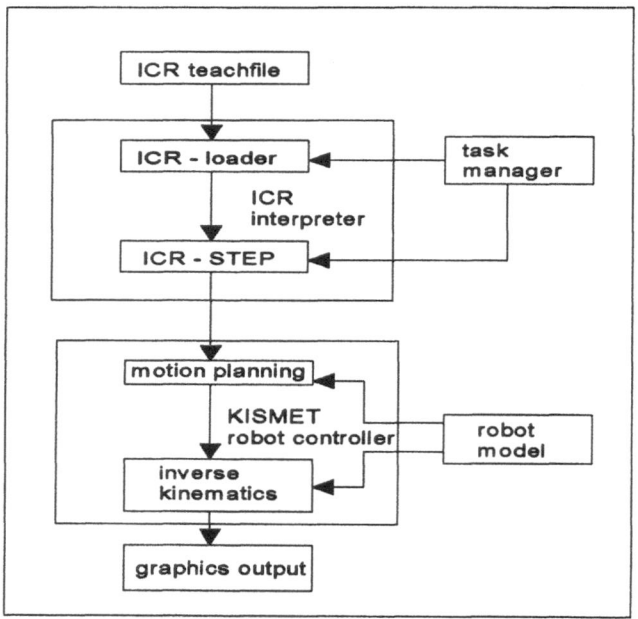

Fig. 7.3.3.-1: Architecture of KISMET teachfile execution

The main difference between the KISMET version of the ICR interpreter and the other applications is that the calling routine and robot dependent part are situated in the same software system. The KISMET robot task manager loads the ICR program for a selected robot and runs the task. In the robot dependent part the desired motion is transmitted to the KISMET path planning module (see Fig. 7.3.3-1).

The task manager assigns the desired teachfile to the actual robot and initiates the loader via the calling interface of the robot independent part. When running the program, KISMET manages the step by step interpretation with the ICR interpreter, recognises user interaction, errors or sensor events and performs the motion demanded by the robot dependent part. To change the duration of the simulation the user can select between real time mode or scaled timebase. Besides ICR, KISMET supports the IRDATA specification and some specific application robot languages.[Kühnapfel 1991A].

The KISMET robot dependent part modifies the robot representation in the virtual robot controller assigned to the loaded program. Here motion target, velocity and acceleration are set or the actual status of the robot is accessed.

During the integration of the ICR interpreter a common gripping procedure has been integrated into KISMET. When an ICR *grclose* command is executed, KISMET looks for the work frame next to the TCP which marks a part to be moved. Similarly a release position is searched when *gropen* is executed. This implementation allows unchanged ICR code to be run both in KISMET and on the robot for the solitaire demo which contains multiple similar gripping operations on different geometric parts. (Kühnapfel 1991A).

7.3.4 Robot Dependent Part for the ABB IRB 2000 Robot

T. Clausen

One of the test implementations of the ICR interpreter is that for the ABB IRB 2000 robot. This Sect. describes the robot dependent part of this interpreter.

The basic implementation of the ICR interpreter

The IRB 2000 robot is not directly prepared for a new robot language. The robot is a stand alone system with two main input facilities: A floppy disk and an RS232 port. Both inputs are basically meant as gates for down loading programs into the robot controller.

In addition, the RS232 interface offers other functions which can be used for:

* Manoeuvring the robot. That is, to command the robot to move to a new position without first creating a program and then executing it.
* Setting digital output ports.
* Reading the value on digital input ports.

These control functions have been used in connection with the implementation of an ICR interpreter for the IRB 2000 robot. The interpreter is implemented on a PC computer external to the robot and every time the interpreter reads an instruction which requires physical action the robot is commanded to do so via the RS232 connection.

The interpreter includes another possibility for running ICR programs on the IRB 2000 robot. Instead of controlling the robot directly via the RS232 channel it is also possible to write the move and digital output instructions in an ARLA program file instead. ARLA is the native language of the IRB 2000 robot. This program can then be downloaded into the robot and executed there. The interpreter is in this case used as a simple compiler.

This way, the interpreter offers two ways of running ICR programs on the IRB 2000 robot which the operator can freely choose between.

The robot dependent part of the interpreter

The robot dependent part can in many ways be regarded as a bridge between the robot independent part of the interpreter and the robot itself. The interface to the robot independent part consists of about fifty functions which reflects the robot dependent instructions in the ICR language. Examples of the robot related instructions are:

* Motion (linear, joint,..)
* TCP control
* Accuracy in the motion
* Velocity, acceleration
* Tool and gripper control
* I/O instructions (digital, analog, text,..)

The interface to the robot consist of two modules. One module manages the communication to the robot via the RS232 channel and another module takes care of writing ARLA programs.

The communication module communicates with the robot by sending and receiving telegrams. The three most central telegrams are:

* Manoeuvring
* Set a digital output port
* Read a value on a digital input port.

These three telegrams can be used for controlling all the above mentioned robot related instruction types.

The Manoeuvring telegram is used to move the robot arm. The telegram carries information about the location to move to and additional parameters related to the motion. The robot starts when it receives the telegram and the robot returns a confirmation telegram when it reaches the end location. The Manoeuvring telegram has the following possibilities:

* Linear and joint motion
* Absolute and relative motion
* Fine and coarse accuracy
* Control of TAP or TCP orientation
* Velocity

The telegrams for digital I/O operations work essentially in the same way. A request for reading or setting a digital I/O channel is send from the interpreter and the robot confirms/answers the request by sending a telegram.

The module for outputting ARLA programs reacts on ICR instructions for motion, setting velocity, and controlling digital output ports. Every time such an instruction is met in the ICR program a corresponding ARLA program line is written.

The following figure illustrates the structure of the ICR interpreter for the IRB 2000 robot:

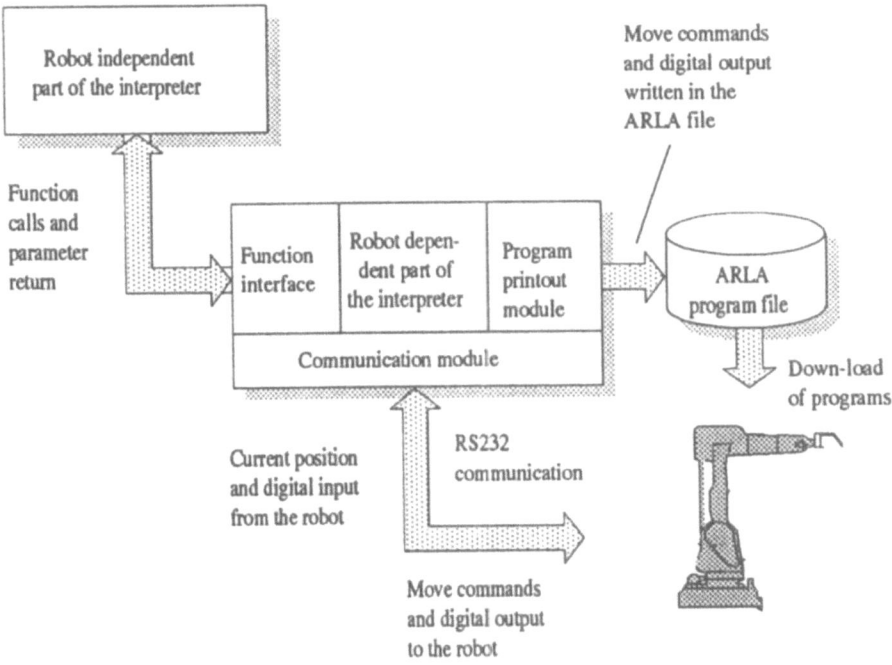

Fig. 7.3.4-1: Structure of the ICR interpreter for the ABB robot

An example

This chapter describes the execution of a robot dependent ICR instruction. The LMOVE instruction is used as an example.

Every time the robot independent part reads a robot dependent instruction it calls the robot dependent part to execute it. This is in practice handled by calling the corresponding function from the interface between the two interpreter parts (refer to Fig. 7.3.4-1). The task of the interface function is then to call other functions which execute the instruction. There is, for example, a function called "rob_lmove" which represents the LMOVE instruction. This function is a part of the interface between the two interpreter parts and it is defined as follows:

```
VOID rob_lmove ( G_INT_TYPE         absrel,
                 CHAR               wait,
                 G_ROBTRT_TYPE      *v_ziel,
                 G_ERROR_TYPE       *fehler)
{
  fehler->error_code =
                move_robot(v_ziel,(absrel-1),wait,0);
}
```

The rob_lmove function takes four arguments:

absrel: Is used to indicate whether the robot should move in absolute coordinates or relative to the present position.

wait: Indicates whether the interpreter should keep on executing the program immediately or wait till the robot has finished its present motion.

v_zeil: Is a data structure containing the next location to move to.fehler:Is a "return variable" holding an error message if the robot did not succeed in executing the LMOVE instruction.

In general, the tasks of the interface functions are not comprehensive, they call other functions from a group of "service functions". These functions implement the real execution of the ICR instructions. For example the "rob_lmove" just calls the "move_robot". The move_robot function completes the execution of the LMOVE instruction by making three new function calls: First a function call which converts from the present orientation format to quaternions (the orientation format of the ABB robot), then a function call which sends a Manoeuvring telegram to the robot, and finally a function call which receives the acknowledgement from the robot. All three function calls again lead to new function calls and the final function calls are sending/receiving bytes to/from the robot.

Evaluation of the implementation

Programs from various sources have been executed on the interpreter for testing the functionality of the ICR interpreter. Examples are hand written programs, programs compiled from IRL, programs generated by the robot programming system GRASP, and programs generated by the CAD system CATIA. The rest of this chapter addresses various aspects related to the evaluation of the interpreter presented here.

First of all, the tests have shown that the ICR format is able to carry a broad range of programs from simple motion programs to programs containing advanced mathematics, and programs with a complex structure.

The implementation has also shown that it is possible to drive a robot with an external ICR interpreter. Quite sophisticated programs have been executed on it. Examples are motion along a spline curve and the Towers of Hanoi problem. The system works satisfactorily in all ways, except for the execution speed. The problem is that the robot makes a short stop between each move (approximately 0.2 seconds) which means that it is impossible to perform continuous motion. The short stop between each motion is due to the way the communication takes place. The robot will not accept a new move command (Manoeuvring telegram) before it has reached the end point of the preceding motion. A short stop is the consequence since it takes a little time to send the next telegram. The problem could be avoided if it was possible to integrate the interpreter closer with the robot controller.

This stop problem can be avoided by using the compiler part instead since ARLA programs can be executed continuously in the robot controller. However, there is one major disadvantage in using the compiler solution: it produces large ARLA programs compared to the ICR programs. Every time the ICR interpreter encounter a move instruction in the ICR program it writes an ARLA move instruction. As a consequence, an ICR program with a loop which is repeated n times containing a move instruction will result in an ARLA program with n move instruction as illustrated by the following program fragments:

ICR program	ARLA program	
REPEAT n TIMES		MOVE
LMOVE	MOVE	n move
END		...
instructions		
		MOVE

This will command the robot to execute the same motion in both cases, but the ARLA program is not as compact as it could be.

Finally, the implemented system is strictly designed for off-line programming which is in line with the fundamental concept for ICR. This forces the user to fully control all parts of the system because it is not possible to make on-line corrections afterwards. It is consequently necessary to go all the way back to the off-line programming tools to correct any errors found in the final execution, and this fact limits the general performance of the ICR implementation. However, in situations where stringent off-line programming is the only correct programming method ICR offers a very general and flexible way of writing programs.

7.3.5 The Robot Dependent Part for the ABB IRB 60 Robot

A. Bautista

The robot dependent part for the ABB IRB 60 robot has been developed by DISEL. The design of the present interpreter for this robot was carried out using two basic concepts. The first one is related to the use of the ASEA software as a communication kernel with the robot to provide a higher degree of "neutrality" and independence from a particular version of the controller (S2 or S3).

Assuming this concept, the type of controller is transparent to the user and no new design is required since this feature is provided by the kernel. This decision was adopted in accordance with the requirements of the final user, CASA.

The second concept refers to the implementation of the move commands where two approaches have been adopted, one directly moves the robot from the external computer while the second creates a compiled ARLA program, loads it into the robots' memory and launches it afterwards. In this sense, the second is not strictly a pure interpreter.

The direct move command provides no interaction with the robot controllers' memory and therefore it is not affected by the number or size of programs already stored. However, when implementng the no_wait option in any move command, control is retained by the robot controller until the movement has almost finished. This can be very inconvenient.

The loading strategy affects the robot controllers' memory since the programs are loaded there and launched afterwards. This might cause some problems if the memory is very "crowded" with pre-existent user programs (the interpreter erases the program it generates after program execution to reduce this problem) and the size of the program (basically the number of points) is too big. The no_wait option is in this case fully implemented.

From the tests performed the second approach provides higher versatility and better movement performance and sensor interface than the first approach, and the space problems do not seem to be relevant under the given test conditions.

The integration between the robot dependent and independent parts of the interpreter is working correctly.

Several of the ICR Interpreter commands cannot be implemented or are just partially implemented (basically, control and acceleration and grasp commands) due to the characteristics of the present robot controller and application.

The following routines have been implemented according to the last version (February 1992, NIRO.WGB.KfK.007.92) of the interpreter:

```
Rob_Lmove_t
Rob_Lmove_tl
Rob_Lmove_j
Rob_Lmove_jl
Rob_Jmove_t
Rob_Jmove_tl
Rob_Jmove_j
Rob_Jmove_jl
```

Rob_r_resdp
Rob_r_resdp1
Rob_w_resdp
Rob_w_resdp1
Rob_r_resip
Rob_w_resip
Rob_check_axes
Rob_w_vel
Rob_r_vel
Rob_limit
Rob_worksp
Rob_prosp
Rob_curpos_r
Rob_curpos_j
Rob_identir
Rob_selectir
Rob_gohome
Rob_cnfgchan
Rob_openchan
Rob_closechan
Rob_resetchan
Rob_statchan
Rob_ain
Rob_wait_le
Rob_wait_ge
Rob_sensstop

Description of routines

A brief description of the routines is now given.

Rob_Lmove_t and Rob_Lmove_t1 represent a linear movement to a cartesian Robtarget in cartesian space both in launching program mode and direct command mode. The main difference between the two modes is the unavailability in the direct command to perform any movement in the No_wait state (do not wait to receive the robots' completion confirmation to process the following ICR instruction) since the control is not returned to the external computer until the movement is almost finished. Movement performance in medium displacement is quite similar in both modes. The Absolute and Relative movement options are fully implemented.

Rob_Lmove_j and Rob_Lmove_j1 represent a linear movement to a joint destination in cartesian space both in launching program mode and direct command mode. This type of command can only be applied to external axes, since joint destination definitions can only be applied to the external axes of the ASEA robot. The Absolute and Relative movement options are fully implemented.

Rob_Jmove_t and Rob_Jmove_t1 represent a linear movement to a cartesian robtarget in joint space both in launching program mode and direct command mode. The movement command is available for all axes since the definition of the final destination is provided in cartesian coordinates and therefore accepted by the ASEA robot. The Absolute and Relative movement options are fully implemented.

Rob_Jmove_j and Rob_Jmove_j1 represent a linear movement to a joint destination in joint space both in launching program mode and direct command mode. This type of command can only be applied to external axes, since joint destination definitions can only be applied to external axes of the ASEA robot. The Absolute and Relative movement options are fully implemented.

Rob_r_resdp and Rob_r_resdp1 represent a reading of the robot's accuracy on the destination pose, and are implemented in the same way for both modes. The joint actuator value cannot be applied to an internal axis of the ASEA robot, since only TCP accuracy is supported.

Rob_w_resdp and Rob_w_resdp1 perform a writing of the robot's accuracy on the destination pose, and are implemented in the same way for both modes. Again, the joint actuator value cannot be applied to an internal axis due to the characteristics of the ASEA robot.

Rob_r_resip and Rob_w_resip perform a reading and writing of the robot's accuracy on the intermediate pose. Both routines have only been implemented for the launching command mode since in direct command mode, intermediate poses are not provided due to the absence of the No_wait state. The joint actuator's accuracy value cannot be applied for the same reasons as in previous paragraphs.

Rob_check_axes is performed reading the Parameter Data Array using the communication kernel.

Rob_w_vel and Rob_r_vel read and write robot's TCP or external axis velocity for every move command.

Rob_limit, Rob_worksp and Rob_prosp are the limit and workspace definition functions. The limit function applies only to external axes since no other control is allowed for a specific axis in the ASEA controller.

Workspace and Prohibited space functions are implemented as minimum and maximum radia and height definitions.

Rob_curpos_r and Rob_curpos_j represent the definitions of the current robot's position in cartesian and joint coordinates. Rob_curpos_j is only used with external axes since information on internal axes is only expressed in cartesian coordinates.

Rob_identir and Rob_selectir are the functions used to identify and select the working robot. Again, a kernel function is used for transparency of use and also to allow the simultaneous (actually fast subsequent) use of the S2 and S3 robot controllers.

Rob_gohome performs a robot and external axis movement from the present position to the synchronization position.

Rob_cnfgchan, Rob_openchan, Rob_closechan, Rob_resetchan and Rob_statchan are the functions used to establish, maintain and close the communication with the robot, and to monitor the information status of such communication.

Rob_ain, Rob_wait_le, Rob_wait_ge represent the function dealing with the analog sensor information aquisition and associated robot command. Since the ASEA robot is not endowed with analog sensor functions these features are implemented in an associated aquisition card. The control of this card is performed by the external computer. The functions perform the reading of the sensor and the control of the robot (waiting) until the value is less_or_equal or greater_or_equal than a fixed value.

Finally and according to documents NIRO.WGB.KfK.007.92 and NIRO.WGB.TGC.04.91 the sensor function Rob_sensetop has been implemented.

The sensing system includes a data aquisition card and two linear sensors.

With the values from the stack including channel, relation and threshold, the corresponding sensor, (indicated by the channel number) value is acquired and compared, with the relation provided, with the threshold value.

The application used to demonstrate the ICR interpreter is the manufacture of carbon fibre ducts using filament winding techniques. Thus if fibre break-down is detected the robot program is interupted.

7.3.6 Robot Dependent Part for Reis ROBOTstar-IV

H. Bruhm

The robot dependent part for the Reis ROBOTstar-IV controller uses the ICR to Reis compiler described in Sect. 7.4.4 of this book in a line by line mode. For this purpose the robot dependent ICR commands are written to an intermediate file (icr.src), padded with PBEG and PEND statements and translated to the Reis Robot Language (RRL) by a call to the icr2reis compiler. The output is an RRL source file, which is further processed and compiled to the binary code of the ROBOTstar IV by the Reis off-line compiler "rrlc". The binary file is downloaded to the robot controller via the DNC interface (options: RS232 or RS422) and immediately executed. The DNC software also allows the ICR interpreter to obtain information on the status of the robot controller, e.g. the current position, and to synchronise its operation with program execution by the robot. For the majority of commands, however, the ICR interpreter is allowed to run ahead of the robots' actions in order to increase its speed. Synchronisation is enforced only at points in the ICR program where the interpreter needs to obtain some information from the robot controller.

The interpreter on the ROBOTstar IV was extended with a special ICR operating mode, which may be described as "external single step mode". In this mode the interpreter will wait at the end of the currently active program and execute any commands transmitted via the DNC interface as they arrive. Initialisation and termination of this operating mode is done automatically by the ICR interpreter using the respective DNC commands.

The ICR to RRL compiler was implemented using the compiler generator tools LEX [Brown, D., Mason, T., 1990] and YACC [Gardner, J., Gisin, E., Retterath,

C., 1985]. The following sections explain how the most important ICR commands have been mapped to RRL and provide some examples.

Tool selection (EFNA)

The ICR command "EFNA" is used for selection of a tool and is called with two parameters. The first one is an integer value which selects the data type by which the tool is specified. Tool selection is executed with the second parameter. This is done either via a string (corresponds to value 5 in ICR) or an integer value (corresponds to integer value 1 in ICR) which is associated with a tool. Up to now, tool selection is only implemented by an integer value and is mapped to RRL as follows:

ICR command	RRL command
1, EFNA,1,#1;	TOOL variable: T_1
1, EFNA,1,#2;	TOOL variable: T_2
.	.
.	.
.	.

The tool command has to be called up in RRL prior to the first movement command. The tool variables have to be defined in RRL before they can be selected in the program. This definition is contained in a program header which is automatically down-loaded to the ROBOTstar IV when the ICR interpreter is started.

Writing movement velocity and acceleration (W_VEL, W_ACCEL)

ICR provides the commands "W_VEL" and "W_ACCEL" for speed and acceleration commanding. The function is determined by two parameters. The first one distinguishes whether the specifications are made for point-to-point or linear movements. The second permits selection between relative (%) and absolute values.

Two further parameters are passed via the stack. Selection of the axis that the command refers to is possible with an integer value. In RRL the speed specification in PTP mode always refers to the fastest moving axis. The second stack parameter indicates the speed.

In RRL these tasks are fulfilled with the command "VELOC_CP", "VELOC_PTP", "ACCEL_CP" and "ACCEL_PTP. In the action routines for "W_ACCEL" and "W_VEL" these are applied according to the parameters. If the selected speed is higher than the maximum admissible speed on the controller, the latter will be used.

ICR command	RRL command
1,PUSHI,I,#0;	

```
2,PUSHR,#50;
3,W_VEL,T,1;                    VELOC_CP (mm/s):1000.0
4,PUSHI,#0;
5,PUSHR,#50;
6,W_ACCEL,T,1;                  ACCEL_CP (%):50
```

Movement control (LMOVE, JMOVE, CMOVE)

Three movement modes are available in ICR. The linear movement is called up with the "LMOVE" command, the point-to-point movement can be selected with "JMOVE", and "CMOVE" commands the circular movement. The parameters are equal for all three commands. Setting of the first parameter indicates whether the position has to be considered as an absolute or a relative one. The second parameter makes a distinction between joint and cartesian coordinates, and with the third parameter further execution of the program can be suspended until the end of the movement. In preparation of a move command, the end position (and an intermediate position for circular movement) must be pushed onto the stack as a "robtarget" data type.

In RRL the corresponding movement modes are activated with the commands INTERPOL #CP-CIRC2, INTERPOL #CP, INTERPOL #PTP. The "POSITION" statement for setting the target of movements uses a position specification in the Reis convention (equivalent to the roll-pitch-yaw convention, but with inverted sign for the pitch angle) and is executed according to the selected movement mode.

ICR command	RRL command
1,PUSHR,#100;	INTERPOL # CP_LIN
2,PUSHR,#200;	POSITION #
	N,X:100.0,Y:200.0,Z:300.0
3,PUSHR,#300;	A:0.0,B:90.O,C:179,VZ1:0,
4,PUSHI,#3;	VZ2:0,VZ3:1,VZ4:0,VZ5:1,VZ6:1
5,PUSHR,#0;	
6,PUSHR,#90;	
7,PUSHR,#179.99;	
8,PUSHR,#0;	
9,GENERR;	
10,PUSHI,#52;	
11,PUSHI,#0;	
12,PUSHI,#0;	
13,PUSHI,#0;	
14,GEND;	
15,GENT;	
16,LMOVE,1,T,W;	

Movement commanding in RRL does not permit specification of the end position in joint coordinates. Furthermore, each move command is completely processed prior to execution of the succeeding command. Special commands ensure the influencing of the robot during movement (see below). Only the corresponding parameter allocations are implemented for the ICR movement commands.

Sensor stop (STOP)

The ICR command "STOP" is used for influencing a movement with external signals, the movement being stopped at the current position after a sensor signal has occurred. Parameters indicate which channel is interrogated and what kind of condition is to stop the movement at what threshold. This command is suitable for the positioning of the end effector depending on information given by the sensors.

In RRL there exists a very similar command "FIND" which, however, supports only the conditions "greater than or equal to" and "less than or equal to".

ICR command	RRL command
1,PUSHI,#1;	FIND #BINARY, Channel:1,
2,PUSHI,#5;	Liminal_Value:1.0,
3,PUSHI,#1;	#NON_INVERT, Label:DUMMY
4,STOP,D;	LABEL "DUMMY"

Stop is an extension to the ICR standard and has been implemented according to the NIRO proposal mentioned in Sect. 6.2.

7.4 ICR Transformation to Robot Vendor Code

7.4.1 Overview

C. Busk

The development of transformation packages for ICR is a way to try to introduce the concept of neutral interfaces into the great population of existing industrial robots. The degree of success for the neutral interface concept could be raised dramatically if it was possible to use the concept in the robots already installed in the industry today. This way the standard interfaces in new robot controllers would be demanded from the end user to ease the integration in the production environment.

ICR is designed for interpretation in the robot controller. As such, fast and well functioning solutions can be developed in new and open robot controllers with ICR as the driving language.

When looking at the installed population of industrial robots it is very seldom we find an open robot controller where a new program execution manager could be incorporated. Therefore transformation packages between the neutral interface language and the native languages of the robot controllers are the only way to apply the concept of neutral interfaces to the existing population of industrial robots.

When using this approach towards the neutral interface concept the result is very dependent on the degree of correlation between the native robot language and the neutral language. The overlap in functionality of the two languages forms a filter through which programs have to pass. It is only possible to express task solutions with functionality covered by both the neutral interface format and the native robot controller language.

This filtering process is one of the major problems when trying to introduce ICR as the neutral interface format for the existing base of industrial robots. ICR is a very strong low level robot programming language and it contains a very broad range of language statements to allow advanced robot programming. Many of these language statements are not available in the native robot languages of the existing industrial robots. Therefore ICR loses very much of its power when the programs have to be transformed into the native language of one of the existing robot controllers.

The existing robot controllers on the other hand are very often specialised for a single robot task such as for example arc welding. This specialisation is the background for a set of dedicated language statements used to define the process. ICR has no direct way of matching these specialised commands, so the task would have to be programmed as an advanced robot task using several of the low level program flow control statements and the low level input/output commands etc. found in ICR. The fact that several statements are needed to express what is found in a very compact way in the existing robot programming languages gives multiple possibilities to solve the same task. This means that it is impossible or at least very cumbersome to translate these program parts from ICR to one of the existing high level robot languages.

The robot programs defined in the neutral language format are transformed into the native robot language. This direction of transformation gives problems for the use of ICR as the neutral language. While ICR is a low level language and therefore excellent for fast interpretation and execution in the robot controller, the native languages of common robot controllers are designed for human interaction directly on the controller itself and so are high level languages. Transformation is then done from a low level neutral language to the high level native language meaning that information gained from a group of commands in the low level language has to be combined into one command in the native high level language on the robot controller.

The fact that several commands in the low level language have to be combined into one or two commands in the high level language gives problems for example with the handling of calls to subroutines. The place where the call is placed in the low level language does not necessarily fit where there is a natural space for a command in the high level language. This has led to problems in the handling of calls to subroutines in the transformation packages which is a major drawback of the solution.

7.4.2 ICR Transformation to the Hirobo Robot

C. Busk

The native language for the Hirobo robot is a very compact special purpose high level language. It covers the following parts of a robot program:

- simple program flow control
- normal movement control
- advanced welding functions
- simple input output control

During the NIRO project the Hirobo robot was used for the demonstration at Odense Steel Shipyard (see Sect. 8.4). Therefore, it was crucial to be able to use most of the functions listed above.

It was not possible to change the behaviour of the controller for the Hirobo robot. It was also not possible to download commands to the controller during run-time. Therefore, the only solution that allowed the use of ICR was a translator. Since ICR was designed for interpretation in the robot controller, this is not the best solution, but during the project the translator was enhanced to a level where it was possible to perform the demonstration task.

The translator was built with the 'C' programming language and the programming tools LEX [Brown, D., Mason, T., 1990] and YACC [Gardner, J., Gisin, E., Retterath, C., 1985]. It runs on a normal DOS based PC.

The translator was built on top of a general ICR scanner/parser developed with LEX and YACC tools. The parser actions for the translator were then programmed in 'C' for the number of ICR statements that are supported in the translator. For the remainder of the ICR commands, error messages are built into the parser actions. This structure allows the possibility of using the scanner/parser part of the translator in other translators for other target languages.

The basic scanner/parser part of the translator, developed by DIS Industrial Consultants a/s, was distributed to the other partners working with translators for different robot languages.

The Hirobo language is very specialised for the welding task, and therefore some of the general purpose functions found in ICR could not be covered by the translator. An example of a Hirobo robot program is shown below. It contains the program to perform the welding of one joint. The process data information needed is not shown in the example. For comparison, the total size of the ICR program with all data files is 10046 bytes.

```
%0
RD1
GC/Added job only
N1M99/1
N2M40/1
N3M50/3D1/10
N4M8X0.0017Y619.9983Z439.9941A89.9996B-44.9996F200.0
N5M8X-235.3853Y404.4006Z1052.3984A134.4532B33.6140F200.0
N6M8X-219.0574Y359.5396Z896.6811A138.1279B43.0103F200.0
N7M50/1D1/10
N8M40/1
N9M50/3D1/10
N10M8X-411.9657Y384.6498Z761.4871A142.7632B32.3073F200.0
N11M9/1TB1F100.0
N12M8X-416.8127Y384.6457Z917.1130A144.3163B31.8814F200.0
N13M9/2TB2F100.0
N14M8X0.0000Y0.0000X0.0000A-9.3109B-10.9582TBTF200.0
N15M72/2
N16M10/1X0.0000Y0.0000Z0.0000A-8.3450B-31.3756TB1S1E1F10.0
N17M8X-401.8198Y366.7877Z761.4889A145.7677B11.0791F200.0
N18M8X-219.0599Y359.5405Z896.6780A138.1279B43.0103F200.0
N19M50/1D1/10
N20M40/1
N21M50/3D1/10
N22M8X-235.3873Y404.4005Z1052.3965A134.4532B33.6140F200.0
N23M8X0.0002Y619.9984Z439.9922A89.9996B-44.9996F200.0
N24M50/1D1/10
N25M3
N26M100
END
```

The use of runtime variables was needed for the demonstration because the welding operations used sensors to determine the location of the joint to be welded. There are special problems with runtime variables when the ICR programs are translated offline. The only data types for runtime variables which it is possible to use when the ICR programs are translated are the data types that the native robot programming language gives access to. In the case of the Hirobo robot language, the number of different data types is limited to poses (position and orientation).

In the translator for the demonstration at Odense Steel Shipyard, the ICR data type ROBTARGET was considered as equal to poses, because of the physical installation. This allowed the storing of sensed poses which afterwards were used as start and end points for the welding operations.

Initially there were problems in the simulation of the search algorithm in the GRASP offline programming system used for the demonstration. Therefore, the translator had to take care of converting all movements corrected by the sensed positions. However the required functionality was subsequently incorporated into GRASP in developments to the 'sense' step.

Other problems encountered originated in the ICR programming language. The handshaking between the gantry and the robot, as an example, demanded the setting of two digital output bits to be set simultaneously. This was not possible for two reasons. Firstly, the ICR specification only allows the values TRUE(1) and FALSE(0). Secondly, it was also not possible to program the output to be simultaneous in the ICR code produced by the offline programming system. This made it necessary to take special actions in the translator to ensure that the handshaking scheme was correct, as needed in the physical demonstration environment.

Furthermore, the demonstration demanded welding operations. Therefore the translator for the Odense Steel Shipyard demonstration was extended with commands for arc-welding and sensing. These commands were defined within the NIRO project and then submitted to the ISO committee working with the standardisation of ICR. The welding commands are only defined for MIG/MAG arc-welding.

7.4.3 ICR Transformation to PDL2 (the Comau Robot)

M. Odifreddi and C. D'Elia

PDL2 is a pascal-like third generation robot programming language developed by COMAU as programming language for all COMAU robots controlled by C3G controller. PDL2 is a symbolic procedural language with standard and robot-oriented data types and instructions. A library of system functions allows geometrical and mathematical computations.

PDL2 programs are translated into an intermediate executable format prior to being run. There is an one-to-one correspondence between PDL2 and intermediate format so that debugging and editing can be done on both PDL2 and the intermediate data format. A cross-compiler is also available on MS/DOS environment allowing program translation and program down loading to C3G.

A PDL2 program has the following structure:

```
PROGRAM program_name
TYPE
   type declaration
VAR
   var declaration
PROCEDURE procedure_name
          VAR ...
```

```
        BEGIN
        . . .
        END procedure_name
    PROCEDURE procedure_name
        VAR ...
        BEGIN
        . . .
        END procedure name
  BEGIN
        . . .
    END program_name
```

The FIAT/SESAM team incorporated a post-processor able to translate ICR statements into PDL2 statements in the NIRO project. The following translations have been implemented:

1) Variable declarations
2) PUSH and POP instructions
3) ADD, SUB, MUL and DIV binary operators
4) AND, OR and XOR binary operators
5) EQ, NE, GT, LE, LT and GE binary operators
6) MIN unary operator
7) NOT unary operator
8) Data type conversion instructions
9) LABEL and GOTO instructions
10) IF instruction
11) JMOVE, LMOVE and CMOVE instructions
12) Logical I/O instructions (including WAIT_B)
13) Hi-level control structures
14) SUBPBEG, SUBPEND and CALL instructions

Arithmetic operators are defined only on INTEGER and REAL data and move instructions are defined only on POSE data.

Software organisation

The ICR to PDL2 post-processor parses an ICR source to generate a PDL2 source through a two-pass parsing with an intermediate step for procedure parameters management.

The ICR to PDL2 post-processor writes the variable list and the statements into two temporary files and merges the two files into a PDL2 program at the end of ICR program translation. When a DECLVAR statement has been parsed a variable declaration is written into the variable list file; when other statements have been parsed the related management procedure is called.

The main post-processor features are the stack management, the procedure management and the high-level control structure management. Move statements have a one-to-one mapping from ICR into PDL2 and are directly translated.

Stack management

The ICR stack is emulated using a binary tree: the top of stack (TOS) is the current node and it is always the second son of an other node; the first son of the father of a generic second son is the element following it on the stack.

PUSH management

When a push instruction has been recognized a new node must be created to hold the push argument. The new node must be the new TOS, so it must be linked to the current TOS with a second new node; the second new node will hold the operand linking the new TOS and the current TOS and it is defined as an unknown operand node. Thus, the push management procedure includes the following steps:

- The 2nd new node takes the place of the current node.
- The current node becomes the 1st son of the 2nd new node.
- The 1st new node becomes the 2nd son of the 2nd new node.

After the push management procedure the 1st new node becomes the current node.

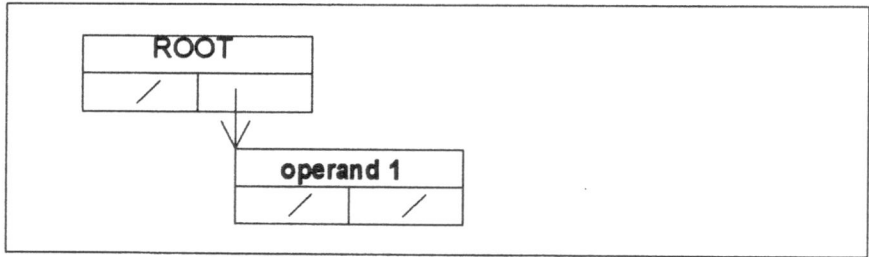

Fig. 7.4.3-1: Stack use "after push operand 1"

Let's consider an empty tree. After a "push operand 1" the instruction tree will look as shown in Fig. 7.4.3-1 with "operand 1" being the current node.

After the "push operand 2" the instruction tree will look as shown in Fig. 7.4.3-2 with "operand 2" being the current node.

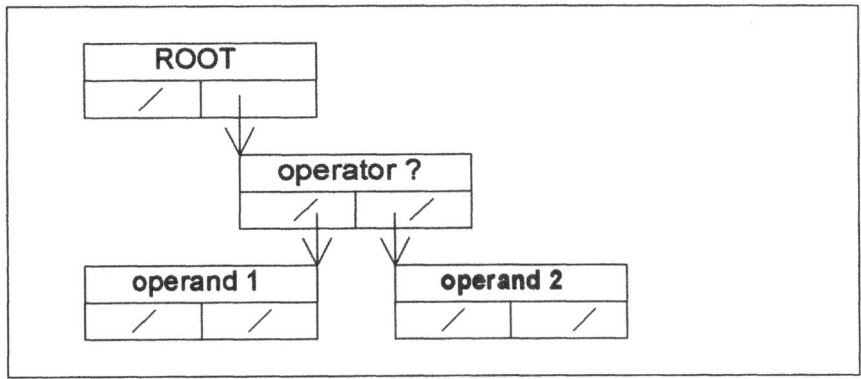

Fig. 7.4.3-2: Stack use "after push operand 2"

After the "push operand 3" the instruction tree will look as shown in Fig. 7.4.3-3 with "operand 3" being the current node.

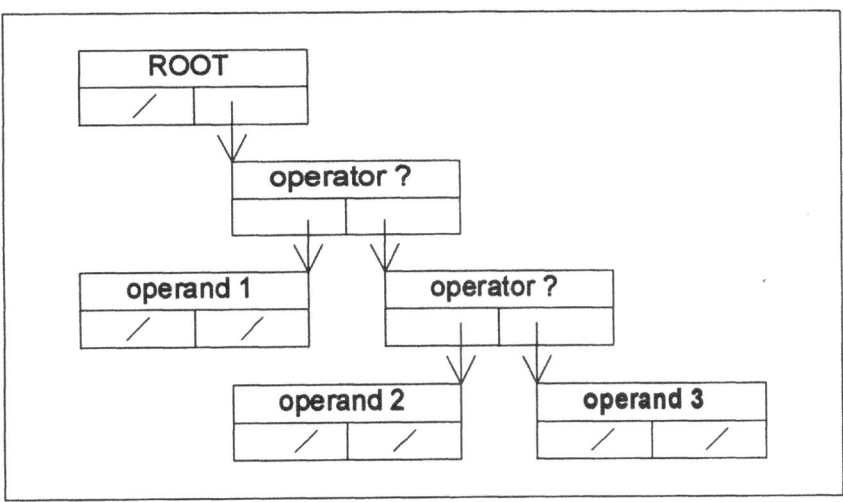

Fig. 7.4.3-3: Stack use "after push operand 3"

BINARY OPERATOR management

When a binary operator has been recognized the TOS and the TOS-1 elements must be its operands and the binary operator must be assigned to the last node father. The last node father is always an unknown operator node. During the stack evaluation the last node father will be the result of the operand application.

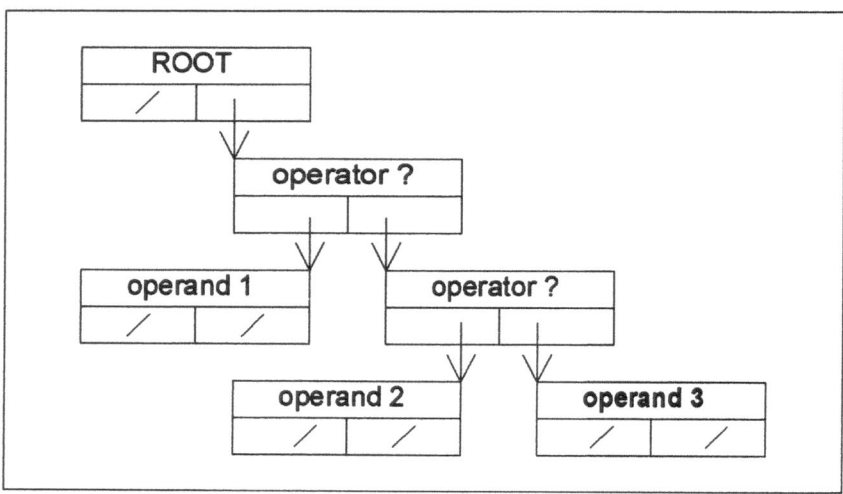

Fig. 7.4.3-4: Binary operator management "after 3 push operands"

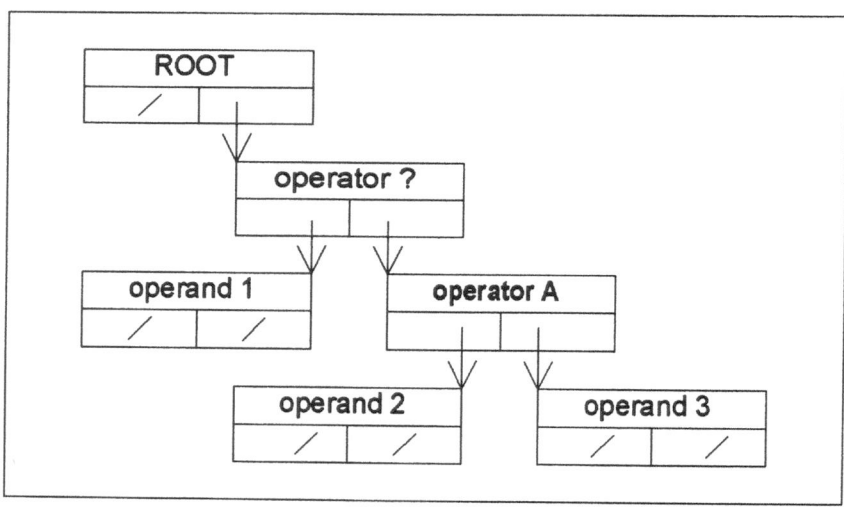

Fig. 7.4.3-5: Binary operator management "after operator A"

The father of the current node is defined with the binary operator. After the binary operator management procedure the father of the current node becomes the current node.

Let's consider an empty tree. After three "push operands", the instructions tree will look as in shown in Fig. 7.4.3-4 with "operand 3" being the current node.

After the "binary operator A" the instruction tree will look as shown in Fig. 7.4.3-5 with "operator A" being the current node.

After the "binary operator B" the instruction tree will look as shown in Fig. 7.4.3-6 with "operator B" being the current node.

Unary operators are managed in the same way

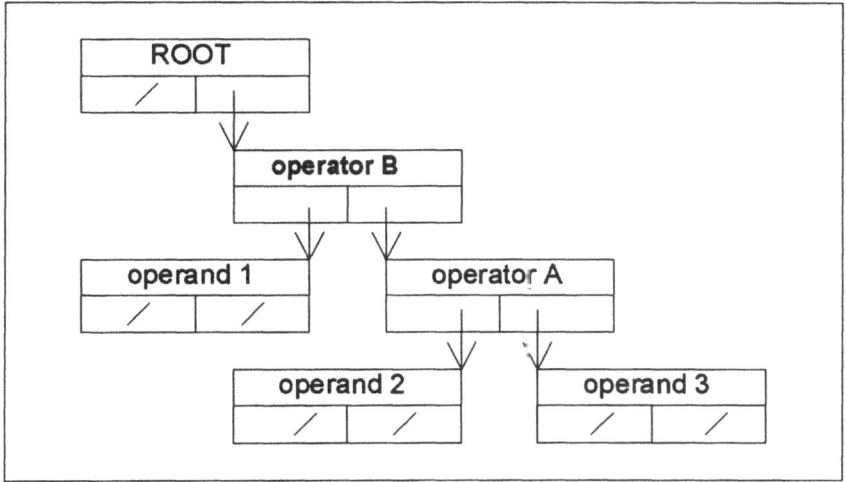

Fig. 7.4.3-6: Binary operator management "after operator B"

POP management

When a pop instruction has been recognized the current tree must be evaluated. This is done by the recursive procedure RecExplor that rebuilds a PDL2 expression starting from its tree representation. The pop management procedure generates the PDL2 assignment:

```
pop_argument =current_expression
```

where current_expression will be evaluated by RecExplor procedure. The pop management procedure also deallocates the expression tree and its father.

To allow the stack to be used to store temporary data the pop management procedure uses the first tree starting with a defined operand, that is the tree starting from the current node (a defined operand can be each operand but the unknown one). Let's look at the following sequence of ICR statements:

```
push expression 1
push expression 2
pop variable A
pop variable B
```

The first pop will find the following tree:

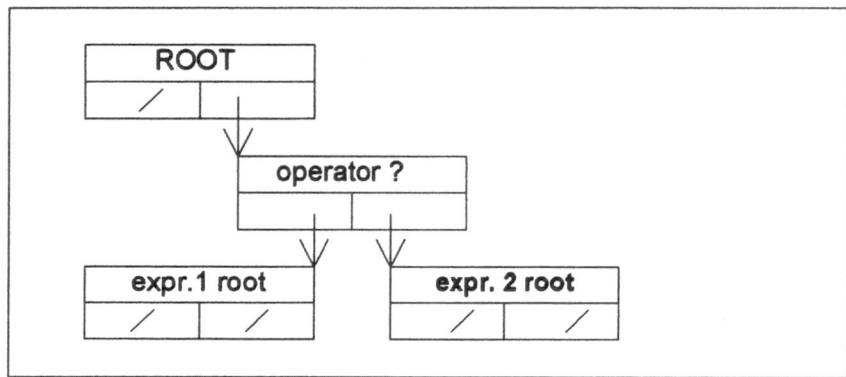

Fig. 7.4.3-7: Pop management "first pop"

It will find the expression 2 tree as the current tree and will generate:

```
A = expression 2
```

cutting "expr. 2 root" and "operator ?" The second pop will find the following tree:

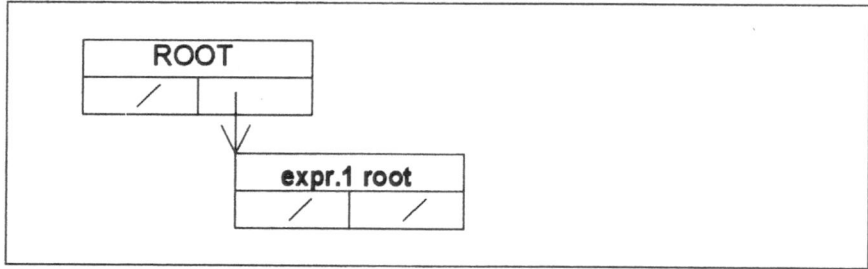

Fig. 7.4.3-8: Pop management "second pop"

It will find the expression 1 tree and will generate:

B = expression 1

cutting "expr. 1 root" and "ROOT" (the empty tree will be generated)

Expression tree evaluation

Evaluating a tree is a recursive activity. A correct tree can include either an operand node or an operator node followed by the correct number of operands; any other tree is wrong; a tree including an operand node has the value of the operand. A tree including an operator followed by its operands has the value of the operator applied to the operands; the operands are evaluated in the same way. If operand generation is required the PDL2 code is generated in two different ways.

If the PDL2 operand is binary infix then:
- evaluating the first operand
- generating the operator
- evaluating the second operand

If the PDL2 operand is unary then
- generating the operator
- evaluating the operand

If the PDL2 operand is functional then
- generate the operator
- evaluate the operator list

Control flow management

In translating the control flow instruction from ICR to PDL2 the hi-level control structures are generally built up.

During the first translation step each jump instruction is related to its target label; during the second translation step the hi-level control structures are built up.

LABEL management

When a label instruction has been recognized it is linked to its jump instruction during the first pass.

If the label closes an IF ... THEN without ELSE branch or the ELSE branch of an IF ... THEN, the label management procedure generates a PDL2 **ENDIF**.

If the label closes an IF ... THEN with ELSE branch the label management procedure generates a PDL2 **ELSE**.

If the label opens a REPEAT ... UNTIL the label management procedure generates a PDL2 **REPEAT**.

A PDL2 lo-level control structure is generated only if no hi-level control structures have been recognized. The label management does not evaluate expressions.

GOTO management

When a label instruction has been recognized it is linked to its target label during the first pass.
If the GOTO is part of an IF ... THEN ... ELSE no code is generated by the goto management.
A PDL2 lo-level control structure is generated only if no hi-level control structures have been recognized. The goto management does not evaluate expressions.

IF management

When an if instruction has been recognized it is linked to its target label during the first pass.
If the if opens an IF ... THEN (with or without ELSE branch) the if management procedure generates a PDL2 **IF ... THEN**.
If the if closes a REPEAT ... UNTIL the if management procedure generates a PDL2 **UNTIL**.
A PDL2 lo-level control structure is generated only if no hi-level control structures have been recognized. The label management calls the RecExplor procedure on the expression tree to get the current logical expression.

Procedure management

The procedure parameter list is not declared by ICR thus the list will be rebuilt by inspecting the source code.
For each missing argument in a procedure a parameter is assumed, when a value has been left on the stack by the procedure code the procedure is declared as a function and the value left on the stack is the function return value.

CALL management

The Call instruction pops a number of operands depending on the function argument list. If the procedure is a routine the PDL2 routine call instruction is generated directly by the call management routine.

```
routine_name (parameter_list)
```

If the procedure is a function the PDL function call instruction will be generated when the related pop will be recognized.

```
pop_argument := routine_name (parameter_list)
```

Software development and validation

The ICR parser has been implemented by means of LEX [Brown, D., Mason, T., 1990] and YACC [Gardner, J., Gisin, E., Retterath, C., 1985] tools. Routines described are the actions called by YACC rules; YACC rules come from the ICR grammar description. The parser has been developed in two steps. During the first step the following features have been developed:

1) Variable declarations
2) PUSH and POP instructions
3) ADD, SUB, MUL and DIV binary operators
4) AND, OR and XOR binary operators
5) EQ, NE, GT, LE, LT and GE binary operators
6) MIN unary operator
7) NOT unary operator
8) Data type conversion instructions
9) LABEL and GOTO instructions
10) IF instruction
11) JMOVE, LMOVE and CMOVE instructions
12) Logical I/O instructions (including WAIT_B)

During the second step the following features have been developed:

13) Hi-level control structures
14) SUBPBEG, SUBPEND and CALL instructions

An ICR program has been written to test the processor and the resulting PDL2 code has been run on a C3G controller connected with a COMAU P-MAST robot.

Gained experience

During post-processor implementation some problems have arisen. Neither the ICR procedure calls nor the ICR procedure declarations have parameter counts thus it is impossible to know how many parameters ICR procedures require. Given that PDL2 is a pascal-like language, PDL2 procedure declarations require a parameter list and a parameter type declaration.

The post-processor tries to evaluate a procedure parameter list examining procedure code with limits, if the use of the parameters inside the procedure is not linear. Adding the procedure parameter number in the ICR procedure call instruction will probably simplify both post-processor and interpreter development.

7.4.4 ICR Transformation to Reis Robot Language

H. Bruhm

The compiler from ICR to the Reis Robot Language RRL was implemented on a PC using the compiler generation tools LEX [Brown, D., Mason, T., 1990] and YACC [Gardner, J., Gisin, E., Retterath, C., 1985]. In order to have an ICR source file *program.icr* compiled to RRL, the compiler must be called with the following syntax:

icrcomp *program.icr*

The compiler will write its output to the file *program.src*. The output file name is generated automatically by stripping off the extension of the input file and replacing it with the new extension *.src*. *program.src* is the compiled program in RRL source format. Conversion to the binary format of the ROBOTstar IV is performed with the Reis off-line compiler *rrlc*:

rrlc program.src

As a result one obtains the binary file *program.bin*, which can be loaded into the ROBOTstar IV via the DNC interface or an attached floppy disk drive.

The rrlc off-line compiler and the DNC software are standard products of Reis.

Commands processed by the ICR to RRL compiler

The commands that the ICR to RRL compiler can process are listed hereafter together with information concerning the parameters supported. It should be noted that data popped from and pushed onto the stack are not considered as parameters, so that some commands do not have any parameters at all.

ADDI	Parameters:	none
ADDR	Parameters:	none
ANDB	Parameters:	none
ANDI	Parameters:	none
BLKBEG	Parameters:	1. block number
		2. block nesting level
	Comments:	For block nesting level 0, the ICR stack on the robot controller is initialized.
BLKEND	Parameters:	none

CALL_S	Parameters:	1. symbol (subprogram name)		implemented

CMOVE	Parameters:	1. abs_rel	1 (=absolute)	implemented
			2 (=relative)	not implemented
		2. pose_type	J (=joint)	not implemented
			T (=cartesian)	implemented
		3. wait	W (=yes)	implemented
			N (=no)	not implemented
	Comments:	Circular movement is mapped onto the RRL command "CP_CIRC2" at which orientation change in the TCP is effected like on a straight line being bent like a semicircle.		

DECLVAR	Parameters:	1. indicator	0 (=global)	implemented
			1 (=local)	not implemented
DELAY	Parameters:	none		

EFNA	Parameters:	1. data_type	1 (=integer)	implemented
			5 (=string)	not implemented
		2. eef_label	integer	implemented
			string	not implemented

GRIPOPEN	Parameters:	none		

GRIPCLOSE	Parameters:	none		

GOTO_SYMBOL	Parameters:	1. symbol (string)		implemented

JMOVE	Parameters:	1. abs_rel	1 (=absolute)	implemented
			2 (=relative)	not implemented
		2. pose_type	J (=joint)	not implemented
			T (=cartesian)	implemented
		3. wait	W (=yes)	implemented
			N (=no)	not implemented

LABEL	Parameters:	1. symbol (string)		implemented

LMOVE	Parameters:	1. abs_rel	1 (=absolute)	implemented
			2 (=relative)	not implemented
		2. pose_type	J (=joint)	not implemented
			T (=cartesian)	implemented
		3. wait	W (=yes)	implemented
			N (=no)	not implemented

NOOP	Parameters:	none		
PBEG	Parameters:	none		
PEND	Parameters:	none		
POPB	Parameters:	int. variable	symbol	implemented
			abs. address	not implemented
			blkrel. addr.	not implemented
POPI	Parameters:	int. variable	symbol	implemented
			abs. address	not implemented
			blkrel. addr.	not implemented
POPR	Parameters:	real variable	symbol	implemented
			abs. address	not implemented
			blkrel. addr.	not implemented
PUSHB	Parameters:	boolean var.	symbol	implemented
			abs. address	not implemented
			blkrel. addr.	not implemented
			constant	implemented
PUSHI	Parameters:	int. variable	symbol	implemented
			abs. address	not implemented
			blkrel. addr.	not implemented
			constant	implemented
PUSHR	Parameters:	real variable	symbol	implemented
			abs. address	not implemented
			blkrel. addr.	not implemented
			constant	implemented
REMARK	Parameters:	comment string		
STOP	Parameters:	1. type	D (=digital)	implemented
			A (=analog)	not implemented
SUBPBEG	Parameters:	none		
SUBPEND	Parameters:	none		
W_ACCEL	Parameters:	1. vel_type	J (=joint)	implemented
			T (=cartesian)	implemented

		2. vel_kind	1 (=%)	implemented
			2 (=m/s)	implemented
			3 (=rad/s)	not implemented
W_VEL	Parameters:	1. vel_type	J (=joint)	implemented
			T (=cartesian)	implemented
		2. vel_kind	1 (=%)	implemented
			2 (=m/s)	implemented
			3 (=rad/s)	not implemented

7.5 Test of Processors

A. Bautista

The test of processors concept is related to the characteristics and quality of the code automatically generated and its interaction with the performance of the equipment.

The first condition to be considered is the scope and extent of the processors. In the present conditions the following group of commands should be included:

* General declaration commands
* Program flow control commands
* Boolean and arithmetic operations
* Check operations
* Move parameter definition
* Movement definition
* Limits related commands
* Robot identification commands
* Tool and gripper commands
* Input/Output operations
* Sensor commands

In some cases miscellaneous commands can also be included.

Once the extent has been established, the quality of the code should then be determined.

Three elements should be taken into consideration :

1) Size of the code

   This is particularly critical in the case of the ICR interpreter on a PC. Since no extended memory is used the maximum size of the program successfully loaded into the interpreter is about two thousand one hundred and fifty lines long (actually 2164 lines usually). The number of lines is therefore limited in certain cases. This represents a problem in point consuming applications such as inspection systems , filament winding systems etc...

2) Speed of code.

To avoid the above mentioned problem the use of data list management functions can be taken into consideration. Therefore where large quantities of points have to be handled the processor should consider a data list management approach. If this resource is too widely used then robot performance problems might show up, in relation with the speed of code processing. Hence a wise combination of data list management use and robot movement strategy is required.

3) User interface quality

Other properties such as readability of the code and automatic initialisation of the program, including open and close procedures and parameter definition, should be analysed.

Finally quality of performance should be tested. In this case a major consideration in checking the code is the close relationship between the quality of the code and the suitability of the code, and between the robot control language and movement strategy of each particular robot. Hence different movement strategies might be required for specific robot controllers and applications in order to produce the best results.

Test example

As an example a set of test programs and movements have been considered for the ICR Interpreter for the ABB IRB 60 robot. The ICR code is generated from a generic movement definition and then tested on the actual robot. Basic features tested are :

* Cartesian movement including
- Type of movement :
+ Absolute movement
+ relative movement
- Type of stop flags :
+ Wait flag (wait until movement finishes)
+ No wait flag (continue while robot is moving)

* Joint movement including
- Type of movement :
+ Absolute movement
+ relative movement
- Type of stop flags :
+ Wait flag (wait until movement finishes)
+ No wait flag (continue while robot is moving)

* Sensor commands such as
- Analog or Digital I/O

- Wait for Analog Input
- Sensstop commands

The example given in Appendix 4 includes a "generic" movement file , with all movement parameters defined in sequential order , and its related ICR code file. The programs included are a "summarised" version of the test programs actually run in the robots' controller where a larger amount of points are actually being used. It reproduces a plane square with and without sensor commands.

Tests at user demonstration sites

The processors for both IRL and ICR described in the preceeding sections have been tested and the results validated at the selected demonstration sites. The diversity of the applications presented show the powerful nature of the neutral file formats in providing a common framework for data transfer in the CIME environment. Results and problems relating to specific processors are discussed in more detail in both the appropriate Sect. in this chapter, and in Chapter 8, "Application of Developments and Results".

8. Application of Developments and Results

Chapter Editor: *H. Bruhm*

8.1 The Rationale

H. Bruhm

The results of the NIRO project have been applied and tested under conditions close to the real production situation by three industrial project partners (CASA, FIAT, and OSS), representing the end users of the technology developed, and additionally by a manufacturer of robots and robot control systems (Reis) in a transportable demonstration set-up, which was shown at the accompanying exhibitions to the CIM-Europe Conference in Odense/Denmark in September 1992 and the ESPRIT '92 Conference in Brussels in November 1992.

The far-reaching potential of the neutral interfacing concept was illustrated by the fact that data transfer via STEP, IRL, and ICR was achieved within existing IT frameworks of a very heterogeneous nature, including a variety of CAD systems, planning and simulation systems as well as robots and control systems from different manufacturers. A lot has been learnt, on the other hand, about requirements for the further development of the interfacing standards. The experience gained was fed back into the respective standardization working groups through direct participation of NIRO members.

It was shown, furthermore, that the approach based on neutral interfaces can help to make CIM concepts economically attractive for industries where this technology has, up to now, achieved only a low penetration. Production in these industries is typically in small to medium size batches, as opposed for instance to the automotive industry, where CIM technology has its traditional stronghold. The introduction of robots into innovative applications is exemplified by the production processes implemented at CASA and OSS: automated fabrication of composite materials parts for the aerospace industry (small batch manufacturing) and welding robots used in ship building (one-off production).

8.2 The Demonstration at CASA/DISEL

A. Bautista

8.2.1 Criteria for Task Selection

Composite materials are increasingly being used in the aerospace industry.

Carbon and glass fibers, epoxy resin matrices, high temperature polymides and other materials were used in creating more than 10000 composite components that went into the B-2 Stealth bomber contured airframe, and citing aerospace industry predictions, composites will account for up to 60 percent of the material used to fabricate next-generation aircraft.

Several processes are used to manufacture carbon fiber components but filament winding is the most widely used to produce revolution pieces. A carbon fiber is impregnated with an epoxy resing and winded with the adecuate position and orientation around the tool to produce the desired shape and strength properties once it is cured .

Since workpieces can widely vary in size, shape and type in the past robot's programs were generated using the robot's teaching mode, but the method required huge amounts of work and adittional risk for the equipments, hence a more integrated solution is required.

In this sense the NIRO project concept engages with the general CIM concept in CASA,also represented by the DNCC system including the control and use of Flexible Manufacturing Cells also connected with CATIA.

CATIA is the most widely used CAD system in CASA and one of the most used in the Aerospace Industry .

8.2.2 Task Description

In general as it has been established, a carbon fiber is impregnated with an epoxy resine and winded with adequate position and orientation around the tool.

Feed rate and orientation of the fiber determine the shape and stregth properties of the piece.

Once the piece has been manufactured a curing process is required.

The conditions under wich this process has to take place greatly vary according to the resine properties, but in general this process requires at least one half hour of air curing.In order to avoid the droppage of the resine towards the lower part of the piece due to gravity effects this process requires an alternative turning process of the external rotary axis.

Finally a subsequent demoulding task (deeply influenced by the concious design of the tool) is followed by a finishing process to provide final shape and tolerances to the piece and in some cases an ultrasonic control quality takes place.

Workpieces widely vary in shape, sice and type, ranging from ducts to fuel deposit covers, but the general case is usually a fiber duct 10 cm to 50 cm long with both ends endowed with characteristic shapes for linking purposes.

To produce these pieces with the flexibility required by the Aerospace Industry, a robot system is actually used.

The system includes :

- An ASEA IRB 60 robot with five axis, with an S2 controller.
- Two external axis one with straight movement another with rotary movement.

The adequate relation between the rotary and straight movements (together with the traslation and orientation movements of the robot) produces the correct orientation of the fiber in the workpiece to provide the desired strength properties.

- An adjustable break to fix the reel to the required winding feed.
- The impregnation system to provide the required amount of resine and catalyst.
- The sensor system including sensors to measure
- The reel level to avoid reel finishing during the winding process.
- The fiber strain / fiber breakdown to stop the process if the fiber is not being winded correctly.
- An IBM system containing CATIA using CATGEO ROBOTICS routines to produce the simulation.

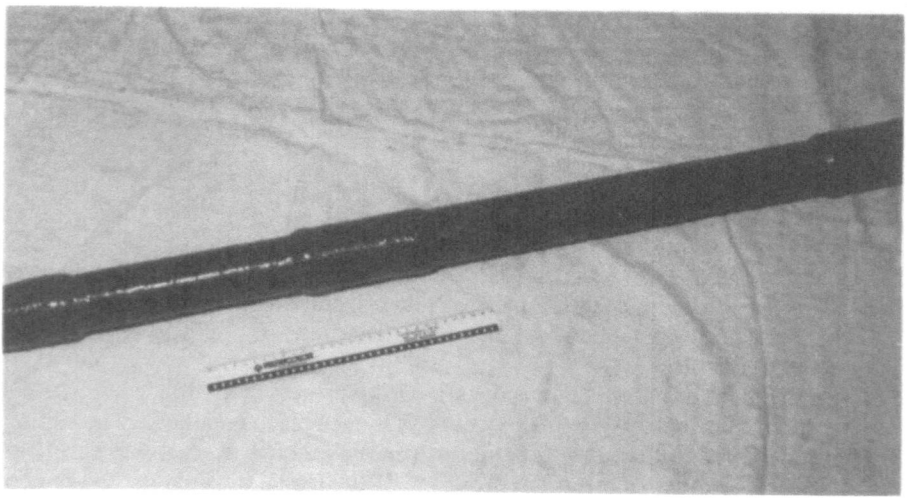

Fig. 8.2.2-1. Carbon fiber work piece

A photo and drawing of the installation are shown in figures 8.2.2-2 and 8.2.2-3.
The sequence of operation follows.

In the general case a file containing the geometric model of the piece will be provided. This file might have been internally generated in CASA, and therefore in CATIA, or it might have been generated in some other company.In this case the STEP-CATIA processor produced in the CADEX project is thought to be used.

Once the model is available a path planning of the robot's trajectory has to be generated.

A process knowledge is required to produce a correct path planning and adequate design of the tool.

Fig. 8.2.2-2: Photo of the Installation

A simulation of this trajectory is performed using CATGEO ROBOTICS routines to check the process. Once the trajectory has been accepted the correspondent ICR code is automatically generated at the CAD station and transferred to the PC. The program is then loaded in the ICR Interpreter.

Running the ICR Interpreter programm manufactures the piece.

The curing process usually also takes place at the station while demoulding and finishing process takes place at adjacent stations.

A photo and a drawing of the simulation process are included.

184

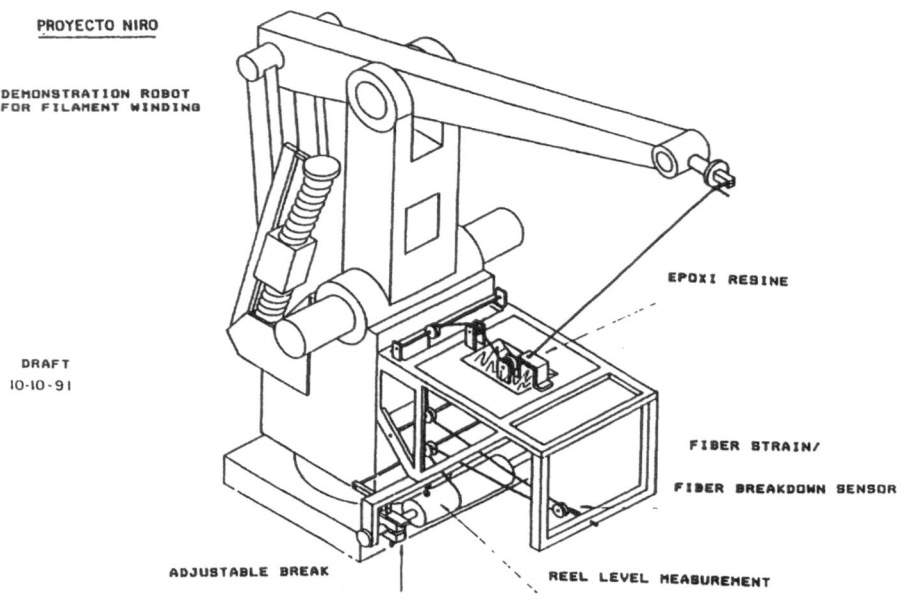

Fig. 8.2.2-3. Drawing of the installation

8.2.3 Information Flow Structure

The general Information Flow Structure, as previously described is represented in Fig. 8.2.3-1.

Again two possible flows are considered. Either the piece is created in an external system were it can be imported using the STEP processors or it can be created inside CATIA.

Since CATIA includes a robotics module the simulation is performed inside CATIA and no processor is required to trasform the information to perform this task.

Once the simulation has been accepted the ICR code is directly generated and loaded into the Interpreter where it will run.

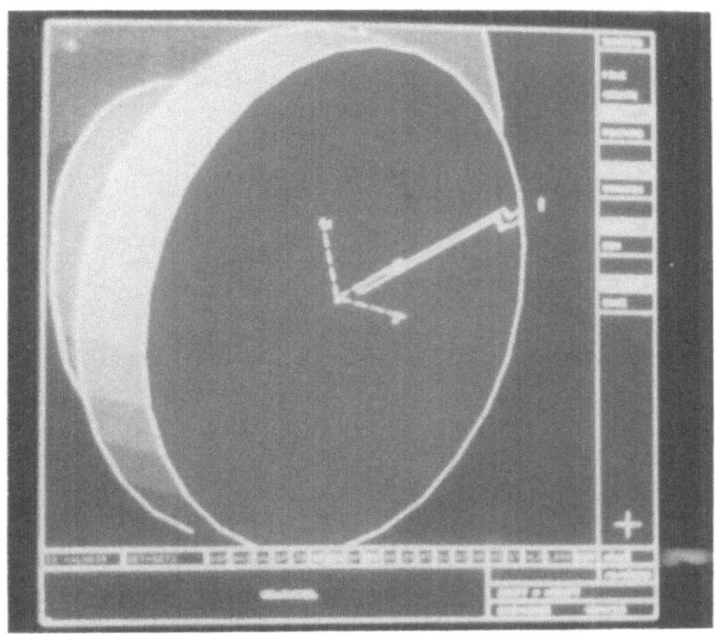

Fig. 8.2.2-4. Photo of the Simulation Process

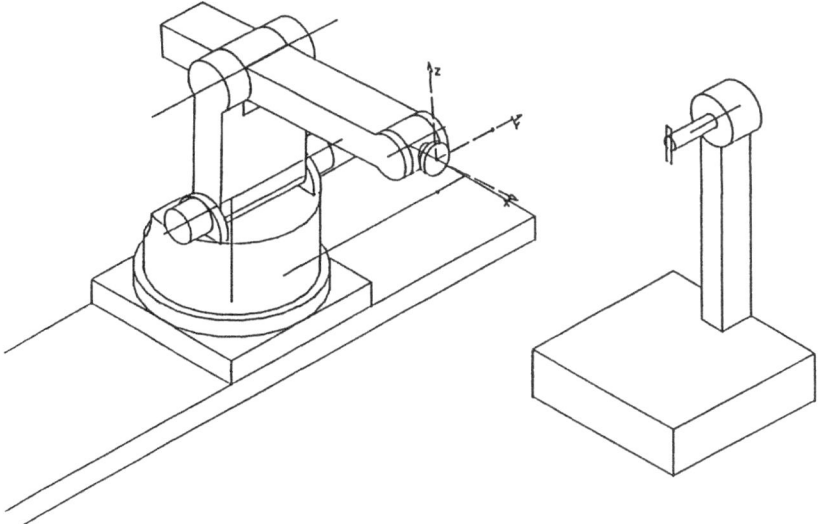

Fig. 8.2.2-5. Drawing of the simulation process

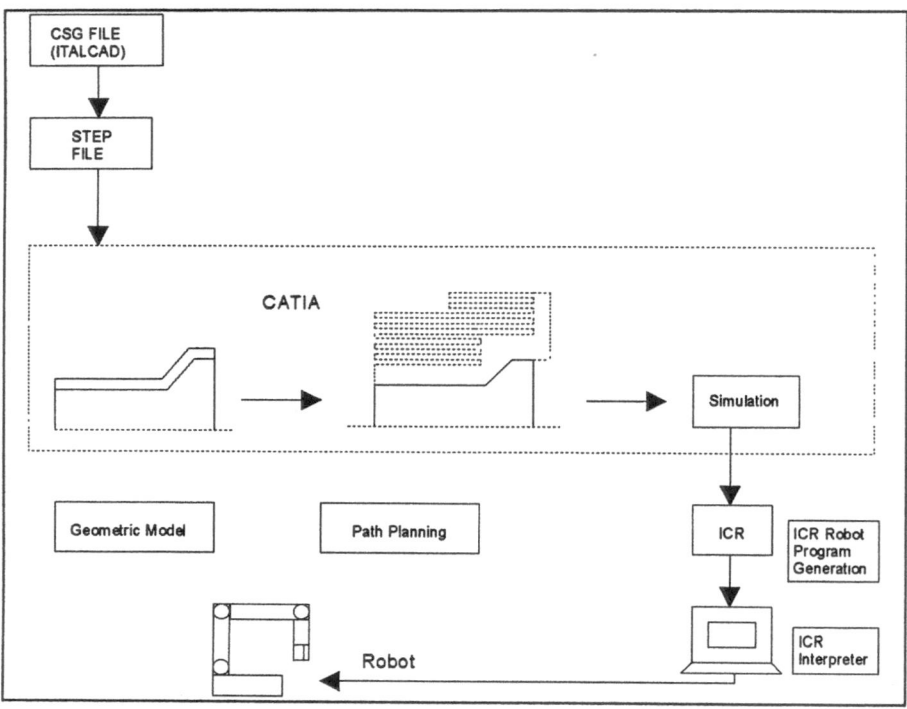

Fig. 8.2.3-1. The general information flow

8.2.4 Implementation Details

With respect to implemntation details two basic aspects have been considered.

The filament winding application is a very high point-consuming process on one hand and also it requires a high degree of continuity.

This leads to aim the two basic problems :

1) Size of the Code

This is particularly critical in the case of the ICR interpreter on a PC. Since no extended memory is used the maximun size of the program succesfully loaded into the Interpreter is about two thousand one hundred and fifty lines long (actually 2164 lines usually).

The number of lines is therefore limited in certain cases.

This represents a problem in applications such as inspection systems, filament winding etc...

2) Speed of the Code

To avoid the above mentioned problem the use of data list management functions can be taken into consideration.

Therefore in case large quatities of points have to be handled the processor should consider a data list management approach.

If this resource is too widely used then robot performance problems might show up, in relation with the speed of performance of the code. Hence a wise combination of data list management use and robot movement strategy is required.

As a conclusion, the use of long programs versus data list management functions has been an issue.

The second basic aspect is related with the way robots are commanded.

One performs a direct command of the robot from the external computer while the second one loads a program in the robot's memory and launches it . This has been the option finally chosen.

The solution of the problem has been addressed in the design of the robot dependet part of the ICR Interpreter.

Finally and just for testing purposes, an automatical generator of ICR code has been developed to avoid the tedious editting of ICR progams from simple move text files. A short example of this files follows.

```
1,PBEG,95,"ROBOT";
2,LABEL,"POSICION";
3,BLKBEG,0,1;
4,PUSHR,#100.000000;
5,PUSHR,#1500.000000;
6,PUSHR,#1600.000000;
7,PUSHI,#6;
8,PUSHR,#1.000000;
9,PUSHR,#0.000000;
10,PUSHR,#0.000000;
11,PUSHR,#0.000000;
12,GENERR;
13,PUSHI,#0;
14,PUSHI,#0;
15,PUSHI,#0;
16,PUSHI,#0;
17,PUSHR,#-1000.000000;
18,PUSHR,#-500.000000;
19,GEND;
20,GENT;
21,JMOVE,1,T,W;
22,PUSHR,#100.000000;
23,PUSHR,#1500.000000;
24,PUSHR,#1400.000000;
25,PUSHI,#6;
26,PUSHR,#1.000000;
27,PUSHR,#0.000000;
```

```
28,PUSHR,#0.000000;
29,PUSHR,#0.000000;
30,GENERR;
31,PUSHI,#0;
32,PUSHI,#0;
33,PUSHI,#0;
34,PUSHI,#0;
35,PUSHR,#-1000.000000;
36,PUSHR,#-500.000000;
37,GEND;
38,GENT;
39,LMOVE,1,T,W;
40,PUSHR,#-100.000000;
41,PUSHR,#1500.000000;
42,PUSHR,#1400.000000;
43,PUSHI,#6;
44,PUSHR,#1.000000;
45,PUSHR,#0.000000;
46,PUSHR,#0.000000;
47,PUSHR,#0.000000;
48,GENERR;
49,PUSHI,#0;
50,PUSHI,#0;
51,PUSHI,#0;
52,PUSHI,#0;
53,PUSHR,#-1000.000000;
54,PUSHR,#-500.000000; 55,GEND;
56,GENT;
57,LMOVE,1,T,W;
58,BLKEND;
59,SUBPEND;
60,LABEL,"INICIO";
61,BLKBEG,2,1;
62,PUSHI,#1;
63,SELECTIR;
64,PUSHI,#1;
65,PUSHI,#0;
66,PUSHI,#0;
67,PUSHI,#0;
68,PUSHI,#0;
69,CNFGCHAN;
70,PUSHI,#0;
71,OPENCHAN;
72,BLKEND;
73,SUBPEND;
74,LABEL,"PARAMETROS";
75,BLKBEG,3,1;
76,PUSHI,#0;
77,PUSHR,#66.000000;
```

```
78,W_VEL,T,1;
79,LIMIT,1,0,#10.000000;
80,LIMIT,1,1,#90.000000;
81,LIMIT,2,0,#10.000000;
82,LIMIT,2,1,#90.000000;
83,PUSHI,#1;
84,PUSHI,#0;
85,W_RESDP,T,2;
86,BLKEND;
87,SUBPEND;
88,LABEL,"CIERRE";
89,BLKBEG,4,1;
90,PUSHI,#0;
91,CLOSECHAN;
92,BLKEND;
93,SUBPEND;
94,REMARK,30,"    ****    MAIN    ****    ";
95,BLKBEG,1,0;
96,DECLVAR,0,VAX,2,0,VAY,2,0,VAZ,2;
97,DECLVAR,0,Q1,2,0,Q2,2,0,Q3,2,0,Q4,2,0,TIORI,1;
98,DECLVAR,0,ROPOSI,8,0,ROORI,9,0,ROPOSE,10;
99,DECLVAR,0,EJEEX,13,0,ERRMOV,1,0,ROFIN,14;
100,CHECK_ORI,6;
101,CHECK_AXES,6,2;
102,CALL_S,"INICIO";
103,CALL_S,"PARAMETROS";
104,CALL_S,"POSICION";
105,CALL_S,"CIERRE";
106,BLKEND;
107,PEND;
```

8.2.5 Results

Some experiences have been carried out . The first one to be presented was shown at the Odense Conference and is descrived in the following paragraph.

The present example is aimed to point out the possibilities of neutral interfaces and the integration of CADEX and NIRO projects.

The model corresponds to a nozzel holder manufactured by FIAT.

ITALCAD generated afterwards a CSG file using an S7 propietary CAD system and transformed it to a STEP file using a pre-processor.

DISEL using the STEP-CATIA processor produced the CATIA model.

Inside CATIA the trajectories are generated and simulated .

Some minor design modifications had to be introduced to allow the nozzle (initially thought to be extruded) to be manufactured by filament winding in carbon fiber.

Once the trajectories have been simulated and accepted the ICR code is generated.

The file is the loaded into the ICR Interpreter and the component is produced in composite material by CASA.

A curing process and some finishing operation are the required to obtain the final component.

Also several test programs have been repeatedly applied . A closer description in the chapter devoted to the test of processors.

From the experiences gained three basic areas should be enhanced:

- A more extense an specific set of sensor commands is required.
- An easy procedure to integrate to integrate specific user libraries should be developed.
- An enhancement of the number of ICR program lines to be loaded in the computer.

8.3 The FIAT Demonstration. Experiences Gained

C. D'Elia

The FIAT demonstration is a robotic workcell in which a P-MAST 25 gantry robot, produced by COMAU, is used for assembling mechanical components. The cell is at the COMAU Beinasco plant and can be characterized as follows:

a) The gantry assembles mechanical power-train components, which are fed to the workstation by a conveyor.
b) The cell is controlled by a C3G COMAU controller.
c) The NIRO concepts are implemented using NIRO-STEP and ICR as standard interfaces. ICR has also been used to controls the conveyor with a separate task labelled "PLC".

The site

Figure 8.3-1 shows the cell. Fig. 8.3-2 shows the cylinder block and the components used for the assembly task. Fig. 8.3-3 shows the P-MAST-25 views. The robot used is equipped with a 4th and 5th axis allowing wrist rotations and a gripper with 2 fingers to perform pick and place operations.

Site information flows

Fig. 8.3-4 describes the data flows and data formats.

191

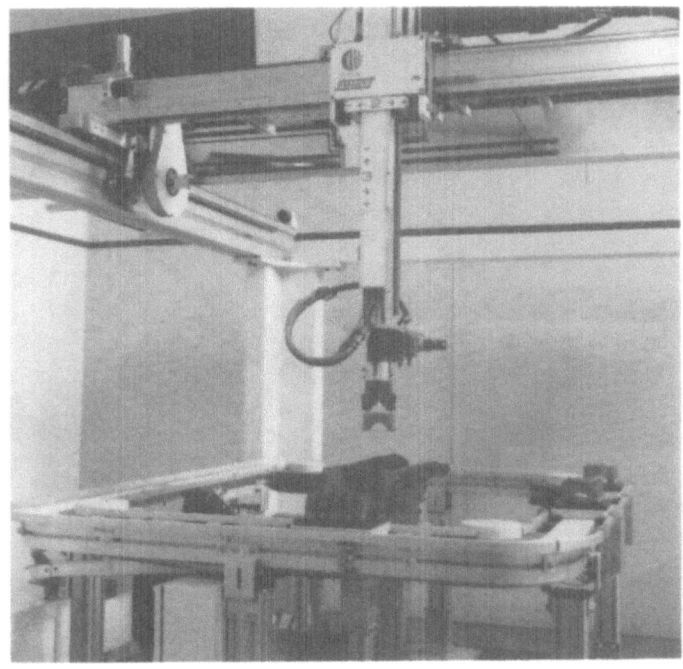

Fig. 8.3-1: Picture of the cell showing gantry robot and conveyor

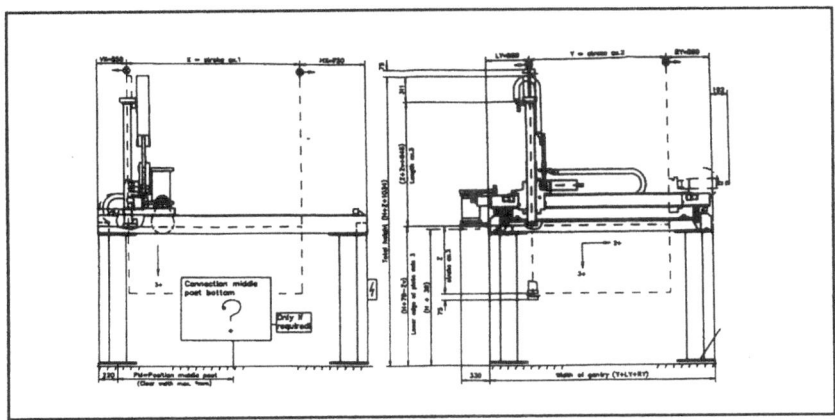

Fig. 8.3-2: P-MAST 25 views

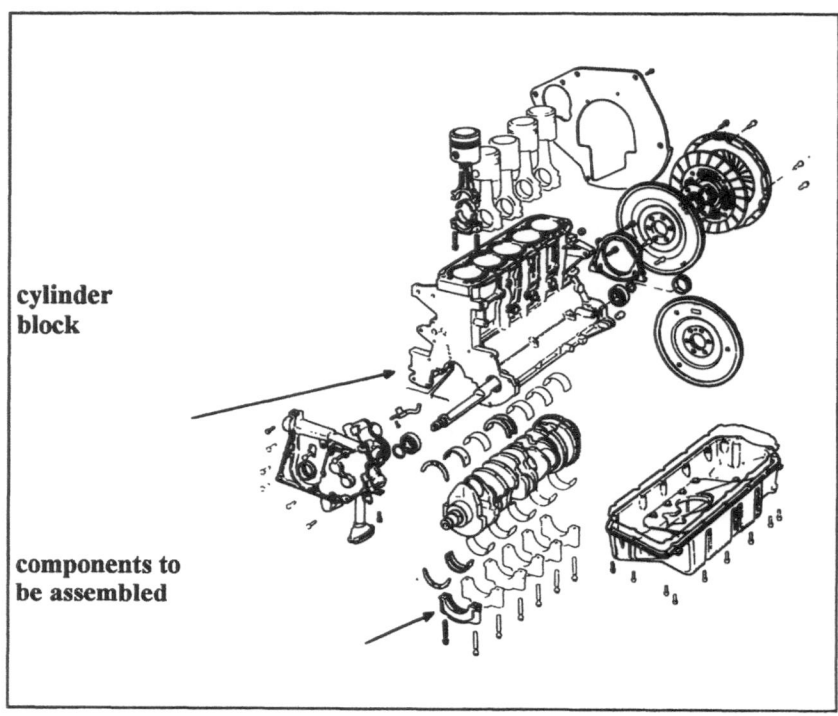

cylinder block

components to be assembled

Figure 8.3-2. Cylinder block and components assembled in the cell

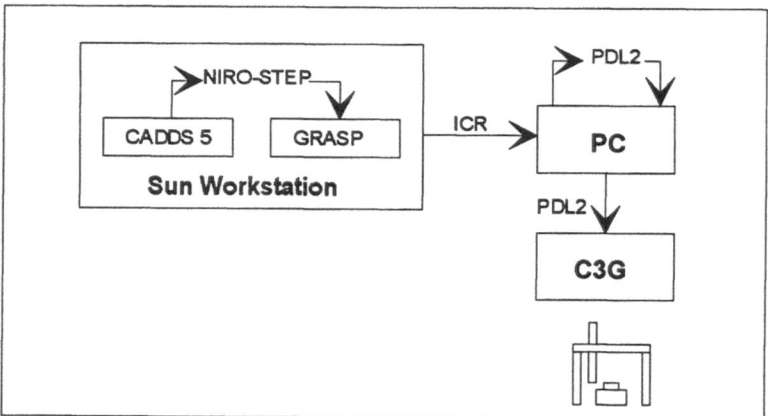

Fig 8.3.4 Information flows in the site

The cell, equipped with the P-MAST-25 robot and the conveyor has been modeled on a CADDS 5 CAD system. Subsequently, external to the CADDS 5 graphic environment, the solid model of the cell has been exported in NIRO STEP neutral format and imported on a GRASP simulation system througth use of BYG post-processor.

Kinematic analysis has been performed at GRASP system level.

ICR code then generated has been translated into PDL2 and then in C3G internal executable format, to accomplish the task of off-line robot programming.

Cell modeling

The Fig. 8.3-5 shows the cell realized on CADDS 5 system.
The Fig. 8.3-6 below shows cylinder block and pallets carrying components.

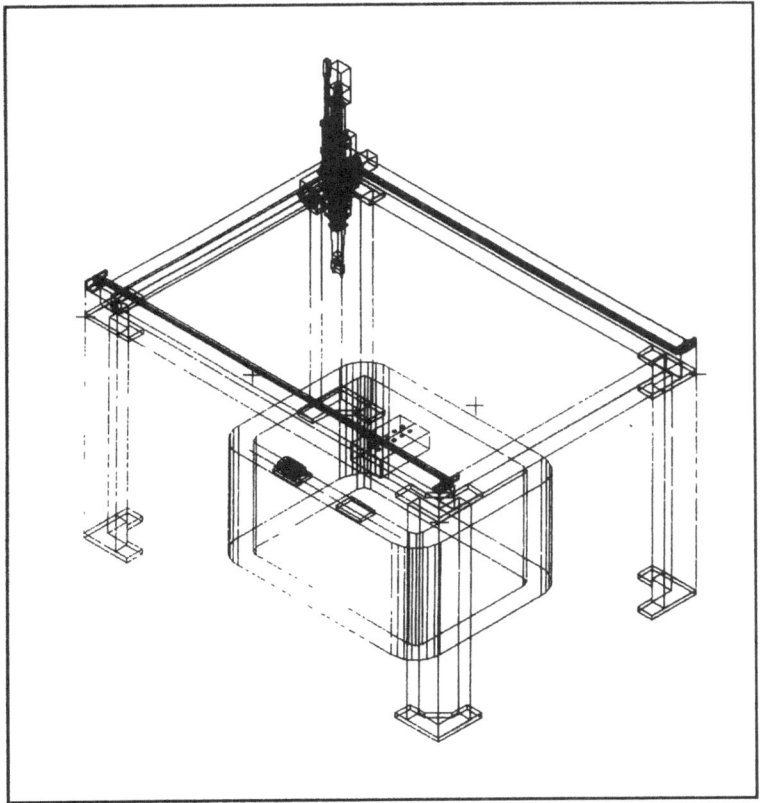

Fig. 8.3-5: CADDS 5 model of the cell

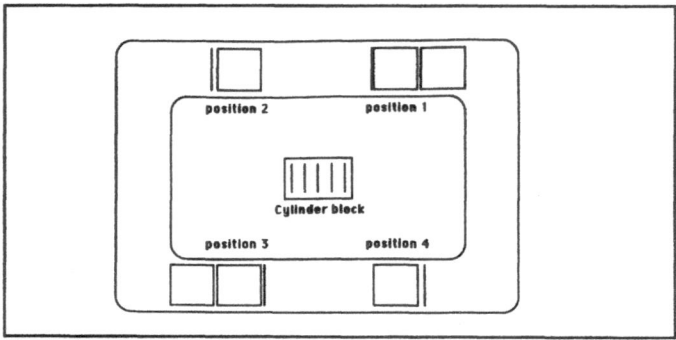

Fig. 8.3-6: Top view of the cell

Cell part-programming

A conveyor belt moves the pallets between four positions. Each pallet contains five components.

The robot arm moves the five supports from the pallet in position 3 to the cylinder block and then from the engine to the pallet in position 4.

The pallet in position 4 consists, therefore of an empty pallet.

The cell is controlled by two interacting programs:

- The Robot Arm Program (Robot Program)
- The Conveyor Belt Program (PLC Program)

The robot arm program controls the 5-axes robot.

The conveyor belt program acts as a PLC program written in ICR and translated into PDL2 (The C3G controller executes both the robot and PLC tasks).

The robot program code

The robot program, after an initialization step, includes the following operations:

1) Waits for a pallet available in position 3
2) Moves the five supports from the pallet in position 3 to the cylinder block
3) Waits for an empty pallet available in position 4
4) Moves the five supports from the cylinder block to the pallet in position 4

The PLC program

The PLC program, after the initialization step, performs the following operations:

1) When position 2 is free it provides a new pallet in position 2
2) When position 3 is free it provides a new pallet in position 3
3) When the supports have been picked from position 3 a new pallet in position 4 is provided
4) When the supports have been dropped to position 4 position 4 is freed (this will provide a new pallet in position 1)

Conclusions

The pre-processor, which is used to extract the solid model of the cell, has been tested by exporting two robot solid models using the NIRO STEP format and subsequently importing them to the GRASP and KISMET simulation systems.

The post-processor used to transate ICR code into PDL2 has been tested on complex ICR programs describing the demo activities.

The post-processor translating ICR code into PDL2 does not allow multitask, while the demo cycle requires two different tasks, one for the robot and one for the conveyor (PLC). As such, minor modifications have to be made on the PDL2 code before it may be executed.

A simplified PDL2 code governing the cell has been appended as an example.

The robot program code

```
PROGRAM robot
... -- variables
ROUTINE move_hand (direct : INTEGER)
BEGIN
  IF direct = 1 THEN -- close the hand
    close the hand
  ELSE -- open the hand
    open the hand
  ENDIF
END
ROUTINE move_arm (open_close : INTEGER; posit:
JOINTPOS)
BEGIN
  move 250.0 over the target position
  move to the target position
  move_hand (open_close)
  move 200.0 over the target position
END
BEGIN -- main program
  -- position on engine
  pos11 := engine
  pos12 := pos11 + displacement
  pos13 := pos12 + displacement
  pos14 := pos13 + displacement
  pos15 := pos14 + displacement
  -- position on pallet1
  pos21 := pallet1
```

```
   pos22 := pos21 + displacement
   pos23 := pos22 + displacement
   pos24 := pos23 + displacement
   pos25 := pos24 + displacement
   -- position on pallet2
   pos31 := pallet2
   pos32 := pos31 + displacement
   pos33 := pos32 + displacement
   pos34 := pos33 + displacement
   pos35 := pos34 + displacement
   ... -- PLC data
CYCLE
   -- pick from pallet1 and drop on engine
   wait for position3 = TRUE
   move_arm (1, pos21)
   move_arm (0, pos11)
   ...
   move_arm (1, pos25)
   stprelgo := TRUE -- make free pallet on
position3
   move_arm (0, pos15)
   -- pick from pallet1 and drop on engine
   wait for position4 = TRUE
   move_arm (1, pos11)
   move_arm (0, pos31)
   ...
   move_arm (1, pos15)
   move_arm (0, pos35)
   stdeptgo := TRUE -- make free pallet on
position4
END
```

The PLC program

```
PROGRAM conveyor
... -- variables
ROUTINE call_pallet (position : INTEGER)
   ... -- unlock called pallet
ROUTINE lock_pallet (position : INTEGER)
BEGIN
   IF position = 1 THEN
     lock position 1
   ENDIF
   IF position = 2 THEN
     lock position 2
     IF pallet on position 2 THEN
       lock position 1
     ELSE
       call_pallet(1)
     ENDIF
   ENDIF
   IF position = 3 THEN
     lock position 3
     IF pallet on position 3 THEN
       lock position 2
```

```
      ELSE
         call_pallet(2)
      ENDIF
   ENDIF
   IF position = 4 THEN
      lock position 4
      IF pallet on position 4 THEN
         lock position 3
      ELSE
         call_pallet(3)
      ENDIF
   ENDIF
BEGIN -- main program
   lock_pallet(1)
   ... -- PLC data
CYCLE
   IF no pallet in position 2 THEN
      lock_pallet(2)
   ENDIF
   IF no pallet in position 3 THEN
      lock_pallet(3)
   ENDIF
   IF stprelgo = TRUE THEN -- pallet free on
position 3
      lock_pallet(4)
   ENDIF
   IF stdeptgo = TRUE THEN -- pallet free on
position 4
      move_pallet(4)
   ENDIF
END
```

8.4 Demonstration at Odense Steel Shipyard

C. Busk and O. Knudsen

The demonstration at Odense Steel Shipyard showed the whole cycle from CAD systems to welding of a ship Sect. with robots. Thereby almost all the interface levels in the NIRO project were present during the demonstration.

The demonstration was performed in close cooperation between Odense Steel Shipyard Ltd, BYG Systems Ltd, Technical University of Denmark and DIS Industrial Consultants a|s. It was presented to the public during the 1st Product Technology Days which took place at Odense Steel Shipyard on 29th and 30th of September 1992.

8.4.1 Criteria for Task Selection

The Odense Steel Shipyard is one of the most modern shipyards of today. The use of information technology spans the whole process from design to production. It includes different CAD systems running on approximately 100 UNIX based work stations. They currently apply approximately 30 industrial welding robots in the production areas together with other automated production equipment.

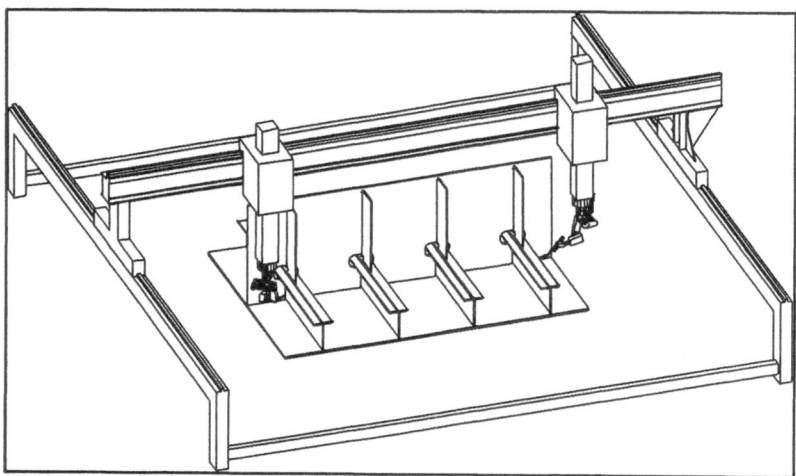

Fig. 8.4.1-1: Demonstration environment at Odense Steel Shipyard

Odense Steel Shipyard's present and future robot applications consist of very complex one-off tasks. For this reason the level of ambition within programming will be high and place heavy requirements on the applicability of the new standards (STEP, ICR, (PLR), MAP/MMS). Because of the difficulty of the automation tasks at the yard, it is crucial that any demo-installation at the Yard reflects at least a fraction of the challenges faced in daily production. Therefore, the choice of demonstration task is based on a normal production situation.

A model of the demonstration environment is shown in Fig. 8.4.1-1. It shows the two Hirobo robots mounted on the great portal gantry spanning the ship Sect. to be welded. This environment is similar to the real production environment and it thereby provides the same structure in the controlling systems.

8.4.2 Task Description

The first part of the demonstration showed the transfer of models from both Hicadec-H and CATIA to the GRASP off-line programming package. From Hicadec-H ship design system, the selected ship Sect. was transferred into STEP format using the Hicadec-H to STEP processor.

In CATIA models of the gantry, the robot and the robot tools had been defined with both geometry and kinematic information contained in the model. These models were also transferred into STEP format using the CATIA to STEP preprocessor developed by DTH and from there read into the GRASP system.

The gantry was modelled as a P-P-P mechanism, that is, a mechanism with three prismatic joints. The robot was modelled as a mechanism with five revolute joints.

The STEP files are converted into GRASP input format by the STEP post-processor developed by BYG Systems Ltd.

When read into GRASP the three models were placed relative to each other modelling the production environment. When this was accomplished, the programming of the robot task was done.

The gantry was programmed to move the robot to different positions for different welding tasks on the workpiece. The tasks for the robot and the gantry are controlled by a robot-controller and a gantry-controller respectively. The coordination is provided by the communication of binary signal in between these two controllers, with the gantry-controller as master.

Programming the task in GRASP contained both the programming of the gantry and the programming of the robot both including the signal handling between the controllers. After programming the robot and gantry task in GRASP, an off-line simulation was executed to secure that no collision was taking place between any of the components in the system at run-time. If collision was encountered, the program was corrected and a new simulation was executed.

When no more collisions were found, the GRASP system generated three ICR files with the programs for the gantry and two ICR files with the programs for the robot. The ICR programs for the robots were then translated into a Hirobo language. The gantry program was loaded into the gantry controller and the robot program was loaded into the robot controller. The programs were then executed and the ship Sect. welding was performed.

8.4.3 Information Flow Structure

The systems used in the demonstration are interconnected as shown in Fig. 8.4.2. Each of the components in this design and production system is briefly described in Sect. 8.4.4.

In the demonstration, the concept of neutral interfaces was used to ensure that each component could easily be exchanged with a similar component without causing changes to the other components in the total system.

This ensures the possibility to change one of the old components to a new and probably more capable system at the lowest price and with a minimum of interface problems. This means that the maintenance cost of the total system is almost reduced to the price of the new components only. The same applies to extending the total system with new functions etc.

For growing companies and companies working in areas with rapid changes, the possibility of easy exchange of system elements in CIM system is vital. It gives the benefit of integrated design and production environments without limiting the ability to respond to changes.

8.4.4 Implementation

Here a short description of each system element in the demonstration system is
presented. For the information flow structure, please refer to Fig. 8.4-2.

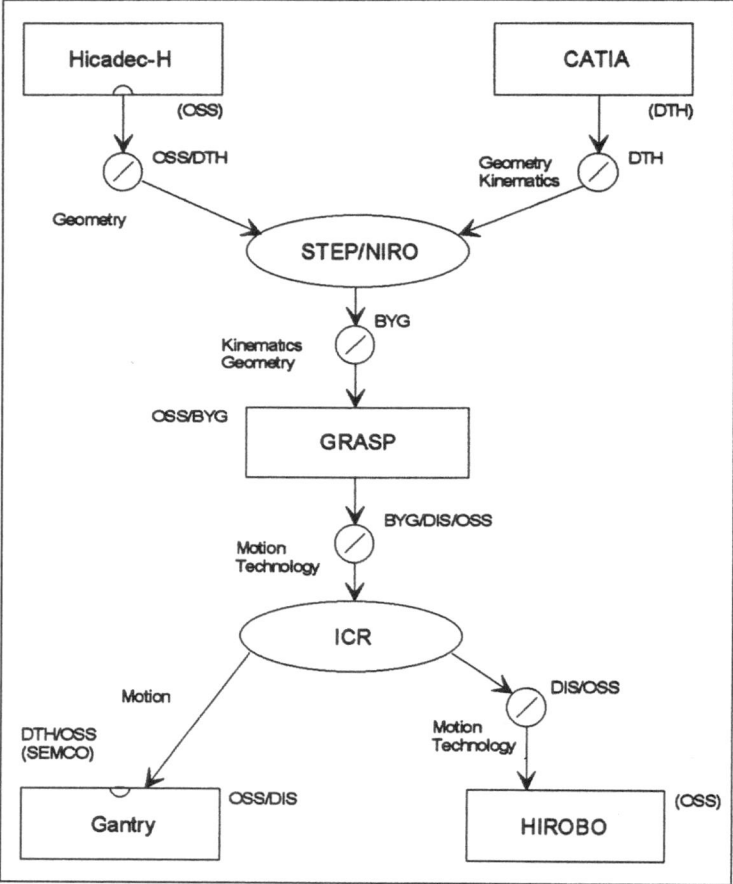

Fig. 8.4-2. Information flow for the demonstration at Odense Steel Shipyard

Hicadec-H ship application system

The Hicadec-H CAD system is a special ship building CAD system used at OSS
for the design and construction of ships. The Hicadec-H works on a database
which contains a model of the total ship and the system offers the naval architects
a set of very advanced features to complete the design of the ship.

In the demonstration, a CAD model of a ship section. was transferred from the database defining the total ship model in Hicadec-H and used in the off-line programming system. The format of the file is the STEP format.

Because Hicadec-H has a closed database it has not been possible to develop a direct Hicadec-H to STEP preprocessor. Instead a IGES to STEP preprocessor has been developed. The IGES to STEP translation is only possible because of the special IGES files produced by Hicadec-H. This IGES to STEP preprocessor has been developed by OSS and DTH in cooperation.

CATIA

CATIA is a CAD system developed by Dassault Systems in France and distributed world-wide by IBM. It features 2D and 3D geometry modules as well as diverse application modules. It has been implemented under the MVS and VM/CMS operating systems for IBM mainframes as well as under UNIX/AIX for IBM's RISC System/6000 series. For the demonstration at OSS, DTH supplied the system with the STEP post-processor running on a RISC System/6000 work station.

CATIA Modules related to robotics

CATIA kinematics- and CATIA robotics modules are two specific modules for dealing with 2D/3D kinematics and robotics tasks. CATIA kinematics, used in conjunction with CATIA Base, CATIA 3D Design and CATIA Solids Geometry, allow the user to simulate the operation of kinematic mechanisms created and stored as models, and to analyze the results of the simulation. It includes the following two functions: The KINEMAT function and the KINEMUSE function. The KINEMAT function is divided into two parts depending on the selected working mode: 2D SPACE or 3D SPACE. It is used to:

- Define the joints, stops, fixed part and commands,
- analyze the kinematic mechanism to control the joint definition,
- erase joints from a mechanism,
- erase, copy or rename a kinematic mechanism,
- isometrically transform a kinematic family,
- unite two kinematic mechanisms.

The KINEMUSE function is used to:

- Simulate the operation of a kinematic mechanism via animation on the graphic screen,
- generate traces and numerical outputs,
- detect automatically a collision between two parts,
- study the velocity range and acceleration of the parts
- define and modify command laws,
- modify the joints, impose stops, fixed parts or commands,
- analyze the kinematic mechanism.

CATIA's role in the demonstration at OSS:

CATIA serves as the geometry- and kinematic- modelling system for the OSS demo.

The kinematic and geometrical model of the gantry and the Hirobo robot, and a geometrical model of the robot tool were generated by the use of the CATIA system.

STEP to GRASP postprocessor

The STEP to GRASP postprocessor processes STEP files and produces files which can be read into the GRASP robot simulation package. These output files are of the GRASP source file format. The solid model and kinematic information held in the STEP file will first be parsed to check that input file syntax is valid.

The STEP to GRASP postprocessor is a complete software package and does not form an embedded part of the GRASP robot simulation package. The postprocessor has been written in the 'C' programming language.

Before starting the offline programming of the robot task, each of the files containing components for the system model had to be converted into the GRASP data format using the STEP to GRASP postprocessor.

GRASP robot simulation package

GRASP is a robot simulation package for the modelling and off-line programming of complex robot cells. 'GRASP' is an acronym for 'Graphical Robot Applications Simulation Package'.

GRASP is used to perform detailed design and planning of robot systems using 3D simulation techniques that allow efficient visualisation and evaluation of the systems. A GRASP robot library is available containing a wide range of robots. GRASP can simulate complete factory systems allowing the planning and evaluation of alternative layouts.

GRASP is also a robot programming system allowing off-line programming of robot cells. The powerful programming commands in GRASP allow the simulation of many robot applications including welding.

GRASP read the source files generated by the STEP to GRASP postprocessor. These files contained the models for the gantry, the Hirobo robot, and the ship Sect.. The operator then used GRASP to program the Hirobo robot and gantry to perform the tasks required for the demonstration. This involved signal communication between the gantry and the Hirobo robot.

GRASP to ICR preprocessor

The GRASP to ICR preprocessor generates ICR robot program files from GRASP robot programs. The preprocessor produces two files when a GRASP robot program is output as an ICR file ; the ICR program file itself, and an associated ICR datalist file.

Once the robot programs controlling the Hirobo robot and the gantry have been written using GRASP, these programs are preprocessed to ICR files. The ICR to GRASP preprocessor was used to create ICR programs for both the Hirobo robot and the gantry. When the ICR files are being produced the user must specify the mapping of GRASP robot program signals used for communication in the simulation to the ICR channels that will be used for actual robot control.

The gantry ICR interpreter

The OSS gantry is controlled by a PLC/PC combination. The PLCs are controlling the gantry's motors, monitoring error conditions, performing calibration, etc. Hence, the PLCs takes care of the gantry's basic motion and it is possible to command the PLCs to move the gantry to a specific position. The PC works on basis of this and features the higher control layers of the gantry controller: It is used for manual controlled motion, initialization of calibration, supervision, man-machine interface, and program execution. The program execution is handled by the ICR interpreter and it is implemented as a state machine. The interpreter reads and executes the program line by line until it meets one of the instructions JMOVE (move the gantry) or EEFM (command the robot to weld). These two instructions involve physical activity, and some special communication procedures to the PLCs are therefore started. The interpreter waits until the gantry or the robot has completed the motion and then continues executing the program.

This Gantry-ICR interpreter accepts a limited instruction set which supports very basic motion and communication with the robot. The interpreters instruction set is illustrated by the following program example which moves the gantry to a new position (0,1710,1800) and commands the robot to start welding a new Sect.:

```
1,PBEG,2,"EXAMPLE";
2,BLKBEG,1,0;
3,PUSHR,#0.0;        Stack the position to move to.
4,PUSHR,#1710;
5,PUSHR,#1800;
6,GENJ;      Convert to JOINT type
7,JMOVE,1,W;        Move gantry to new position
8,REMARK,0,"AGT";
9,EEFM,1;    Command the robot to start welding
10,BLKEND;
11,PEND;
```

If the interpreter reads an unknown instruction or an instruction which it is impossible to execute it stops immediately and changes the controllers mode from "execution" to "ready".

ICR to Hirobo translator

The ICR to Hirobo translator was used to transform the ICR robot programs into the robot language defined by the robot vendor. This transformation is done off-line and cannot be changed during execution of the program. The transformation of the ICR programs was necessary because it has not been possible to interface a PC based on-line interpreter with the robot controller.

The use of a transformation package as the interface tool between the off-line programming system and the robot controller gives certain limits on which parts of ICR can be used. These limitations arise because the robot controller has no commands for handling many of the basic ICR commands (e.g. arithmetic operations). Because of the lack of commands on the robot controller, it is difficult to translate ICR programs into Hirobo language. ICR programs are based on many very simple operations whereas the Hirobo language is equipped with highly integrated commands to handle the special tasks for the Hirobo robot.

In the demonstration, the ICR to Hirobo translator read a ICR file produced with GRASP and transformed it into a file containing a Hirobo program. This file was then copied into a bubble cassette and loaded in the robot controller. When all the programs were loaded and started the gantry controller took control over the total system and synchronized the operations.

8.4.5 Experiences and Recommendations

The demonstration at Odense Steel Shipyard used most of the interface formats handled in the NIRO project. It is therefore well suited to work as a basis for gaining experiences with all the neutral interfaces in the project.

The interfaces in the design phase of the system all worked perfectly. The STEP format made it possible to use different CAD systems for the design without causing additional interface costs for the offline robot programming system.

As the STEP format becomes more and more widely spread in the different systems, the use of this interface format could be expected to give easier integration between the different systems.

In the production phase of the demonstration system the interface problems were harder to overcome. The ICR format lacked some of the vital process functions and even with these functions added to the ICR format, the interface still was a bottleneck in the programming of the robots. The use of ICR as the programming language for the gantry controller was far more successful due to the fact that the gantry controller was designed to run with ICR as the programming language.

Therefore, the result of the demonstration regarding the ICR language would be that the language still has to undergo further revisions before having the real impact on the robot installations. It should also be used as the direct programming language of the robots. Because of the difference in language level that will often be the case between the ICR language and a robot vendor language, the use of ICR for transformation to robot vendor code is not the best approach. The PLR language (earlier called IRL) is another neutral robot language under development within the scope of the ISO. This language is at a higher language level than the ICR language and could therefore become a better solution in the long run but PLR is a very new language and the versions that have been released until recently have not been stable enough to become the basis of work like the work done in the NIRO project.

8.5 The Reis Demonstration

H. Bruhm

8.5.1 Criteria for Task Selection

The selection of the task for the Reis demonstration was based on the following criteria:

- Portability of the whole demonstration set-up. The system was to be shown at the CIM Europe Conference in Odense (Denmark) in September 1992 and at the ESPRIT Conference in Brussels in November 1992.
- Ability to demonstrate interfacing between CAD on one side and robot control and simulation on the other side via the neutral languages IRL and ICR.
- Inclusion of the following products of project partners in the demonstration: ICR interpreter for Reis ROBOTstar IV, BYG's robot programming system GRASP, the IRL to ICR compiler by PSI, and the KISMET simulation system from KfK.
- Demonstration of the capability of the system to deal with geometric uncertainty in the task using an external sensor. The necessary operations had to be expressible in IRL and ICR.
- Interactive nature of the task: Some kind of user (or exhibition visitor) intervention was desired, proving that the whole task was not completely pre-programmed.
- Demonstration of information flow from the GRASP workstation to the robot controller (robot commands) and vice versa (sensor information about the environment).
- Industrial relevance of the task: Despite the limitations imposed by the required presence at two major public events, the characteristics of the task should be similar to real industrial situations.

8.5.2 Task Description

The task that was selected for the Reis demonstration is a special sequence of palletizing operations: The robot is used to move the figures in a Solitaire game.

Fig. 8.5.2-1.Photograph of Solitaire demonstration cell

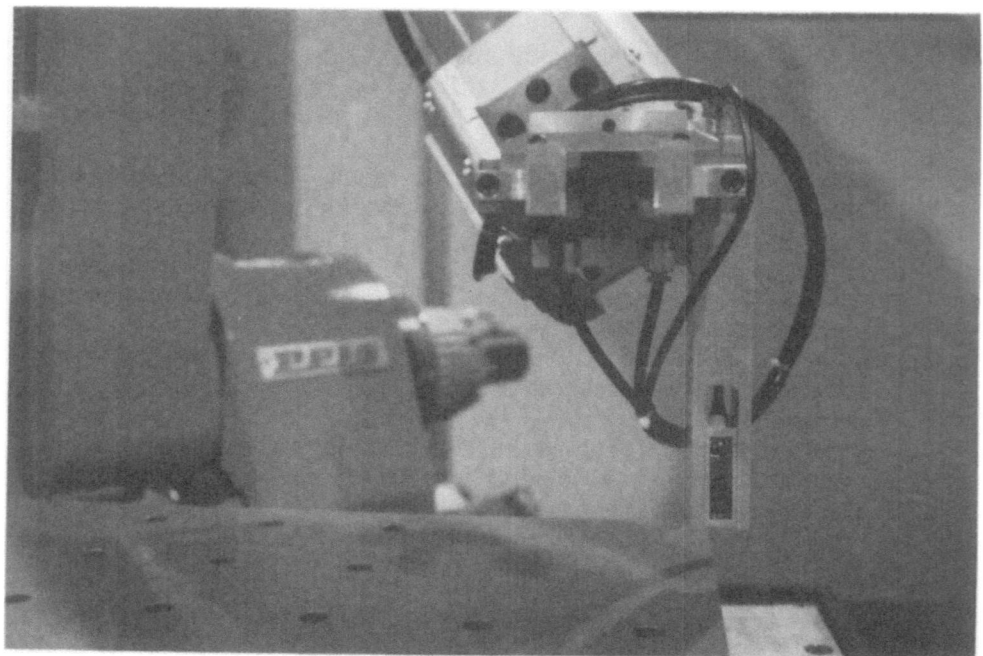

Fig. 8.5.2-2: Close-up photograph of sensor over edge of board

The configuration of the demonstration cell is shown in the photo of Fig. 8.5.2-1. It comprises a Reis RV15 robot with ROBOTstar IV controller, a work table mounted on a common base plate with the robot, a Solitaire board and corresponding figures (pawns) of suitable size, and a magazine where the unused pawns can be stored.

As an interactive element in the task and in order to visualize the positional uncertainty that is typical of real industrial applications, it is possible to move the game board in three degrees of freedom: two translations in the plane of the working table plus rotation about the vertical axis. The change in position and orientation is constrained by mechanical stops at the edges of the table. Also this is typical for real world tasks: the magnitude of the inevitable uncertainties is limited to values which, in all cases of practical relevance, are much smaller than the range of motion of the Solitaire board in this demonstration.

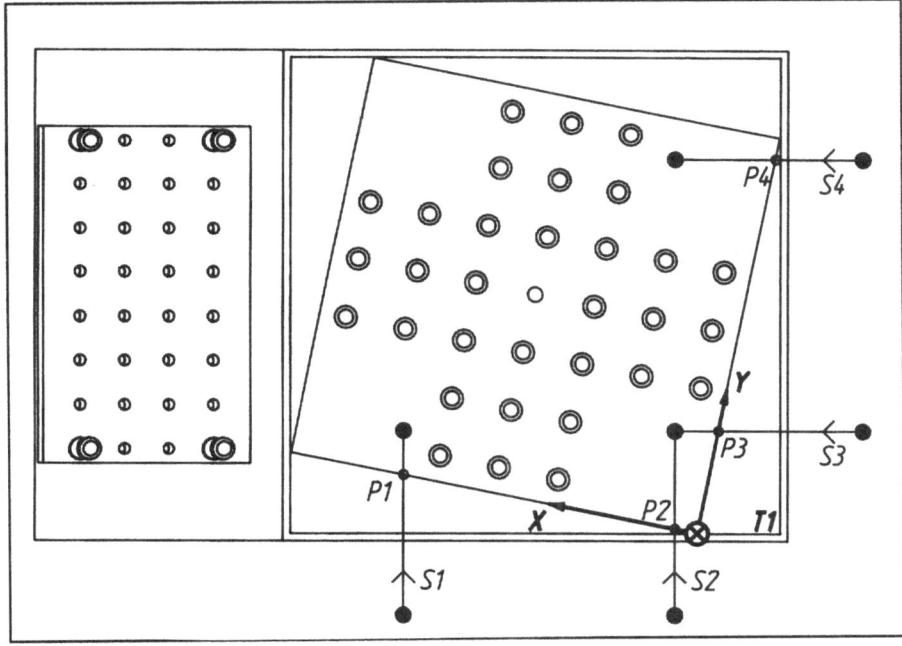

Fig. 8.5.2-3. Diagram of scan paths.
  S1-S4:   scan lines
  P1-P4:   edge points found
  T1:   board reference frame

An optical proximity sensor attached to the robot gripper (see Fig. 8.5.2-2) is used to determine the exact initial position and orientation of the pallet. The procedure used for this purpose requires four scanning motions of the robot along the paths shown in Fig. 8.5.2-3. The position data of the four points where the sensor detects the edges of the board are recorded and used to calculate the board reference coordinate frame T1.

It should be noted that three points would be sufficient to determine the board position and orientation and that the four measurements performed lead to an over-determined problem. The equations are resolved with a least squares algorithm programmed in IRL or ICR. This solution method is rather complex, but it has the advantage that it makes the sensing procedure more robust with respect to measurement errors. What happens is comparable to the well-known technique of averaging over repeated measurements.

The complexity of the task is further increased by the fact that the magazine is tilted 10° with respect to the surface of the table. This means that the poses in the robot program must be different not only in position but also in the orientation part. In the simpler case of a completely planar problem, all the robot motions could have been effected with constant orientation.

Apart from the task execution with the real robot there is a simulation running in parallel on the KISMET system, which has been developed by KfK. Two Reis robots with different kinematics - the RV15 of the physical set-up and an RH15 as shown in Fig. 8.5.2-4 - have been modelled in the simulation and may be used for the task interchangeably. Only the foot point of the RH15 had to be shifted to another location due to different workspace configurations.

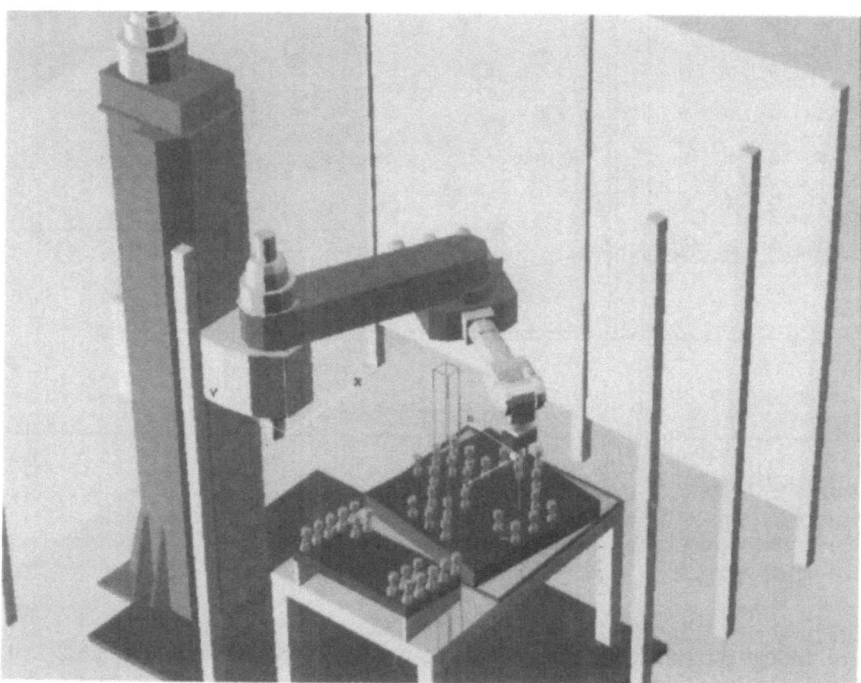

Fig. 8.5.2-4: Execution of Solitaire game with Reis RH15 robot simulated on KISMET workstation

The demonstration cell with the Reis robot and the KISMET simulation constitute two stations on the execution level. The processors developed in the NIRO project make it possible that both stations can execute the same IRL and ICR programs. These programs are generated off-line using the GRASP system from BYG. GRASP generates an IRL program which is converted to the corresponding ICR code by the PSI compiler. The two execution stations are controlled by the ICR interpreters that KfK and Reis have developed for their respective systems. All the computer systems involved in the demonstration are connected by a network (see Fig. 8.5.2-5) so that programs can be downloaded directly from the GRASP workstation to the KISMET simulation and to the robot in the demonstration cell.

The execution of the Solitaire game programmed in IRL on systems from different vendors shows impressively the advantages of the neutral language interfaces: No extra programming effort is necessary when the application is transferred from one system to another. In the industrial world this would also mean that non-productive time for pose teaching is avoided. The main advantage for the vendors of programming systems such as GRASP is that they need to have just one unique language interface on the output side of their systems; there is no need any more to develop a specific post-processor for each target system.

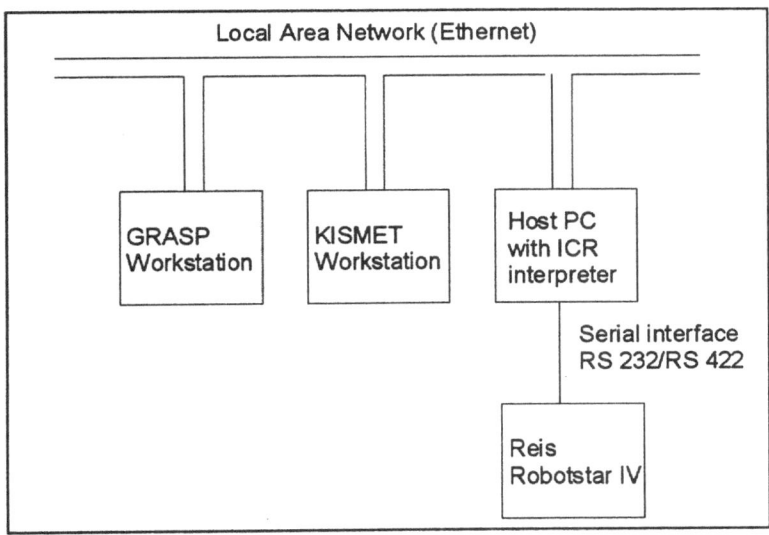

Fig. 8.5.2-5: Computer network for the Solitaire demonstration

Although this is not visible any more in the final result, it should be noted that the STEP interface was also used with advantage in setting up the Solitaire demonstration: CAD data originating from the Bravo3 system have been transferred to GRASP and KISMET via STEP files.

8.5.3 Information Flow Structure

J. Reim

Close to the developed processors of the NIRO Project, model and movement information flow starts at the mechanical CAD system and arrives at the execution of the programmed task on the robot respectively its simulation (see Fig. 8.5.3-1).

The origin of the Solitaire demo is at the Bravo3 CAD workstation where the engineer designs the game containing table, game board, magazine and figures. The fence surrounding the robot's workspace and the robot itself is added. The robot kinematics is defined using the ROBOT application package to Bravo3. Both, robot and solid scene are dumped on an intermediate format and passed through the BRAVO3/KISMET Preprocessor to the STEP neutral file.

The GRASP and KISMET systems process the STEP file and retrieve the model information, the former for off-line programming the latter for robot simulation. Inside GRASP the model is customized for the programming task. The figures are told to be movable parts, table and game board position are calibrated. Main activity of the GRASP simulation is path planning and the generation of the robot program. Finally GRASP converts the generated program from its internal representation to IRL.

The generated IRL program is compiled to the intermediate code ICR and downloaded to the Reis robot for execution and to KISMET for simulation. Both target systems - KISMET and the host PC to the Reis RV15 robot - are running the ICR interpreter from KfK with robot dependent parts for Reis ROBOTstar IV and KISMET, respectively.

Figure 8.5.2-5 visualizes task and data flow of the Solitaire demo. A second version of the Solitaire program supports a sensing task at the beginning. This allows to play the game with a changed board position. The measured board position and the changed program are transferred to KISMET before the game itself begins.

8.5.4 Implementation Details

J. Reim

Transfer of the measured game board position

As mentioned above, the sensing task for searching the game board is not supported by the simulation systems KISMET and GRASP. Therefore the procedure is split up into a sensing and a playing task and executed in two parts.

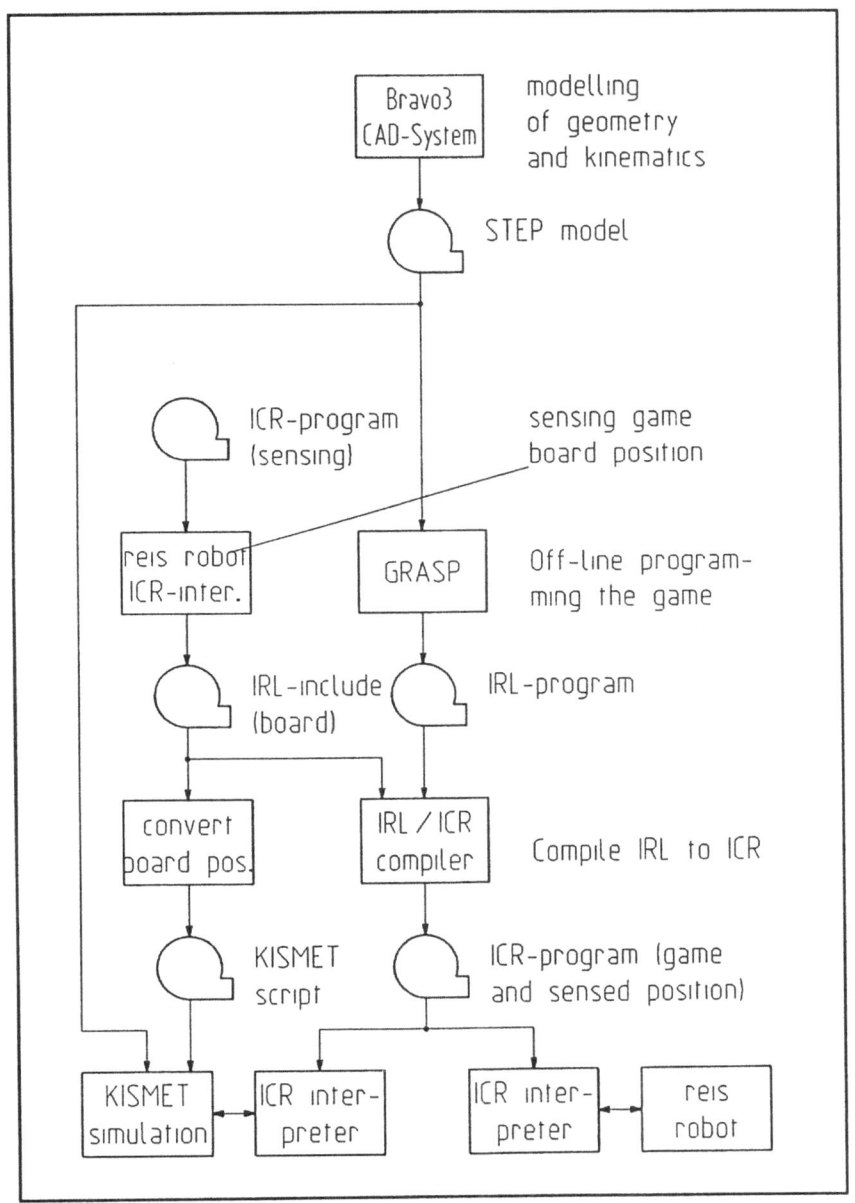

Fig. 8.5.3-1: Information flow structure in the Solitaire demonstration

In the first part the ICR program for the sensing task is run on the robot. The found board position is written to an IRL include file utilizing a newly implemented file output facility of ICR. To manage disk files with ICR, the robot dependent parts for Reis and KISMET got pre-defined input and output channels which are assigned to disk files with names optionally specified in the program or otherwise fixed by default. IRL input and output functions produce ICR I/O statements which access the specified channels.

When compiling the IRL game procedure the created include file with the actual board position and orientation parameters is integrated by the IRL compiler. The resulting ICR code contains the correct measured board position and is distributed to the KISMET and Reis ICR interpreters. Before running the game on KISMET the game board must get the measured position. Therefore the IRL include file which was produced by the sensing task is transferred to the KISMET workstation and interpreted by a special transformation routine. Its result is a KISMET macro (called script file) which modifies the position of the game board during start-up phase.

After these preparations the second part of the demonstration - the actual game - can be started. The ICR interpreter is invoked on both target systems and runs the game. First the game is set up. All figures necessary for the game are taken from the magazine and set on the game board. Then the game is played with a fixed solution strategy.

End effector selection

The Reis robot dependent part of the ICR interpreter references tool center point definitions which are predefined on the robot controller. For this functionality the ICR command EFNA is used. To run the same ICR code on the KISMET simulation, the predefined end effectors were added to the KISMET robot dependent part.

Communication between robot controller and ICR interpreter

The ICR interpreter for Reis ROBOTstar IV is implemented on a host PC and communicates with the robot controller via a serial interface, which is operated at a transmission rate of 19200 baud in this case. It was realized during the initial tests that the on-line transmission of the Solitaire program in single commands created a great amount of overhead which could not be tolerated, because it would lead to a very slow program execution.

The problem was solved by the introduction of a block transmission mode together with a looser synchronization between host interpreter and robot controller. Now the ICR interpreter runs ahead of program execution on the robot for most of the time so that block transmission of buffered commands becomes possible. The two processes are synchronized only at points in the program where the interpreter needs to get back some information from the robot controller. With this new operating mode the ICR interpreter has shown a very satisfactory performance. The speed of execution for the Solitaire program is now determined by the robot (as it should be expected) and not any more by the interpreter.

8.5.5 Results

H. Bruhm and J. Reim

Model consistency

During the preparation of the demo, some inconsistencies between CAD model and reality needed special attention:
-   On the way from design to the manufacturing of the game some details were changed without updating the CAD model. The errors were not easy to trap because they first appeared during the test procedure for the ICR interpreter. For the next project it will be necessary to insert a verification procedure for the CAD model. For example some key dimensions of the scene which are transferred with the model will help to trap inconsistencies.
-   The second topic is the calibration of the model. When the scene is assembled, not all the dimensions of the CAD model fit exactly. Therefore the IRL source programs calculate all move targets relative to the table, the game board and the magazine position. Table and magazine are calibrated with a special procedure on the Reis robot controller. The game board position results from the sensing task.
-   The calibration data for these three elements are written down and changed in the IRL source program and the Bravo3 CAD model. The updated model is transferred to KISMET. Inside GRASP a special calibration facility is fed with the measured points and calculates the updated model.

Workframes

Running the simulation of the solitaire game on KISMET and GRASP, a common problem of the STEP specification arose: How can I specify that a part can be moved away ?

Both simulation systems - GRASP and KISMET - allow the exchange of tools and gripping operations. But the parts to be moved and the settle down positions must have some additional information because the simulation system has to decide which part is to be moved. To manage this problem the tool entity of the STEP specification was applied. All movable parts which do not belong to the kinematic part of the model (in this demonstration the figures of the game) are treated as tools. The tool attachment point of the robot is set at the tool center point of the gripper. With this definition, gripping and releasing a of figure is modelled as mounting or, respectively, dismounting a tool.

The Bravo3 source model of the Solitaire scene contains predefined KISMET workframes. Their type code, which is important for the KISMET gripping strategy, is stored in the INDEX_ENTRY name string belonging to its placement information.

Tree of solid instances

The three utilized systems Bravo3, GRASP and KISMET store static models in a tree or tree-like data structure. For example, Solitaire figures are placed on the game board which can be moved on the table. The STEP implementation only distinguishes between solids belonging to the kinematics or solids fixed on the ground. Two restrictions were necessary for the solitaire demo:

- The game board must not move when figures are placed on it. They will not follow in the simulations.
- The above mentioned use of the tool entity is the only way to grasp the figures with their additional frames.

Benefits of neutral interfaces

Apart from the problems discussed above, the expected benefits of interfacing between systems of different vendors via neutral representation formats for CAD data and robot programs could be verified. The task performed in the demonstration - though designed to be easily transportable and to serve as an eye-catcher for exhibition events - was characterized by a complexity level similar to real industrial applications. The sensing strategy employed to resolve positional uncertainty of the objects that the robot is to work on is definitely at the forefront of currently established procedures in industrial practice. Altogether the task in the Solitaire demonstration is very similar to typical small part handling tasks in industry.

8.6 Other Applications Outside the NIRO Project

S. Trostmann, E. Trostmann, and L.F. Nielsen

This chapter describes a Robot Off-line Programming and real-time SIMulation system called ROPSIM currently under development [Trostmann, S., Nielsen, L.F., 1992] at the Control Engineering Institute, Technical University of Denmark. This system is, among others, based upon the validity of the concept and results of the NIRO project.

8.6.1 Introduction

Robots offer a high degree of automation, programmability, and repeatability. This makes robots suitable as production components in a computer integrated manufacturing and engineering (CIME) system.

In order to develop reliable robot programs and as a result of the ambition to produce at a high speed and with high repeatability a real-time simulation, including the true dynamic behaviour of the total robot system and the use of external sensors is beneficial.

The ROPSIM system must in order to be a true CIME-subsystem, facilitate the exchange and reuse of robot system definition data and robot program definition data with systems of other origin or different functionality. For that reason the STEP neutral interface is utilised in ROPSIM as an interface for exchange robot manipulator models and the ICR neutral interface is utilised as an interface for exchange of robot programs.

8.6.2. Architecture and Information Flow

The ROPSIM software is implemented using the object-oriented programming language C++. The software presently runs on a Silicon Graphics GTB80 computer. The overall architecture of the ROPSIM software system is illustrated in Fig 8.6.2-1.

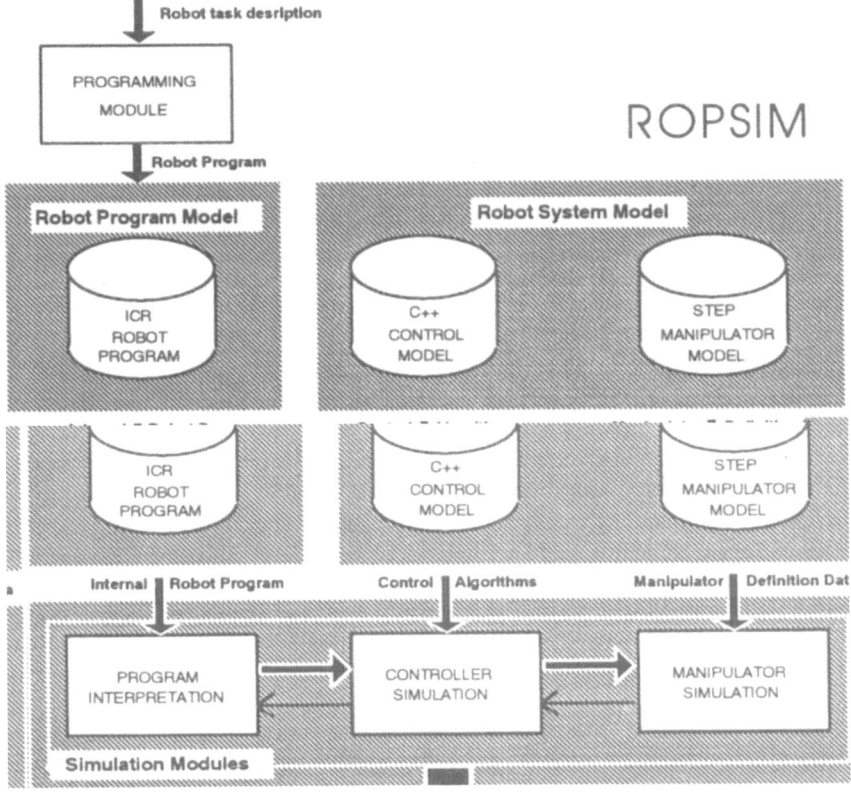

Fig. 8.6.2-1: The ROPSIM architecture

The components of ROPSIM can be categorised into models and modules where the models carry essential robot system information inclusive the robot program and the modules carry out the necessary data processing. The figure shows the models (represented as "cylinders"), the modules (represented as "rectangles"), the interfaces between the modules (represented as horizontal arrows), and the information flow between the models and modules (represented as vertical arrows). The thick vertical arrow represents the information flow from the three simulation modules (program interpreter, controller simulation, and manipulator simulation) to the presentation module. In the following the individual models and modules are described.

8.6.3 Models

As can be seen in Fig. 8.6.2-1 two different models are applied in the ROPSIM system. The first is a robot program and the second is a robot system model.

Robot program model. Generally speaking the information in a robot program is a description of actions the robot must perform in order to accomplish a desired task. The instructions of a robot language can be divided into two categories. One part is dealing with the data manipulation and another part is dealing with the robot control. The data manipulation instructions implement the logical structure of the program. The robot related instructions implement the control of the robot. Also instructions for external sensor feed-back is included in this category.

In ROPSIM it has been decided to use the ICR robot programming language. ICR is a suitable robot programming language which include facilities for implementing both categories of instructions. Further the application of the ICR programming language facilitates the simulation performed by ROPSIM to be driven by the same robot program which is later applied for real-time control in the task execution. Also robot programs can be exchanged via the ICR neutral interface with other sub-systems in the CIME system.

The ICR robot program serves the information needed in the program interpretation module, refer to Fig 8.6.2-1.

Robot system model. As illustrated in Fig. 8.6.2-1, the robot system model can be divided into a manipulator model and a control model.

Manipulator model. The manipulator model must carry information about the kinetics of the robot manipulator. Kinetics is the study of all aspects of motion, and as a result the manipulator model may include information about the kinematic structure, the dynamic relationship and the geometric shape of the manipulator. The kinematic structure describes the relationship between the links and the joints in the manipulator, and defines the realisable manipulator movements. The kinematic structure is the essential core for simulation of manipulator movement. The dynamic relationship may be described by mass properties (concentrated parameters defining the mass distribution) of the manipulator links and the behaviour of the servo motor drives. The dynamic relationship is essential for simulation of manipulator movement under the action of torque/force. The geometric shape describes the shape of the individual links in the manipulator. This geometric information of the manipulator is essential for performing collision tests and for visualisation via a 3D animation of the robot movements.

In ROPSIM it has been decided to use the STEP neutral interface for manipulator models. This means that the manipulator model defined in the NIRO project is a true subset of the manipulator model utilised by ROPSIM.

Since the ROPSIM system is designed to simulate the dynamic behaviour of the robot manipulator, the manipulator model utilised by ROPSIM is extended compared to the NIRO manipulator model. These "user-defined" extensions are mainly addressing the following two subjects:

- Definition of mass-properties of the manipulator links.
- Definition of the dynamic properties of the manipulator servo motor drives.

The user-defined extensions was implemented in Express and mapped to the exchange file format using the formal methods defined within the STEP proposal. Owing to the formal approach for definition of neutral interfaces provided by the STEP proposal, the process of defining the extensions is straight forward and consistent.

The STEP manipulator model describing the manipulator kinetics provides the information for the manipulator simulation module, refer to fig. 8.6.2-1.

Control model. At present time there exist no neutral control model interface. This means that the control model must be defined using the internal control model defined in ROPSIM. The ROPSIM control model holds a description of the robot controller and is presently capable of describing motion control. The motion control model is implemented in C++ and defines a fundamental structure of the robot controller. The model classifies six functions which are cartesian path planner, cartesian interpolator, inverse kinematic transformation, joint path planner, joint interpolator, and a joint controller. Also the control model specifies a specific interconnection of these six functions, establishing the information flow between the functions. Also internal and external sensor feed-back are considered and defined by the structure of the ROPSIM control model. The sensors in the servo-loops are considered as internal sensors while additional sensors are considered as external.

The only freedom when creating a control model of a specific robot system lies in the definition control algorithms used in the six functions. This makes sense since the six functions and the specified interconnection are found in almost all industrial robot controllers.

The control model provides information for the control simulation module.

8.6.4 Modules

Programming module. The function of the programming module is to transform a given robot task description into a robot program. The approach taken in ROPSIM is to base the simulation on imported ICR robot programs generated in available off-line programming systems, refer to Fig 8.6.5-1.

Program interpretation module. The program interpretation module can be viewed as a functional unit. The input is an ICR robot program. The ICR robot program is a description of the actions the robot system should perform in order to accomplish the desired robot task. The program interpretation module perform an interpretation of the ICR robot program on basis of a syntactic and semantic model of the ICR programming language. The interpretation takes one statement at the time and perform the semantic actions of the statement (e.g. changes the value of a variable or send a move instruction to the Robot controller). The actions will depend on the internal state of the interpreter (e.g. the values of variables). The output is a flow of move instructions passed to the robot control simulation module.

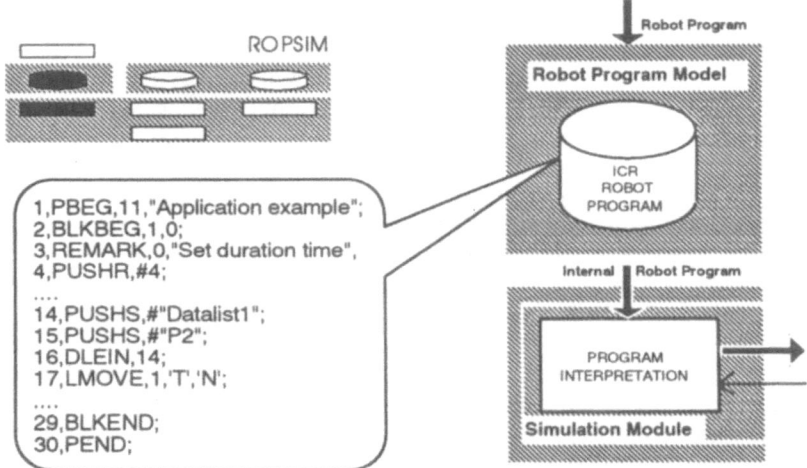

Fig. 8.6.4-1: The input and output model of the interpretation module

As some robot tasks need information about the present state of the robot system in order to perform the succeeding actions, various kind of information may be fed back from the controller simulation module on requests. These input/output relations and the program interpretation are illustrated in Fig. 8.6.4-1.

Controller simulation module. The controller simulation module calculates the servo motor control signals on basis of the control model. The input to the controller is the flow of robot move instruction generated by the robot program interpretation module. The controller model includes algorithms that can effectuate the instructions and accordingly generate servo motor drive control signals as output.

As most robot controllers use sensors on the manipulator this sensor information must be fed back. The controller algorithms include internal control states which are updated during the simulation. These input/output relations and the controller simulation are illustrated in Fig. 8.6.4-2.

The accuracy of the control simulation is heavily depending on the controller model which at present time must be implemented by the user in C++, presumably using the structured generic controller model. Because the control simulation must reflect the true behaviour of the robot controller working in the real-world, the same algorithms used for real-time control of the robot should be applied for the simulation.

The definition of such a control model is not always possible for a ordinary user. For this reason the ROPSIM software has been prepared for a future definition of a neutral interface for control models. Such a neutral interface would allow users to receive an exact control model from the robot vendor and apply this model for simulation purposes. Also the control models could be exchanged with software systems of other origin and different functionality.

The ROPSIM system is prepared for application of such a neutral interface for control model exchange, refer to Figs. 8.6.2-1 and 8.6.5-1.

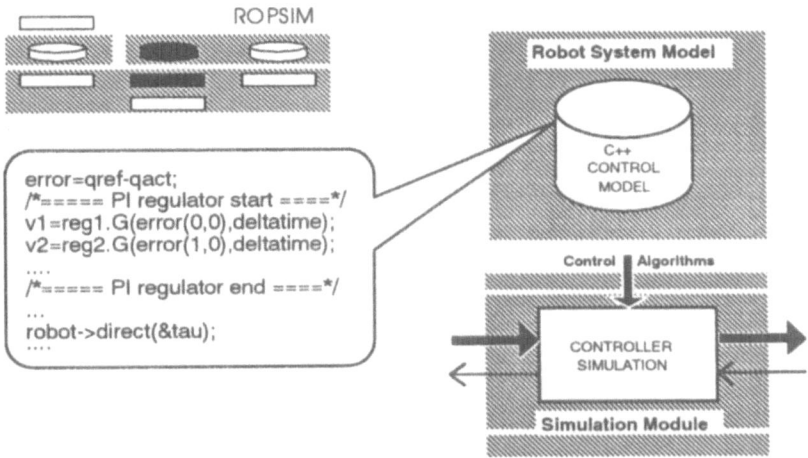

Fig. 8.6.4-2: The input and output model of the control simulation module

Manipulator simulation module. The manipulator simulation module calculates the kinetics of the manipulator on basis of the STEP manipulator model. Because the applied STEP manipulator model can describe both the mass-properties of the robot arm system and the dynamic properties of the servo motor drives the manipulator simulation reflects the dynamic behaviour of the manipulator.

The input to the manipulator simulation are the servo motor control signals generated by the controller simulation. These control signals will effect the manipulator through the servo motor drives (e.g. a moment) and the internal states of the manipulator (e.g. the position, velocity and acceleration of the robot joints) will be updated. The result of the manipulator simulation is an updated manipulator model. These input/output relations and the manipulator simulation are illustrated in Fig. 8.6.4-3.

Presentation module. The presentation module is the module where selected states of the system model (robot program and robot system) are presented. The states of the total system are generated and updated by the program interpretation, the controller simulation and the manipulator simulation modules. The presentation of the current states can take different forms depending on the application. This is illustrated in Fig. 8.6.4-4

Visualisation through animation has shown to be a very effective way to illustrate the geometric relations between objects. Animation is very useful to illustrate the movements of a robot in some surroundings, which might include other moving objects. In Fig. 8.6.4-4 the animation is illustrated by the right rectangle with the "moving" robot.

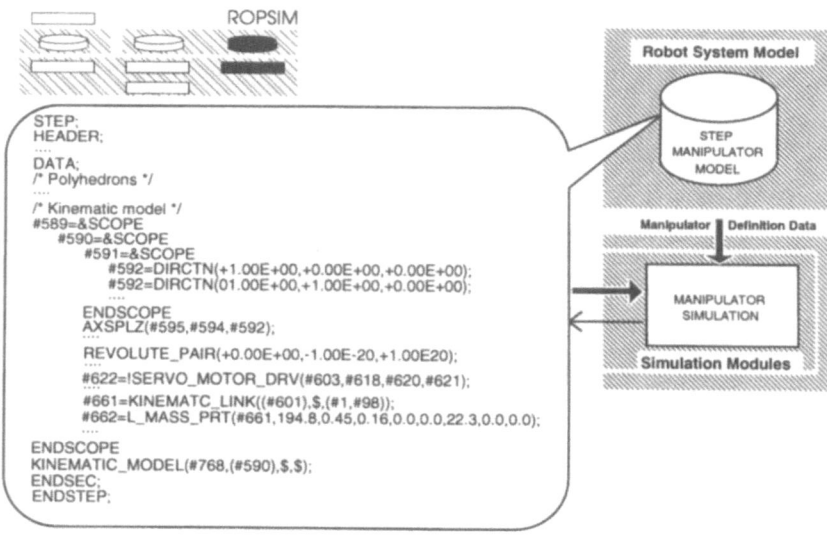

Fig. 8.6.4-3: The input and output model of the manipulator simulation module

Diagrams have been applied for presentation of relation ships between variables in many domains. Within robot simulation many relationships might be of interest and diagrams are often a useful way to present these relationships. Transient diagrams can be used to evaluate state variables of the robot system over time. This is illustrated by the lower middle rectangle in Fig. 8.6.4-4.

To indicate the status of variable or for instance program status the program execution text are often applied. A pointer to the actual program instruction is often applied to shown the current program execution. This is illustrated by the left rectangle in Fig. 8.6.4-4.

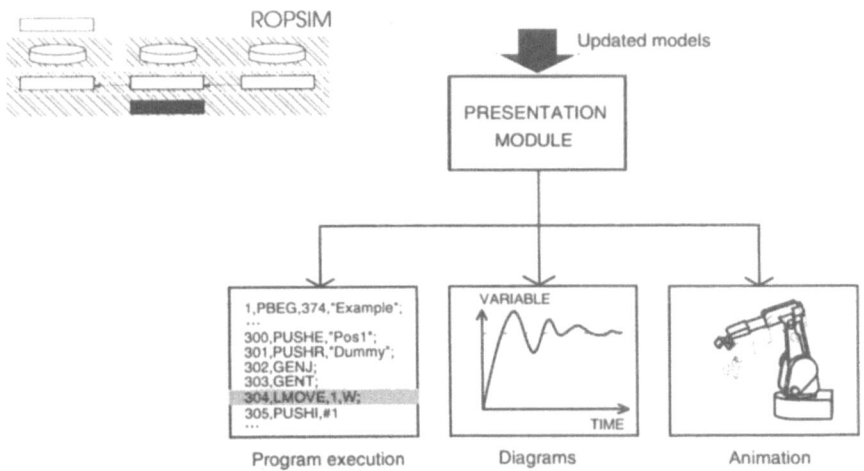

Fig. 8.6.4-4. The input and output model of the presentation module

8.6.5 The Ropsim Model Interface

The architecture for exchange of robot programs and robot system models between ROPSIM and other CIME sub-systems are illustrated in Fig 8.6.5-1.

8.6.6 Example

To illustrate one of ROPSIM's capabilities a simulation example of the dynamic behaviour of a two-link hydraulic test robot called MULTIDOS is given. The robot illustrated in fig. 8.6.7.a is approximately man size and very powerful. The robot has been designed, developed, and constructed at the Control Engineering Institute.

A kinematic and geometric model of the robot has been designed in CATIA and transferred into ROPSIM via the STEP interface. The STEP manipulator model was manually extended with definitions of the mass properties of the links and definition of the servo motor drives.

222

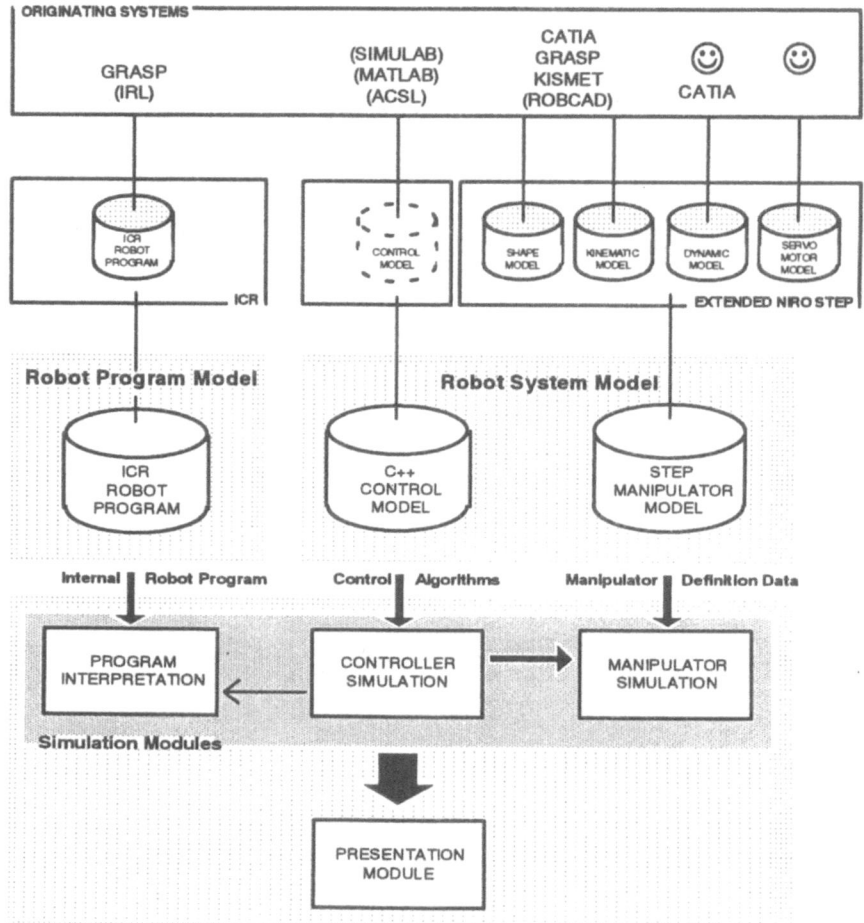

Fig. 8.6.5-1.: The integration of ROPSIM in the CIME system presently operative at the Control Engineering Institute.

The control model was designed using the internal ROPSIM motion control model. The control model used in this example is relatively simple. It consists of two independent PI-regulators. No optimisation of the control model has been carried out. The example has been chosen to illustrate the capabilities of the ROPSIM system rather than the behaviour of a specific control model.

In the example the robot was programmed in ICR to follow a 0.1 meter times 0.1 meter square path at four different locations in the robot workspace, refer to Fig. 8.6.7.a. The horizontal displacement between the squares is 1.0 meter and the vertical displacement is 0.6 meter. The programmed velocity was 0.4 m/s.

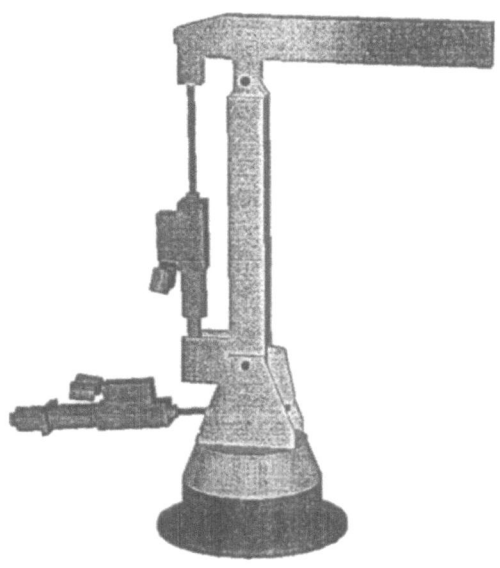

Fig. 8.6.5-2. The MULTIDOS hydraulic test robot

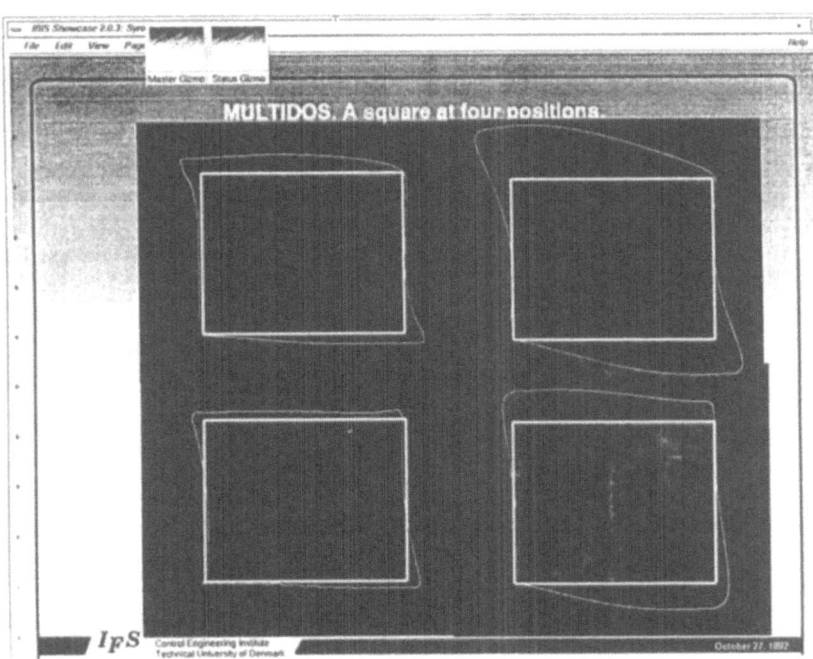

Fig. 8.6.5-3 . The simulated TCP path error along the four squares

The programmed and simulated paths are illustrated in Fig. 8.6.7.b. As expected the TCP errors varies with the position of the path.

Such a simulation performed by the ROPSIM system can be used to evaluate the different parts in the robot system model. On this basis the different parts of the robot system model can be optimised. This could be the control algorithms in the control model or it could be the whole manipulator construction. Also the ICR program could be optimised on basis of the simulation.

8.6.7 Conclusion

The simulation algorithms in ROPSIM modules are designed and implemented using a model driven approach. Because ROPSIM is based on these general and model driven algorithms it is possible to avoid dependence on dedicated models. As a result only the models representing the robot system must be developed when a robot system is simulated in ROPSIM.

In view of modern technology, a simulation system like ROPSIM must be prepared for integration into a more comprehensive CIME environment. This is accomplished by the ROPSIM system by means of the ICR and STEP neutral interfaces for exchange of robot programs and robot system models respectively. Further the modular structure of the ROPSIM architecture prepare the system for application of a future neutral interface for control model exchange.

The main results achieved by development of ROPSIM are

ROPSIM is a true CIME sub-system. ROPSIM is prepared for integration in a larger CIME system environment due to the application of STEP interface for manipulator models and ICR interface for robot programs.

The ICR robot program interface. Due to the ICR robot program interface ROPSIM can simulate execution of very advanced robot programs. Further ROPSIM can exchange robot programs with other off-line programming and simulation systems as well as real-world robot control units.

The STEP manipulator model interface. Robot manipulator models can be exchanged and shared with other systems in the CIME system via the STEP manipulator model interface.

Controller model. With the implemented control model a wide range of robot controllers (independent joint control, multi-variable control, adaptive control)can be modelled. As already mentioned a neutral interface for exchange of robot controller definition data would be extremely applicable in this context. Such an interface would facilitate the exchange of robot controller models with other systems in the CIME system.

Model driven simulation algorithms. All the simulation algorithms implemented in ROPSIM are model-driven. This means that the simulation performed by ROPSIM is driven by the above described models and thereby supports the utilised neutral interfaces.

Presentation of simulation results. The ROPSIM presentation module is capable of presenting any state in the robot system model.

Acknowledgement

The work reported in this Sect. (8.6) has been sponsored by Statens Teknisk-Videnskabelige Forskningsroed, the Danish Technical Research Council, under the research program for Integrated Manufacturing Systems, "Integrerede Produktionssystemer (IPS)", under contract no. 5.26.09.01. We gratefully acknowledge the support which we have received.

9. Conclusions of the Project

9.1 Interface Between CAD, Robot Programming and Simulation Systems

E. Trostmann

9.1.1 General

The investigation of feasibility and practicality of product model data exchange using the STEP interface proceeded along the following main lines:

- Description of concepts and information pertaining to the robotics area in a neutral medium. This included the geometry of robots and scene, the kinematics of mechanisms, and several robot specific entities.
- The actual implementation of processor software for the exchange of this information between systems of different functionality as demonstrated in real industrial applications.

Within the limitations established on the scope of the project, and inherent in an assessment project, the use of STEP as a neutral interface proved feasible and practical.

Seen from the vendor partners perspective, the project had the strategic value of

a) gaining experience in formulating product data exchange specifications for particular application areas (e.g. kinematics and robotics).
b) gaining experience in interfacing techniques for commercial software packages and processor coding.
c) Achieving a first hand knowledge of the weak points to be resolved.
d) Participating in international co-operative work opening the possibility of potential contacts and markets.

From the user partners perspective, several advantages can be mentioned.

a) Getting aquatinted with new tools, methods and techniques that may help to accelerate the automation of their enterprises.
b) Becoming aware of weaknesses and limitations as well as the potential benefits of introducing standard interfaces.
c) Being in a much better position to specify requirements and evaluate proposals for interfacing tools in CIME.

One of the most important conclusion of the project is that a thread of automated data transfer from design to production via interfaces is feasible practical, and convenient.

As expected there are still problems to be solved. Most of the problems experienced in the realisation of system independent interfaces concern the mapping between entities among CAD and Robot programming and simulation systems, as well as the neutral description. A compromise must always be met, sacrificing some generality and flexibility in order to be able to transfer as much information as possible between the sender and receiver.

The disparity of capabilities in different CAD systems only allow for the correct transfer of those features that have an exact counterpart in the neutral description. Some times the CAD system has capabilities not accounted for in the STEP information model. These are irreversibly lost or replaced by other, more elementary representation that result in an equivalent but degraded model. Other times, features in the neutral description have no counterpart in the receiving system. In this case, the recovery of the information is partial. Yet other times, approximations in terms of specific supported capabilities are introduced in the transaction programs to interpret unsupported features in a certain way as to simulate specific behaviours.

The problems experienced with model degradation in transfer events were described in the former chapters. Some examples are:

Closed kinematic chains, as found in many industrial robots can be modelled in sending systems and correctly mapped onto the STEP description. However, none of the post-processors was capable to recover these kind of structures into the receiving systems.

The (redundant) information of 3 co-ordinates for points that must lie on a certain plane gives rise to numerical inaccuracies. In some occasions the point is slightly off the plane in the sending system and consequently in the STEP description. If the receiving systems tolerances are too strict, the system fails to recognise the point belonging to the plane and degrades a (polyhedral) solid model to a wire-frame.

The sub-assembly hierarchy of objects intended by the sending system is not properly covered in the STEP description. Once recovered in the receiving system, a robot trying to pick a sub-assembly component, picks the entire subassembly instead.

These problems belong in the category of solvable ones, and were to some extent solved within the group of systems employed in the project, by resorting to ad-hoc solutions. Nevertheless they will reappear and have to be taken into consideration for every new system and the corresponding processors.

These and other problems will have to be faced in the stabilisation and maturing process of STEP evolution.

There are great expectations around STEP. A great effort has been already been invested in this international enterprise to provide usable CIME tools for industry. However, the migration to a new standard in a slow process. The main point is to have an engagement of industry to commit itself to accept and support the standard.

The NIRO contribution in this context has been to test a small but important part of product description and to demonstrate that STEP can be extended into application areas like robotics.

9.1.2 Contributions to STEP Standard

The main conclusion to be drawn regarding contributions to the STEP standard is that progress in the development of STEP is pretty unpredictable. The reason is that the standardisation process is not a development project that is run according to certain plan and with control of certain resources. It is rather a consensus-building process whose resources are provided by voluntary contributions from the various countries involved. Thus, in the course of events the persons involved changes as well as their priorities and the methodologies.

During the lifetime of the NIRO project the most important changes were:

- the switch from bottom-up to top-down approach: originally resources information models in well-established domains were to be build up first, application protocols were intended to be build on top. Now the approach is to define application protocols first and to derive from them resources models and the most recently invented application interpreted constructs (AICs).

- the introduction of the integration process: this process originally was intended to insure that compatibility was achieved between different information models. However, integration did not restrict itself to this task but rather imposed a specific style of information modelling (based on a specific school of thought and very much influenced by relational database modelling) to all models which was given higher priority than the consensus of the field experts on the suitability of the resulting data structures.

As a result of this development the predictions made on the basis of the prototype processors which were developed in various ESPRIT projects (such as CAD*I, CADEX, NIRO, as well as several US developments in PEDES Inc.) and which all predicted that STEP can be implemented efficiently, are now less reliable. In fact, it is predictable that STEP implementation (e.g. due to the fact that EXPRESS made the ANDOR construct a default, and due to the way how part 41 and 43 developed) will be much less efficient in both development and use than compared to the findings of the various ESPRIT projects.

In the future, ESPRIT projects should try to increase their mutual support even in areas where they do not directly cooperate technically, in order to improve the continuity of the influence from ESPRIT projects on the development of STEP. Furthermore, it is strongly recommended to seek similar strong support as the NIRO project has received from the German DIN, not just from one national standardisation body but from the standardisation bodies of several nations.

9.2 Interface Between Robot Off-Line Programming and Control Systems

H. Bruhm

The NIRO project was mainly focused on the use of ICR as the input format of application programs for robot control. IRL was partly used as an intermediate step, but was always compiled to ICR before execution; the direct program transfer in the IRL format was not within the scope of NIRO.

Robot programs generated off-line in the GRASP simulation system were successfully mapped to both IRL and ICR. In the case of IRL output, a compiler developed in the project was used to transform the programs from IRL to ICR. ICR interpreters, and also compilers to the native language of some controllers, were developed for a variety of execution systems: six different robot controllers and the KISMET simulation system. All the interfacing processors to IRL and ICR were tested individually and as part of the information transfer chain from off-line programming to robot control, with test scenarios based on complex application tasks from the automotive, aerospace and shipbuilding industry.

A comparative assessment of IRL and ICR as alternative neutral interfacing formats for off-line programming and simulation systems such as GRASP showed clearly that IRL is more suitable if the interface is to work in both ways, i.e. for output of robot programs generated in GRASP and also for retrieval of programs from other sources back into GRASP. The reason is that the internal representation in GRASP is a high level language similar to IRL. Retrieving data from a low level language such as ICR introduces many problems which are described elsewhere in this book.

The development of ICR interpreters and compilers revealed that similar mapping problems occur when going from ICR to (modern) native robot languages, which are, though designed for interpretation, on a much higher level than ICR and partly even more powerful than IRL. The latter is true mainly for commands for general geometric program transformation (shifting, rotation, scaling, mirroring), parametric program instantiation (palletizing function, multi-layer welding function) and on-line path "modulation", e.g. with superimposed weaving patterns (welding!) or feed-back from task-related external sensor information (contour tracking etc.). Native robot languages are also very strong when it comes to commands controlling not the robot itself but the application process into which the robot is integrated.

Both ICR and IRL have limitations in their ability to handle information needed to control application processes in a consistent and user-friendly way. These issues need to be addressed if the neutral file formats are to be used in real production environments and commercial vendors are to take full advantage of the results.

Despite the problems detected, the concept of using standard neutral interfaces for data transfer (here: IRL and/or ICR) is very promising for vendors of systems (here: off-line programming systems) which produce input data for other systems (here: robots). Reduction in development and support of proprietary processors allows resources to be concentrated in other areas of valuable research. Little or no additional work is required to support new systems which already have processors to the neutral file format, thus increasing competitiveness. However, the benefits depend directly on the clear definition of the neutral file formats, preferably by the way of international standards, and even more importantly on their acceptance by vendors and users.

There is still a considerable amount of work to be done before this may be achieved: standardization proposals for IRL and ICR were rejected by ISO in the first balloting round. A partial success has been achieved, however, with the decision of ISO to issue the two proposals in the form of ISO Technical Reports. NIRO has continuously supported and influenced the standardization process over the whole duration of the project, through participation in national and international working group meetings, revision and practical testing of the concepts, and elaboration of proposals for further enhancements. Much of the material from the NIRO project is still awaiting its integration into the official documents, however, due to the slow nature of the standardization process.

10. Outlook

O. Knudsen

The major goals of ESPRIT can be grouped into at least three different areas. First are the aim to give companies from different EC-countries the possibility of joining efforts and use advanced technology to produce results which can be brought to the marketplace. Second is the encouragement to produce tools, methods and processes which enhance the development or performance of industrial manufacturing processes, leading to shorter development times, higher quality, better yields and reduced costs. Third is the work of making the EEC contributing with new improvements and proposals to the international standardization work which are going on world-wide.

NIRO has made important contributions to these goals by focusing on the different aspects in applying international standards to data exchange between different modules of complete CIM applications and the results of this work can be divided into three main groups:

1) Each of the partners in the project has achieved a knowledge about where and when standards are applicable. A knowledge which will grow and spread during the coming years with the rising needs for higher integration on the shop-floor and with the higher experience in implementing these standards in real production.

2) The project has developed a number of processors which can produce or transform data to one of the standards STEP, ICR or IRL (PLR). These processors are available and for sale through the partners which have done the main work in developing them.

3) The NIRO project has contributed to extensions and refinements of standards which are or are going to be ISO-approved. These contributions have mainly focused on enhancing the industrial use and applicability of the standards.

ESPRIT project NIRO has developed neutral interfaces (neutral data formats and associated processors) which can now be used for the exchange of geometrical, technological, and programming functions and data between CAD, CAM-modules (robot planning, programming, and simulation), and robot control. Also the user oriented robot programming language has been treated as an interface, e.g. between the programmer and the programming system. These developments are based on formal techniques for specifying the interfaces that are relevant for the robotic application area.

The NIRO project has developed processors, software tools and methods that enables multi-vendor systems to be composed to the benefit of both users and vendors. An important topic of the project has been to demonstrate the feasibility of the approach by prototypes (distributed over the partners) by proving the intercommunication capabilities between systems built from multi-vendor modules.

The contribution of NIRO to the goals of ESPRIT

The single partners in the NIRO project give their own important contribution to the goals of ESPRIT by making very accurate and ambitious exploitations plans for the results of the NIRO work in the future. These exploitation plans are listed below.

BYG Systems Ltd. has developed a number of processors during the NIRO project. The size of the potential market for these products will depend upon the extent of acceptance of both the NIRO-STEP and IRL neutral interfaces by other vendors of CAD and robot control systems.

BYG Systems are going to market the processors which are developed in-house. These will be supplied in conjunction with the existing GRASP simulation and off-line robot programming systems in Europe which has a large and fast growing user base. It is from this user base that it is envisaged the initial customers for the exploitations of the results of this project would come.

Marketing of the processors will be directly through BYG Systems existing Sales Department and also, through-out Europe using BYG's established network of distribution partners.

In addition to the above, there is potential for marketing agreements between BYG Systems and other partners in the project. BYG Systems would be interested in supplying other products, through existing sales network, under licence from the developers. This could be, for example the IRL/ICR interface developed by PSI.

CASA has medium and long term production plans that mean commitments in production rates and supplying dates. These commitments are important, not only regarding to the internal plans but also because of CASA's growing participation in European projects as Airbus or EFA. The exact fulfilling of them are of critical importance. To automate processors is one of the ways to achieve this accuracy in the planning. To make automatic data transfer between different departments is part of the work which currently does not have an open solution in the case of design and robot manufacturing.

CASA are going to exploit the new standards in three different kind of production equipment:

1) Ultra sonic inspection system.
2) Food processing system
3) Filament winding machine

DIS's expertise within the CIM-area and the business interest consists first of all in the ability to use CIM and CIM-technologies to increase competitive strength, to act as integrators and to design and implement architectures. DIS has a thorough background in process technology and knowledge of the exploitation of IT for programming controls and integration of technical and administrative systems in industrial enterprises.

Through the project DIS has increased their competitive strength as total system designers and integrators within the CIM-domain, and by this gained a stronger and enlarged position on the market of small and medium sized companies in Denmark and in other European countries.

This has been achieved by improving the ability to integrate off-line programming and simulation facilities in CIM-projects especially in the area of robotics and FMS's.

DISEL has two commercial objectives in the area of CIM standardization. One is the commented system integration under the important aspect of product data transfer. The other is the robotic, a technology of powerful emergence in Spain, and the whole Europe, that claims for the integration more than others, but has lesser development in open solutions than the classical manufacturing technologies. Beside of this, the project has had a clear actual manufacturing orientation, which is the normal environment of DISEL's current works.

DISEL has developed pre- and post-processors between a CAD package of wide use in Europe and the robotics standards to deal with in the project. It has also solved the problems of a prototype implementation in an actual manufacturing line.

The interface and prototype will be the basis for a new tool to be commercialized among the customers of DISEL. The implementation in a manufacturing company will be the best reference to sell them.

DTH, The Technical University of Denmark (Danmarks Tekniske Højskole) is a government owned university. The general purpose of DTH is to educate engineers at M.Sc. and Ph.D. level and to carry out technical and scientific research. Results of research projects like NIRO are transferred to industry directly through publishing or consultative activities as well as indirectly through improvement of the engineering curricula. More especially the concept of using neutral interfaces in CIM and implementation rules for this will be adopted in running courses. DTH will grant licences on developed software to private companies and other institutions.

Firstly, DTH participates through four of its institutes in a national research program "Integrated Production System", where one of the major efforts is to develop a realtime simulation system for robotic systems and where one of the basic ideas is the application of neutral interfaces between subsystems such as extended STEP and ICR formats. Further in this project issues on introducing STEP standards in Danish industry are considered.

Secondly, at the IFS (Instituttet for Styreteknik, the Control Engineering Institute at DTH) an educational system for teaching mechanical engineering students is being developed. One of the purposes of this system is to teach engineering students (at Master Degree and Ph.D levels) in the understanding and operation of neutral interface based systems for CAD, off-line programming and real-time control of robotic systems. This development emphasizes a theoretical as well as experimental approach and will culminate in formulation of formal courses in robotics.

Thirdly, activities within exchanging CIME data between subsystems are being developed in relation to industrial development projects and consulting. These activities are run within the Institute for Product Development, IPU, an institute at DTH, driven by four scientific institutes within the mechanical engineering department at DTH. Licences concerning DTH developed software for the exchange of product definition data in CIME are granted to external partners.

FIAT/SESAM will exploit the results according to:

- Utilization of the software in a pilot environment and its gradual diffusion.
- Introduction to the FIAT group a common understanding and an operation environment (language, model and software) for the exchange of the technical data from CAD systems to robots planning systems.

- Strong influence of significant software vendors in the adoption of the same common understanding
- Education of users on common understanding and environment

CADDS 5 is a Prime Computervision (now Computervision) new CAE/CAD/CAM product that has been released during 1992. CADDS 5 that replaces the CADDS-4X product, performs 2D and 3D CAE/CAD/CAM analysis.

COMAU owns the NIRO-Step processor interfacing CADDS 5. The processor has been developed by SESAM with support of Computervision (Italian R&D Group), which does not have any exploitation or commercial right on the processor.

COMAU owns the ICR-PDL2 processor developed by SESAM. Companies of the FIAT Group interested to use the processor have free access.

The exploitation of the CADDS 5 to NIRO-Step processor is directly related to the diffusion of Step in the CAD/CAM community and in particular in the FIAT Group. Following the CADEX NIRO activities, FIAT is planning an internal event in order to present Step Technology and the CADEX NIRO results to FIAT Divisions.

ICR to PDL2 processor will be marketed by COMAU as addition features of COMAU robot controller.

KfK will, as previously with the results of ESPRIT project 322 (CAD*I), exploit the project results in two ways:

1. direct application of the project results in the main areas of technical development at KfK. This includes particularly the European projects JET (Joint European Torus) and the European Fusion Technology project including NET and the European contribution to the world-scale ITER project.
2. KfK will grant licences to companies for commercial exploitation of the project results.

OSS is an end user in the NIRO project consortia. It is therefore obvious that OSS' exploitation of project results is basically to be realized as internal utilization of results, that is internal adoption of the developed standards as well as actual utilization of standardized processors developed during the project to the improvement of the functionalities of the yard's CIM system for robot applications.

It should however be mentioned that the solutions in NIRO are of a fundamental value to a broader application of IT based robot systems in heavy industry. It is also the intention of OSS as a pioneer within this field to stimulate other heavy industry companies not only to apply more IT, but also to base the application on emerging standards.

The major reason for ship-builder's excessive needs for standardized interfaces are the following:

- A ship is a multi-system product witch in itself calls for utilization of a number of CAD-systems in the design phase.
- Production of a ship contains discrete parts processing, continuous processing, onsite construction and so on which means that all kinds of CAM tools are required in order to enable automation.
- A major part of known manufacturing processes are put into action during the production cycle of a ship and all kinds of geometrical condition with respect to robot utilization are experienced. No robot manufacturer can cover every demand, which makes need for standardized interfaces between simulation/programming tools and robot controllers very obvious.

PSI is interested in commercializing an IRL programming system in two versions: as a PC based product and as a portable software package ready to be integrated in larger programming or control systems.

Based on a market study which was done by PSI in the second half of 1988 it can be stated that neutral interfaces are the key point for the more general introduction of off-line programming: it must be possible to interface the off-line programming system to existing CAD systems as well as to robots without the need for two different programming languages.

From the PSI participation in the national standardization work it is known that the most important German car manufacturers are strongly interested in an internationally standardized programming language for industrial robots. This is about 50% of the German user market. It can be assumed that there is no big difference in other European countries. This guarantees the acceptance of the new interfaces on the market, because of the user demands the competing producers and vendors of robot controllers must enhance their products through the neutral interfaces.

The big push for off-line programming is only possible if the new interfaces IRL and ICR become more or less widespread within a relatively short amount of time. This can only be achieved if an easily portable software package for IRL and ICR is offered at a reasonable price. It must be possible

- to integrate this package into appropriate existing robot controllers to enhance them by using IRL and/or ICR INTERFACES
- extend it to a stand-alone PC based system in order to connect an IRL programming system to ICR and perhaps an ICR interface

This can be offered only by a neutral developer and vendor who has no interest in specific robots, controllers or other equipment. This has been the role of PSI. PSI is now offering a PC based product and a portable package "IRL programming system" to the market of producers and vendors of robot controllers and of graphical programming systems. According to the experience of the PSI market study it can be expected that especially those companies which do not count to the big ones in the market will be of most importance. As they cannot invest in a development of their own they want to buy software from outside to improve their products.

REIS will exploit the results of the NIRO ESPRIT project inside the project with other participants as well as outside of the project to penetrate new market segments, for instance the automotive industry or segments where robots are programmed with high CAD support.

The interest of the market in ICR is presently much lower than expected at the outset of the NIRO project. Lacking contribution from the robot users to the standardization work is just one indication of this. Furthermore, there is still a lack of IT products on the market that can generate ICR code.

Despite these obstacles to commercialisation. Reis is prepared to offer the ICR interpreter for the ROBOTstar IV controller to interested customers.

Reis is also investigating the technical feasibility of direct compilation from the neutral high level robot programming language IRL to the Reis Robot Language, using the existing front end of the PSI compiler for IRL. A decision on the development of such a product will be made in the near future.

The role of interfaces for CIM in the future

Standard interfaces in robotics will increase the productivity and competiveness of the European industry applying new methods and systems. The introduction of neutral interfaces will have its main impact on the small batch manufacturing industries and on those vendors of robots whose expertise lies specifically in the application domains of the project. Especially the small and medium size vendors who can supply individual robot system modules but cannot supply the whole robot system, including all modules for robot planning, programming, testing and control. They will benefit from the neutral interfaces.

Robotics is an important part of computer integrated manufacturing solutions and solutions in this area include a great number of components situated at different steps in the design and production cycle, at different locations, and with different levels of functionality. Neutral interfaces can connect these components to form an integrated solution and still provide manufacturing system flexibility by allowing rapid replacement of any functional module by its more advanced and better suited successor. Neutral interfaces allow for gradual evolution from specialized and stand alone components to integrated system structures. With standardized interfaces, small and medium sized European vendors in robotics can more easily penetrate the market, if their small or medium sized robot systems or components can be plugged into complete CIM structures via standardized interfaces.

Neutral interfaces will have an impact on the future of European industry only if they are implemented in international standards. Otherwise, robot users and small or medium sized CIM-vendors have to accept the CIM systems supplied by large robot system vendors with private (non-neutral) interfaces which inhibit rather than permit competition. Hence, influencing the international standardisation is essential for a project concentrating on neutral interfaces.

References

Aho, V.A., and Ullman, J.D. (1979): Principles of Compiler Design. Addison-Wesley, Reading, MA.

ANSI Y 14.26 M. American National Standards Institute. Digital Representation of Product Definition Data. (September 1981)

Blume, C., W. Jakob (1986): Programming Languages for Industrial Robots. Springer-Verlag, Berlin.

Blume, C., Jakob, W., Schlechtendahl, E.-G. (1990): New Standards for Neutral Interfaces for Robotics, IFS Publications, UK, 271-279.

Brown, D., Mason, T. (1990): Lex & Yacc. UNIX Programming Tools, O'Reilly and Associates Inc., Sebastopol 1990.

Craig, J.J. (1989): Robotics: Mechanics and Control. Addison-Wesley, Reading, MA.

Denavit, J., and Hartenberg, R.S. (1955): A kinematic notation for lower-pair mechanisms based on matrices, Journal of Applied Mechanics, Volume 77, pp. 215-221.

DIN 66 312 Teil 1, Beuth Verlag, Berlin (1992).

DIN/89 - Industrial Robot Language (IRL) - Language Description (Draft)(1989) NAM in the DIN committee working paper N 96.2.2/15 - 89.

DIN/90 - DIN 66 313: IRDATA - Interface between programming system and robot control. Beuth Verlag, Berlin (1990).

Feuer, A. and Gehani, N. (Eds.): Comparing & Assessing Programming Languages, Ada C Pascal. Prentice Hall, New Jersey.

Fu, K.S., Gonzalez, R.C., Lee, C.S.G. (1987): ROBOTICS: Control, Sensing, Vision, and Intelligence. McGraw-Hill, New York.

Gardner, J., Gisin, E., Retterath, C.(1985): MKS Lex & Yacc Tutorial and Programming Guides, Mortice Kern Systems Inc., Waterloo, Canada.

ISO (1990): Industrial Automation Systems - Exchange of Product Model Data (STEP) - Part 105: Kinematics Information Model - Committee Draft (ISO 10303, St. Louis 1990).

ISO (1990): Industrial Automation Systems - Exchange of Product Model Data - (STEP) Part 11: The EXPRESS Language, ISO/TC184/SC4/WG1/ N496, March 1990.

ISO (1991): Industrial Automation Systems - Exchange of Product Model Data (STEP) - Part 42: General Resources, Representation and Format Description. ISO/DP 10303-42.

ISO (1992): Industrial Automation Systems - Product Data Representation and Exchange (STEP) - Part 21: Clear Text Encoding of the Exchange Structure, ISO/TC184/SC4/WG7, 1992.

ISO/89 - External Representation of Product Definition Data (STEP). ISO DP 10303, (1989).

ISO/89-1 - ICR - Intermediate Code for Robots. ISO DP 10562-1. 1989.

ISO: Manipulating Industrial Robots - Intermediate Code for Robots, ISO DP 10562.2, (1991).

Kroszynski, U., et al. (1989): Geometric Data Transfer Between CAD Systems, IEEE J. of Computer Graphics and Applications 9 (5), pp. 57-71.

Kroszynski, U., et al. (1991): Driving Robots Via Neutral Interfaces, Proc. ESPRIT '91 Conf. Brussels, 25-29 Nov. 1991, pp. 646-660.

Kroszynski, U., Sorensen, T. (1992): Library of Test Models and Exchange Suites for the Transfer of Kinematics and Robotic Models, June 1992. NIRO.WGA.DTH.006.92, 126 pp. Unpublished report. Danmarks Tekniske Hojskole.

Kroszynski, U., Sorensen, T., and Trostmann, E. (1993): Issues in Modelling Techniques for the STEP Based Exchange of Robotics Data. Paper submitted for the IFIP Workshop on Interfaces in Industrial Systems for Production and Engineering, March 15-17, 1993, IGD, Germany.

Kroszynski, U., Sorensen, T., and Schlechtendahl, E.G. (1991): NIRO-STEP Specification for the Transfer of Robotic Models, Final Version: L02V00 "December 91". NIRO.WGA.DTH.010.91, 86 pp. Unpublished report. Danmarks Tekniske Hojskole.

Kroszynski, U.I., Soerensen, T., Schlechtendahl, E.G., (1991) STEP Specification for Kinematics and Robotics CAD Data Exchange.

Kühnapfel, U. (1991A): "Grafische Realzeitunterstützung für Fernhandhabungsvorgänge in komplexen Arbeitsumgebungen im Rahmen eines Systems zur Steuerung, Simulation und Off-Line-Programmierung". Dr.-Ing. Thesis, Universität Karlsruhe. Bericht KfK-5052 (1992).

Kühnapfel, U. (1991B): "KISMET User Data Specification". Unpublished report. Kernforschungszentrum Karlsruhe.

McCarthy, J.M. (1990): Introduction to Theoretical Kinematics. MIT Press, Cambridge, MA.

Model Driven Simulation of Robot Systems. Models and Algorithms. Ph.D thesis. S. Trostmann and L.F. Nielsen. IFS Rep. No. S92.66.

Newmann, W.M., Sproull, R.F.(1981): Principles of Interactive Computer Graphics, McGraw-Hill, New York.

Paul, R.P. (1981): Robot Manipulators - Mathematics, Programming and Control, MIT Press, Cambridge MA.

Raflik, M., Pätzold, B. (Eds.) (1990): CAD*I Database. An Approach to an Engineering Database. Version 4.0. Research Reports ESPRIT. CAD Interfaces (CAD*I) Vol. 5. Springer-Verlag, Berlin.

EXPRESS Language Reference Manual (N466) (1990) Schenck, Douglas. McDonnell Aircraft Company.

Schenck, D. (N466) (1990): EXPRESS Language Reference Manual. McDonnell-Douglas Aircraft Company

Schlechtendahl, E.G. (Ed.) (1988): Specification of a CAD*I Neutral File for CAD Geometry. Version 3.3. Third edition. Research Reports ESPRIT. CAD Interfaces (CAD*I) Vol. 1. Springer-Verlag, Berlin.

Schlechtendahl, E.G. (Ed.) (1989): CAD Data Transfer for Solid Models. Research Reports ESPRIT. CAD Interfaces (CAD*I), Vol 3. Springer-Verlag, Berlin.

Sørensen, T., et al. (1992): Neutral Interfaces and Emerging Standards for Robotics, CIRP '92 Conference, June 11-12, 1992, Danmarks Tekniske Hojskole.

Thomas, D., Van Maanen,, J., Mead, M. (Eds.) (1989): Specification for Exchange of Product Analysis Data. Version 3. Research Reports ESPRIT. CAD Interfaces (CAD*I) Vol. 2. Springer-Verlag, Berlin.

Trostmann, E. (1990): Robotics Research within ESPRIT. Proc. 21st. Int. Symposium on Industrial Robots (ISIR), Copenhagen, pp. 377-382.

Trostmann, E. (1991). Intelligent Interfaces in Robotics. The 1st EuroCIM-conference, "Mechatronics and Robotics, I" 8-10 July 1991, Aachen. ISO Press, Amsterdam, Pp. 231-239.

Appendix 1

STEP Specification
for Kinematics and Robotics CAD Data Exchange.
Final Release: L02V00

U. Kroszynski, T. Sørensen, and E. Schlechtendahl

Preface

This specification is intended to serve as a basis for the implementation of STEP processors to communicate information from CAD systems to robot planning, programming, and simulation systems. Although based upon STEP, this document is not intended to act as a STEP Application Protocol. Hence, it does not always conform to the rules established by ISO for such type of documents.

Rather than using the latest releases of ISO Drafts, this specification is based mostly upon the versions available at the beginning of 1991. In particular, the EXPRESS data definition language version employed in this document is not the latest ISO Committee Draft, but rather an older version. Since then, modifications have occurred. For instance, the MAP construct, frequently used in this document for the purpose of mapping an entity specification from an original STEP schema, is no longer supported in EXPRESS, where it has been replaced by the USE and REFERENCE constructs.

Moreover, this document is not self contained, in the sense that explanatory texts in the STEP documents have not been copied here for obvious space reasons.

Therefore, the reader is strongly advised to refer to the versions of the STEP documents listed in Sect. 1.1 for a proper understanding of the semantic interpretation of entities and attributes mapped into the present specification.

According to project decisions the content is restricted to polyhedral solid models (i.e. boundary representations with implicit planar surfaces and straight edges) for the representation of shape, and to simple kinematic topology (i.e. only revolute and prismatic pairs) for the representation of kinematics. Motion definitions are not included.

This document has its origin in the L01V03 specification of February 1991, (Copenhagen revision, 11 March 1991). The reasons for this final release are:

- The publication of a modified STEP specification for kinematics (STEP part 105, Sapporo version, of July 12, 1991), with a description that allows to define the base of a mechanism on the ground or on a link of another mechanism.
- Some modifications that permit the definition of kinematic skeletons without attached geometry.
- The introduction of new user-defined entities, not supported in STEP, but considered necessary in the representation of robotic models.

The document is the result of the effort of all WGA members in the NIRO project, as well as other persons in the partner organisations:

> H.P. Lorenz, S. Haas, R. Lutz, and E.G. Schlechtendahl, of KfK (D),
> P. Sorenti and G. Barker, of BYG (UK),
> C. D'Elia and F. Rusina, of SESAM (I), and
> E. Trostmann, L.F. Nielsen, S. Trostmann, T.G. Clausen, J.C. Jensen,
> T. Sørensen, and U.I. Kroszynski, of DTH (DK)

This specification will serve as reference for the remaining period of the project. Since the L01V03 version proved stable under the critical scrutiny of the first pilot processor programmers, the format and general structure of that document was maintained. However, explanations and examples were introduced, together with the additions and modifications, in order to make this document more readable to the uninitiated. We certainly hope that, in this L02V00 version, this goal is achieved.

Specific NIRO entities as well as a simple representation of the schema and examples are presented in Sects. 8 to 12.

1. STEP Specification for the Transfer of Robotic Models: General Issues

1.1 Definition Basis

Since the last release (L01V03) in February/March 1991, several changes in (and new releases of) the source STEP documents occurred, and will continue to occur after this (L02V00) final release. As the L01V03 version proved stable in most aspects, rather than following up these developments, it was decided to keep this final specification based upon the following documents:

[1] ISO DP 10303 Industrial Automation Systems - Exchange of Product Model Data - Representation and Format Description. Part 42: General Resources: Shape Representation. Proposed CD Edition. January 1991.

[2] NIRO-STEP Specification for the Transfer of Robotics Models. Version: L01V03 "February 91" revised after the Copenhagen meeting 91.03.11, Uri I. Kroszynski & Torben Sørensen, DTH, and Ernst G. Schlechtendahl, KfK.

[3] ISO/TC184/SC4/WG1/N466 Industrial Automation Systems - Exchange of Product Model Data - Part 11: The EXPRESS Language (March 90).

[4] Industrial Automation - Specification for the Exchange File Implementations of the Standard for the Exchange of Product Model Data (STEP Part 21) Version 24 July 1990. (Jan van Maanen)

[5] Industrial Automation Systems - Exchange of Product Model Data - Part 105: Kinematics Information Model - Committee Draft. (ISO 10303 - St. Louis 1990)

[6] ISO CD 10303-105. Product Model Data Representation and Exchange - Part 105: Kinematics. Version "Sapporo 91", July 1991, only in what concerns the mechanism entity.

[7] Information received about changes to the exchange file specification, in particular with respect to the position of the scope section and the default token (document [4]), and the removal of the local co-ordinate system reference from cartesian point and direction entities (doc. [1]).

1.2 Chapter Structure

The specification is structured according to the entity types supported by this level and version of STEP. For each entity type the reader will find all, or most of, the following sub-headings:

1. *The original STEP specification in EXPRESS.* The formal part of the entity (or type) specification is repeated from the respective STEP document in the form [ref, page] . Thus, [1, 31] refers to the STEP part 42 document page 31. Where no corresponding STEP entity exists, no such heading will appear.

2. *The adopted specification for NIRO.* For entities which are mapped from a corresponding STEP entity, the formal MAP definition is given. (See remark in the Preface concerning the use of the MAP construct).

3. *NIRO specific restrictions.* Under this heading informal explanations and restrictions are presented.

4. *Mapping to the physical file.* This is the formal syntax of the respective entity on the physical file in WSN (Wirth's Syntax Notation).

5. *Usage.* This heading gives recommendations on the use of the entity or more detailed restrictions about its appearance on the

physical file. In particular, any context dependencies for the file which cannot be handled by the formal parts of the specification are given under this sub-heading.

1.3 Deviations from the STEP Specification

1.3.1 New Entities

New entities have been defined which are specific to the NIRO project and do not exist in the STEP specification. These entities are defined in Sect. 8 and include, among others:

- the TYPE *aliasable-entity* describing which kinds of entities can be addressed by the index-entry mechanism

- the *render-face* entity for rendering of faces in solid models

- the *point-direction-pair* entity

- the *solid-mass-properties* entity

- the *link-mass-properties* entity

- the *actuator* entity

- the *rgb-colour* entity

- the *rgb-colour-table* entity

- the *colour-attribute* entity that assigns a colour to a solid model

- the *tool-attachment-frame* entity

- the *tool* entity

- the *mount-tool* entity.

1.3.2 Inconsistencies Found in the STEP Specification

Shape_model was removed as an entity although e.g. *solid_model* is still a SUBTYPE OF (shape_model) [1, 179].

The *solid_instance* was defined as a SUBTYPE OF (solid_model) [1, 193] although solid_model is not a SUPERTYPE OF (solid_instance) [1, 179].

Several inconsistencies were found in document [4], particularly in what concerns the representation of the HEADER section on a STEP file.

NOTE: In the meantime, the corresponding STEP documents have been updated and reached the status of Committee Drafts. In these latest versions, the inconsistencies mentioned above have been resolved.

1.3.3 Deviations with Respect to the STEP Specifications

In the STEP specification for geometry, the *cartesian point* [1, 30] and *direction* [1, 32/33] have an optional parameter indicating a reference to a local co-ordinate system. Since the publication of the document, this parameter has been removed again [7]. Therefore, point co-ordinates and direction components are given relative to co-ordinate systems implied in the context in which they are used.

- *Point co-ordinates* and *direction* components for points and directions within *facetted_breps* are with respect to the *absolute co-ordinate system*.
- The same applies for points and directions used in the *transformation* entity in *solid_instances*, and the ones used in the *frame* for the *ground* of a *kinematic_model*.
- *Point co-ordinates* and *direction* components for points and directions within *base frames* of *mechanisms* are given with respect to a *reference frame*. This reference frame can either be the *ground frame* or one of the *"additional frames"* in a *kinematic link* of another *mechanism*.
- *Point co-ordinates* and *direction* components for points and directions within *frames* expressing the *placements* of *kinematic pairs* or the *"additional frames"* within a link are with respect to the *"link frame"*.
- *Point co-ordinates* and *direction* components for points and directions within a frame expressing the *unmounted position* of a tool are given with respect to the *absolute co-ordinate system*. This frame indicates the *position* and *orientation* in space of the tool origin if the tool is not mounted.
- Point co-ordinates and direction components of points and directions within frames representing the *tool-centre-points* are relative to the tool origin (i.e. the absolute co-ordinate system). When the tool is mounted, the tool origin is understood to coincide with the specified *tool-attachment-frame*.
- The co-ordinates of the *centre of mass* in the *solid-mass-properties* entity are relative to the *absolute co-ordinate system*. The *axes of inertia* pass through the *centre of mass* and have the same orientation as the *absolute co-ordinate axes*.
- The co-ordinates of the *centre of mass* in the *link-mass-properties* entity are relative to the *"link co-ordinate system"*. The *axes of inertia* pass through the *centre of mass* and have the same orientation as the *"link co-ordinate system"* axes.

In the STEP specification for kinematics, *SU-parameters* are employed to express the transformation between the "link co-ordinate system" and each of the joints in that link [5, 11]. Since commercial robot simulation systems are more likely to provide matrices as their internal representation for such transformations, we have chosen to employ "frame" as the general placement method. The frame entity is equivalent to the axis2_placement entity. The reference co-ordinate system for the placement is implied by context.

The STEP specification concerning the mapping to a physical file [4], is unclear in some respects. The HEADER section has therefore been specifically tailored for NIRO files.

2. Syntactical and Semantical Matters

In the following chapters we use token names as they are specified for the STEP file syntax in the relevant STEP documentation. These token names are recognisable from the fact that they are capitalised such as STRING, INTEGER, REAL, and ENTITY_NAME.

In addition, we use some token names which are defined as follows:

```
default_token  = "$" .

reference_list =
     "(" [ ENTITY_NAME { "," ENTITY_NAME } ] ")".

string_list = "(" [ STRING {"," STRING } ] ")".

real_list    =    "(" [ REAL {"," REAL } ] ")".

     where
             { } means 0 or more
             [ ] means 0 or once.
```

Since optional lists in the STEP specification for kinematics were specified to have at least one element, the case of empty lists, which should normally be represented by "( )", is mapped as non-existent, i.e. by the default token "$". This is used for the link-shape type (a list of references to solids) in case no solids exist, and for the list of references to "additional-frames" in a kinematic link, in case no additional frames exist.

Just after the publication of [4], the STEP committee decided that the keyword for scoped entities should appear following the scope section [7]. This is therefore the method adopted in NIRO.

It was decided for the STEP specification that comments should be ignored by the post processors.

It was decided for the STEP specification that instanced facetted_breps should be considered as auxiliary bodies. This means that upon recovery at the receiving system after post-processing, the source polyhedron should not to be visualised and only its instances should be seen.

3. Specification of the STEP_file for NIRO

This section provides the target symbol for the grammar file syntax definition and some auxiliary tokens used in the subsequent chapters.

```
STEP_file      = "STEP;"
                       header_section
                       data_section
                  "ENDSTEP;" .

header_section = "HEADER;"
                       file_identifier_occurrence
                       file_description_occurrence
                  "ENDSEC;" .

data_section = "DATA;"
                  { [ rgb_table_occurrence ]            |
                    [ kinematic_model_occurrence ]      |
                    solid_model_occurrence              |
                    solid_mass_properties
                                        _occurrence |
                    colour_attribute_occurrence         |
                    tool_occurrence                     |
                    index_entry_occurrence }
                  "ENDSEC;" .
```

- NOTE - The optional occurrences may appear only once in the DATA section.

4. Data Transfer Application

4.1 File_Identifier

4.1.1 The Original STEP Specification in EXPRESS

Refer to [4, 39].

```
ENTITY file_name;
  name                             : STRING(256) VARYING;
  time_stamp                       : STRING(256) VARYING;
  author           : LIST [1:#] OF STRING(256) VARYING;
  organization     : LIST [1:#] OF STRING(256) VARYING;
  step_version                     : STRING(256) VARYING;
```

```
   pre_processor_version          : STRING(256) VARYING;
   originating_system             : STRING(256) VARYING;
END_ENTITY;
```

4.1.2 The Adopted Specification for NIRO

```
MAP file_identifier FROM file_name;
WHERE
  step_version = 'STEP L02V00';
END_MAP;
```

4.1.3 Mapping to the Physical File

```
file_identifier_occurrence = file_identifier_keyword
                       "("   STRING        ","
                             STRING        ","
                             string_list ","
                             string_list ","
                             STRING        ","
                             STRING        ","
                             STRING   ")" ";".
```

4.2 File_Description

4.2.1 The Original STEP Specification in EXPRESS

Refer to [4, 39].

```
ENTITY file_description;
  description : LIST [1:#] OF STRING(256) VARYING;
  implementation_level:       STRING(256) VARYING;
END_ENTITY;
```

4.2.2 The Adopted Specification for NIRO

```
MAP file_description FROM file_description;
END_MAP;
```

4.2.3 Mapping to the Physical File

```
file_description_occurrence = file_description_keyword
                  "(" string_list "," STRING ")" ";".
```

4.3 Index_Entry

4.3.1 The Original STEP Specification in EXPRESS

Refer to [4, 42].

```
ENTITY index_entry;
  alias        : STRING (64) VARYING;
  step_entity  : aliasable_entity;
END_ENTITY;
```

4.3.2 The Adopted Specification for NIRO

```
MAP index_entry FROM index_entry;
END_MAP;
```

4.3.3 NIRO Specific Restrictions

- "aliasable_entity" is not a formally defined entity type in STEP so far. It is defined as a NIRO specific type (SELECT type) in "ALIASABLE_ENTITY" (see Sect. 8).

4.3.4 Mapping to the Physical File

```
index_entry_occurrence =
      ENTITY_NAME "=" index_entry_keyword
                  "(" STRING "," ENTITY_NAME ")" ";" .
```

4.3.5 Usage

The ENTITY_NAME corresponding to the aliasable entity is a reference to an entity defined in the file. The allowed types of aliasable entities are defined in Sect. 8.

4.4 Export_List

Refer to [4,17]

Referencing entities defined within the scope of other entities is not possible unless the referenced entities are made more globally known. This is achieved via the "export list" defined in [4, 17]. The "export list" is used in this specification for the ground frame, for additional frames of links, for tool-centre-point frames in tools, and tool-attachment-frames in mechanisms.

4.4.1 Mapping to the Physical File

```
export_list_occurrence = ENTITY_NAME "="
                          export_list_keyword
                    "(" reference_list ")" ";" .
```

An export_list_occurrence appears in the same scope of the entities to be exported. The entities included in the reference_list are thus recognized in the same scope as the enclosing entity.

5. Shape

5.1 Solid_Model

5.1.1 The Original STEP Specification in EXPRESS

Refer to [1, 179].

```
ENTITY solid_model
SUPERTYPE OF(ONEOF( csg_primitive, csg_solid,
                    facetted_brep, manifold_solid_brep,
                    swept_area_solid ))
SUBTYPE OF (shape_model);
END_ENTITY;
```

NOTE: There is no entity shape_model having solid_model as a SUBTYPE. Moreover, solid_instance should also be in the list.

5.1.2 The Adopted Specification for NIRO

```
ENTITY solid_model
SUPERTYPE OF (ONEOF( facetted_brep, solid_instance ));
END_ENTITY;
```

5.1.3 NIRO Specific Restrictions

    o   The solid_model is a facetted_brep or an instance of a facetted_brep.

5.1.4 Mapping to the Physical File

```
solid_model_occurrence = facetted_brep_occurrence |
                         solid_instance_occurrence .
```

5.2 Solid_Instance

5.2.1 The Original STEP Specification in EXPRESS

Refer to [1, 193].

   The original STEP document is inconsistent here. It defines solid_instance as a SUBTYPE OF (solid_model) while solid_model is not a SUPERTYPE OF (solid_instance) See Sect. 1.3.2.

```
ENTITY solid_instance
SUBTYPE OF (solid_model);
  solid_to_be_copied : solid_model;
  location           : transformation;
END_ENTITY;
```

5.2.2 The Adopted Specification for NIRO

```
MAP solid_instance FROM solid_instance;
WHERE
   facetted_brep IN TYPEOF( solid_to_be_copied );
END_MAP;
```

5.2.3 NIRO Specific Restrictions

- The solid instance shall be a replica of a facetted_brep with new location (position and orientation).
- Whenever an instance of a facetted_brep is encountered, the interpretation upon post-processing is that, rather than being duplicated, the source polyhedron is an auxiliary, non-visible solid. There might be several instances of the same source facetted-brep.

5.2.4 Mapping to the Physical File

```
solid_instance_occurrence =
   ENTITY_NAME "=" solid_instance_scope_section
                   solid_instance_keyword
             "(" ENTITY_NAME "," ENTITY_NAME ")" ";".

solid_instance_scope_section =
                   "&SCOPE"
                       transformation_occurrence
                   "ENDSCOPE".
```

5.3 Facetted_Brep

5.3.1 The Original STEP Specification in EXPRESS

Refer to [1, 182].

```
ENTITY facetted_brep
SUBTYPE OF (solid_model);
   outer   : shell_or_logical;
   voids   : SET [0:#] OF shell_or_logical;
WHERE
   WR1 : dimensionality(facetted_brep) = 3;
   WR2 : qualitative_extent(facetted_brep) =
                                 finite_non_zero;
   WR3 : manifold(facetted_brep);
   WR4 : constraints_topology_brep(facetted_brep);
   WR5 : constraints_geometry_brep(facetted_brep);
END_ENTITY;
```

5.3.2 The Adopted Specification for NIRO

```
MAP facetted_brep FROM facetted_brep;
WHERE
   shell IN TYPEOF (outer);
   0 = extent( voids );
END_MAP;
```

5.3.3 NIRO Specific Restrictions

- The facetted_brep has no internal voids in NIRO. Shell is the type selected from shell_or_logical.

5.3.4 Mapping to the Physical File

```
facetted_brep_occurrence =
  ENTITY_NAME "=" facetted_brep_scope_section
                  facetted_brep_keyword
              "(" ENTITY_NAME "," default_token ")" ";".

facetted_brep_scope_section =
     "&SCOPE"
       { render_face_occurrence              |
         direction_occurrence                |
         point_direction_pair_occurrence     |
         shell_occurrence                     |
         face_occurrence                      |
         loop_occurrence                      |
         point_occurrence                     }
     "ENDSCOPE" .
```

- The ENTITY_NAME shall reference the shell occurrence inside the scope section.
- At least four points in three loops corresponding to three faces in one shell shall be defined within the scope of a facetted _ brep, for a valid polyhedron.

5.3.5 Usage

All entities which constitute a facetted_brep occurrence shall be in the scope section of that facetted_brep.

5.4 Topology for Facetted-Breps

5.4.1 The Original STEP Specification in EXPRESS

Refer to [1, 122].

```
ENTITY topology

SUPERTYPE OF (ONEOF(vertex, edge, path, loop, face,
                    sub-face, shell,connected_edge_set,
                    connected_face_set ));
END_ENTITY;
```

5.4.2 The Adopted Specification for NIRO

```
MAP topology FROM topology;
SUPERTYPE OF (ONEOF( loop, face, shell ));
END_MAP;
```

5.4.3 NIRO Specific Restrictions

- The topology represents only those entities that are relevant for a facetted_brep.

5.4.4 Mapping to the Physical File

- No mapping.

5.5 Shell

5.5.1 The Original STEP Specification in EXPRESS

Refer to [1, 135].

```
ENTITY shell
SUPERTYPE OF (ONEOF( vertex_shell, wire_shell,
                     open_shell, closed_shell ))
SUBTYPE OF (topology);
WHERE
  WR1 : { 0 <= dimensionality (shell) <= 2 };
  (* this proposition uses functions
              not completely coded in EXPRESS *)
  WR2 : (qualitative_extent (shell) = finite_non_zero)
              OR (qualitative_extent (shell) = zero);
  (* this proposition uses functions
              not completely coded in EXPRESS *)
  WR3 : genus (shell) >= 0;
  (* this proposition uses functions
              not completely coded in EXPRESS *)
END_ENTITY;
```

5.5.2 The Adopted Specification for NIRO

```
MAP shell FROM shell;
SUPERTYPE OF (closed_shell);
END_MAP;
```

5.5.3 NIRO Specific Restrictions

- A shell is always a closed shell in the facetted_brep context. Moreover, a facetted brep has only one shell.

5.5.4 Mapping to the Physical File

```
shell_occurrence = closed_shell_occurrence .
```

5.5.5 Usage

- The shell entity does not occur on a STEP file from NIRO.

5.6 Closed Shell

5.6.1 The Original STEP Specification in EXPRESS

Refer to [1, 142/143].

```
ENTITY closed_shell
SUBTYPE OF (shell);
  shell_boundary  : SET [1:#] OF face_or_logical;
WHERE
  WR1 : manifold (closed_shell);
(* this proposition uses functions not completely coded
in EXPRESS *)
  WR2 : dimensionality (closed_shell) = 2;
  (* this proposition uses functions
              not completely coded in EXPRESS *)
  WR3 : genus (closed_shell) >= 0;
  (* this proposition uses functions
              not completely coded in EXPRESS *)
  WR4 :constraints_topology_closed_shell(closed_shell);
  (* this proposition uses functions
              not completely coded in EXPRESS *)
  WR5 :constraints_geometry_closed_shell(closed_shell);
END_ENTITY;
```

5.6.2 The Adopted Specification for NIRO

```
MAP closed_shell FROM closed_shell;
```

```
face IN TYPEOF (shell_boundary);
END_MAP;
```

5.6.3 NIRO Specific Restrictions

* Face is in the type selected from face_or_logical [1, 121].

5.6.4 Mapping to the Physical File

```
closed_shell_occurrence =
    ENTITY_NAME "=" closed_shell_keyword
                "(" reference_list ")" ";" .
```

* The reference_list consists of references to the type "face".

5.7 Face

5.7.1 The Original STEP Specification in EXPRESS

Refer to [1, 133].

```
ENTITY face
SUBTYPE OF (topology);
  outer_bound  : OPTIONAL loop_or_logical;
  bounds       : SET [1:#] OF loop_or_logical;
  face_surface : OPTIONAL surface_or_logical;
WHERE
  WR1 : dimensionality (face) = 2;
  (* this proposition uses functions
              not completely coded in EXPRESS *)
  WR2 : qualitative_extent (face) = finite_non_zero;
  (* this proposition uses functions
              not completely coded in EXPRESS *)
  WR3 : manifold (domain (face));
  (* this proposition uses functions
              not completely coded in EXPRESS *)
  WR4 : arcwise_connected (face);
  (* this proposition uses functions
              not completely coded in EXPRESS *)
  WR5 : constraints_topology_face (face);
  (* this proposition uses functions
              not completely coded in EXPRESS *)
  WR6 : constraints_geometry_face (face);
  WR7 : (NOT EXISTS(outer_bound) OR
```

```
                                    (outer_bound IN bounds));
END_ENTITY;
```

5.7.2 The Adopted Specification for NIRO

```
MAP face FROM face;
WHERE
  NOT EXISTS (outer_bound);
  loop IN TYPEOF (bounds);
  NOT EXISTS (face_surface);
END_MAP;
```

5.7.3 NIRO Specific Restrictions

- In this STEP specification the outer bound does not appear.
- Loop is in the type selected from loop_or_logical [1, 121].
- The face_surface does not appear as it is implied to be a plane defined by the points generating the loops on the face.

5.7.4 Mapping to the Physical File

```
face_occurrence =
      ENTITY_NAME "=" face_keyword
                  "(" default_token ","
                      reference_list ","
                      default_token ")" ";" .
```

o   the reference_list consists of references to the type "loop".

5.7.5 Usage

- The implied orientation of a face is away from the material.

5.8 Loop

5.8.1 The Original STEP Specification in EXPRESS

Refer to [1, 128].

```
ENTITY loop
SUPERTYPE OF (ONEOF(vertex_loop, edge_loop, poly_loop))
SUBTYPE OF (topology);
```

```
WHERE
  WR1 : {0 <= dimensionality (loop) <= 1};
  (* this proposition uses functions
                not completely coded in EXPRESS *)
  WR2 : (qualitative_extent (loop) = finite_non_zero)
                OR (qualitative_extent (loop) = zero);
  (* this proposition uses functions
                not completely coded in EXPRESS *)
  WR3 : closed (loop);
  (* this proposition uses functions
                not completely coded in EXPRESS *)
  WR4 : genus (loop) >= 0;
  (* this proposition uses functions
                not completely coded in EXPRESS *)
END_ENTITY;
```

5.8.2 The Adopted Specification for NIRO

```
MAP loop FROM loop;
SUPERTYPE OF (poly_loop);
END_MAP;
```

5.8.3 NIRO Specific Restrictions

- As subtypes only the poly_loop is allowed.

5.8.4 Mapping to the Physical File

```
loop_occurrence = poly_loop_occurrence .
```

5.8.5 Usage

- The loop entity does not occur on a STEP file from NIRO.

5.9 Poly_Loop

5.9.1 The Original STEP Specification in EXPRESS

Refer to [1, 131].

```
ENTITY poly_loop
SUBTYPE OF (loop);
  polygon : LIST [3:#] OF UNIQUE point;
WHERE
  WR1 : dimensionality (poly_loop) = 1;
  (* this proposition uses functions
                 not completely coded in EXPRESS *)
  WR2 : qualitative_extent (poly_loop) =
finite_non_zero;
  (* this proposition uses functions
                 not completely coded in EXPRESS *)
  WR3 : manifold (poly_loop) AND
                          arcwise_connected (poly_loop);
  (* this proposition uses functions
                 not completely coded in EXPRESS *)
  WR4 : planar (poly_loop);
  (* this proposition uses functions
                 not completely coded in EXPRESS *)
  WR5 : genus (poly_loop) = 1;
  (* this proposition uses functions
                 not completely coded in EXPRESS *)
  WR6 : constraints_geom_polyline (poly_loop);
END_ENTITY;
```

5.9.2 The Adopted Specification for NIRO

```
MAP poly_loop FROM poly_loop;
END_MAP;
```

5.9.3 NIRO Specific Restrictions

- The point sequence defining the poly_loop must comply with the orientation away from material.

5.9.4 Mapping to the Physical File

```
poly_loop_occurrence =
              ENTITY_NAME "=" poly_loop_keyword
                        "(" reference_list ")" ";" .
```

- The reference_list consists of references to the type "point".

6. Geometry

6.1.1 The Original STEP Specification in EXPRESS

Refer to [1, 29].

```
ENTITY geometry
SUPERTYPE OF (ONEOF( coordinate_system, point, vector,
                     axis_placement, transformation,
                     curve, surface ))
END_ENTITY;
```

6.1.2 The Adopted Specification for NIRO

```
MAP geometry FROM geometry;
SUPERTYPE OF (ONEOF( point, vector, axis_placement,
                     transformation ));
END_MAP;
```

6.1.3 NIRO Specific Restrictions

• The geometry entities are restricted to those relevant in NIRO.

6.1.4 Mapping to the Physical File

• No mapping

6.2 Point

6.2.1 The Original STEP Specification in EXPRESS

Refer to [1, 30].

```
ENTITY point
SUPERTYPE OF (ONEOF( cartesian_point, point_on_curve,
                     point_on_surface ))
SUBTYPE OF (geometry);
END_ENTITY;
```

6.2.2 The Adopted Specification for NIRO

```
MAP point FROM point;
SUPERTYPE OF (cartesian_point);
END_MAP;
```

6.2.3 NIRO Specific Restrictions

- The point has only the cartesian point as subtype.

6.2.4 Mapping to the Physical File

```
point_occurrence = cartesian_point_occurrence.
```

6.2.5 Usage

- The point entity does not occur in a STEP file from NIRO.

6.3 Cartesian_Point

6.3.1 The Original STEP Specification in EXPRESS

Refer to [1,30].

```
ENTITY cartesian_point
SUBTYPE OF (point);
  local_coordinate_system : OPTIONAL coordinate_system;
  x_coordinate            : length_measure;
  y_coordinate            : length_measure;
  z_coordinate            : OPTIONAL length_measure;
DERIVE
  dim : INTEGER := count_dimensions([x_coordinate,
                                     y_coordinate,
                                     z_coordinate]);
END_ENTITY;
```

- NOTE - The local_coordinate_system has been removed from the cartesian_point entity in the version of part 42 of STEP following [1] and in this STEP specification as well (refer to 1.3.3.).

6.3.2 The Adopted Specification for NIRO

```
MAP cartesian_point FROM cartesian_point;
WHERE
  EXISTS (z_coordinate);
  length_measure = REAL;
END_MAP;
```

6.3.3 NIRO Specific Restrictions

- Coordinates are with respect to the absolute coordinate system, except when used in the scope of axis2_placement, where the coordinates are with respect to a coordinate system determined by context.
- The z-coordinate must be present.
- Coordinates are real numbers representing meters.

6.3.4 Mapping to the Physical File

```
cartesian_point_occurrence =
        ENTITY_NAME "=" cartesian_point_keyword
                    "(" REAL "," REAL "," REAL ")" ";" .
```

6.4 Transformation

6.4.1 The Original STEP Specification in EXPRESS

Refer to [1, 38].

```
ENTITY transformation
SUBTYPE OF (geometry);
  axis1        : OPTIONAL direction;
  axis2        : OPTIONAL direction;
  axis3        : OPTIONAL direction;
  local_origin: OPTIONAL cartesian_point;
  scale        : OPTIONAL REAL;
DERIVE
  dim  : INTEGER := space_dimension
(local_origin,axis1,axis2,axis3);
  scl  : REAL := NVL (scale, 1.0);
  u    : LIST [2:dim] OF direction :=
build_axes(dim,axis1,axis2,axis3);
```

```
origin : cartesian_point :=
base_origin(dim,local_origin);
WHERE
  WR1: scl > 0.0;
  WR2: (NOT EXISTS(local_origin) OR
       (coordinate_space(local_origin) = dim)) AND
       (NOT EXISTS(axis1) OR
           (coordinate_space(axis1)= dim)) AND
       (NOT EXISTS(axis2) OR
           (coordinate_space(axis2)= dim)) AND
       (NOT EXISTS(axis3) OR (coordinate_space(axis3)=
dim));
END_ENTITY;
```

6.4.2 The Adopted Specification for NIRO

```
MAP transformation FROM transformation;
WHERE
  EXISTS (axis1);
  NOT EXISTS (axis2);
  EXISTS (axis3);
  EXISTS (local_origin);
  NOT EXISTS (scale);
END_MAP;
```

6.4.3 NIRO Specific Restrictions

- No scaling is allowed (scale := 1.0).
- The **axis3** represents the z_axis of the transformation.
- The vector product **axis3** x **axis1** represents the y_axis of the transformation.
- The vector product **y_axis** x **axis3** represents the x_axis of the transformation.

This excludes mirroring transformations though still covers rigid rotation and translation.

6.4.4 Mapping to the Physical File

```
transformation_occurrence =
        ENTITY_NAME "=" transformation_scope_section
                        transformation_keyword
                 "(" ENTITY_NAME "," default_token ","
                     ENTITY_NAME "," ENTITY_NAME ","
                     default_token ")" ";".
```

```
transformation_scope_section =
                    "&SCOPE"
                    { direction_occurrence |
                      cartesian_point_occurrence }
                    "ENDSCOPE" .
```

6.5 Vector

6.5.1 The Original STEP Specification in EXPRESS

Refer to [1, 32].

```
ENTITY vector
SUPERTYPE OF(ONEOF( direction, vector_with_magnitude ))
SUBTYPE OF (geometry);
END_ENTITY;
```

6.5.2 The Adopted Specification for NIRO

```
MAP vector FROM vector;
SUPERTYPE OF (direction);
END_MAP;
```

6.5.3 NIRO Specific Restrictions

- Only the direction is relevant as sub-type of vector.

6.5.4 Mapping to the Physical File

- None.

6.6 Direction

6.6.1 The Original STEP Specification in EXPRESS

Refer to [1, 32/33].

```
ENTITY direction
SUBTYPE OF (vector);
  local_coordinate_system : OPTIONAL coordinate_system;
```

```
x  :  REAL;
y  :  REAL;
z  :  OPTIONAL REAL;
DERIVE
  dim : INTEGER := count_dimensions([x,y,z]);
WHERE
  WR1 : vector_magnitude(direction) > 0.0;
END_ENTITY;
```

- NOTE - The local_coordinate_system has been removed from the direction entity in the version of part 42 of STEP following [1] and in this STEP specification as well (refer to 1.3.3.).

6.6.2 The Adopted Specification for NIRO

```
MAP direction FROM direction;
WHERE
  EXISTS (z);
  vector_magnitude (direction) > 0.0;
END_MAP;
```

6.6.3 NIRO Specific Restrictions

o Direction components are given in terms of the absolute coordinate system, except when used in the scope of axis2_placement, where the components are with respect to a coordinate system determined by context.

o The z component must be given.

6.6.4 Mapping to the Physical File

```
direction_occurrence =
        ENTITY_NAME "=" direction_keyword
               "(" REAL "," REAL "," REAL ")" ";".
```

6.6.5 Usage

- Note that the direction data on the physical file must be normalized in the receiving system.

6.7 Axis_Placement

6.7.1 The Original STEP Specification in EXPRESS

Refer to [1, 34].

```
ENTITY axis_placement
SUPERTYPE OF(ONEOF( axis1_placement, axis2_placement ))
SUBTYPE OF (geometry);
END_ENTITY;
```

6.7.2 The Adopted Specification for NIRO

```
MAP axis_placement FROM axis_placement;
SUPERTYPE OF (axis2_placement);
END_MAP;
```

6.7.3 NIRO Specific Restrictions

- The axis_placement has only axis2_placement as subtype.

6.7.4 Mapping to the Physical File

```
axis_placement_occurrence = axis2_placement_occurrence.
```

6.7.5 Usage

- The axis_placement does not occur in a STEP file from NIRO.

6.8 Axis2_Placement

6.8.1 The Original STEP Specification in EXPRESS

Refer to [1, 35].

```
ENTITY axis2_placement
SUBTYPE OF (axis_placement);
  location       : cartesian_point;
  axis           : OPTIONAL direction;
  ref_direction  : OPTIONAL direction;
```

```
DERIVE
  dim : INTEGER  := coordinate_space (location);
  p   : LIST [2:dim] OF direction :=
                   place_axis (dim, axis, ref_direction);
WHERE
 WR1 :(NOT EXISTS(axis)OR coordinate_space(axis) = dim)
            AND(NOT EXISTS (ref_direction) OR
                coordinate_space(ref_direction) = dim);
 WR2 :((NOT EXISTS (location.local_coordinate_system)
      AND(NOT EXISTS (axis.local_coordinate_system))OR
        ((location.local_coordinate_system =
         axis.local_coordinate_system) AND
         (location.local_coordinate_system =
         ref_direction.local_coordinate_system));
END_ENTITY;
```

   o NOTE - The local_coordinate_system has been removed from the cartesian_point and direction entities (refer to paragraph 1.3.3, 6.3 and 6.6).

6.8.2 The Adopted Specification for NIRO

```
MAP axis2_placement FROM axis2_placement;
WHERE
  EXISTS (axis);
  EXISTS (ref_direction);
END_MAP;
```

6.8.3 NIRO Specific Restrictions

- All attributes must be given.

6.8.4 Mapping to the Physical File

```
axis2_placement_occurrence =
        ENTITY_NAME "=" axis2_placement_scope_section
                        axis2_placement_keyword
                "(" ENTITY_NAME "," ENTITY_NAME ","
                    ENTITY_NAME ")" ";" .

axis2_placement_scope_section =
                        "&SCOPE"
                        { cartesian_point_occurrence |
                          direction_occurrence }
                    "ENDSCOPE" .
```

6.8.5 Usage

- The axis2_placement is employed to represent frames in kinematics models, as well as in tools.
- It represents a general rigid rotation and translation with respect to a coordinate system implicit in the context in which the transformation is used.
- The first reference is to a cartesian_point within the scope section, representing the origin location of the placement.
- The second reference is to a direction within the scope section, representing the z-axis of the placement.
- The last reference is to a ref_direction.
- The vector product **axis** x **ref_direction** represents the y-axis of the placement.
- The vector product **y-axis** x **axis** represents the x-axis of the placement.

7. Kinematics

Very long original descriptions in EXPRESS have been omitted concerning WHERE and DERIVE clauses. The reader is referred to STEP document [5]. Concerning the mechanism entity, the reader is referred to STEP document [6].

7.1 Link_Shape

7.1.1 The Original STEP Specification in EXPRESS

Refer to [5, 4].

```
TYPE link_shape = SET [1:#] OF shape_model;
END_TYPE;
```

7.1.2 The Adopted Specification for NIRO

```
TYPE link_shape = SET [1:#] OF solid_model;
END_TYPE;
```

7.1.3 NIRO Specific Restrictions

- The solid_model shall be a facetted_brep, or an instance of a facetted_brep.

7.1.4 Mapping to the Physical File

```
link_shape_type = reference_list.
```

In the STEP specification for kinematics, the link-shape appears as an optional type. According to the definition, the list, when existent, must have at least one reference. As any optional list, when not existent, it is represented by the default token "$".

In this case: `link_shape_type = default_token`.

7.1.5 Usage

- link-shapes refer to solid models.

7.2 Kinematic Model

7.2.1 The Original STEP Specification in EXPRESS

Refer to [5, 5/6].

```
ENTITY kinematic_model;
  ground     : ground;
  mechanisms : LIST [1:#] OF mechanism;
  control    : OPTIONAL kinematic_control_model;
  result     : OPTIONAL kinematic_result_model;
END_ENTITY;
```

7.2.2 The Adopted Specification for NIRO

```
MAP kinematic_model FROM kinematic_model;
WHERE
  NOT EXISTS (control);
  NOT EXISTS (result );
END_ENTITY;
```

7.2.3 NIRO Specific Restrictions

- This STEP specification does not include control or result information.

7.2.4 Mapping to the Physical File

```
kinematic_model_occurrence =
        ENTITY_NAME "=" kinematic_model_scope_section
                        kinematic_model_keyword
                "(" ENTITY_NAME ","
                    reference_list ","
default_token
                "," default_token ")" ";" .

kinematic_model_scope_section =
                "&SCOPE"
                { solid_model_occurrence              |
                  colour_attribute_occurrence         |
                  solid_mass_properties_occurrence    |
                  tool_occurrence                     |
                  mount_tool_occurrence               |
                  ground_occurrence                   |
                  mechanism_occurrence                |
                  index_entry_occurrence              }
                "ENDSCOPE" .
```

o NOTE - The ground appears only once within a kinematic model.

7.3 Ground

7.3.1 The Original STEP Specification in EXPRESS

Refer to [5, 6].

```
ENTITY ground;
  ground_frame               : frame;
  geometric_representation : OPTIONAL link_shape;
END_ENTITY;
```

7.3.2 The Adopted Specification for NIRO

```
MAP ground FROM ground;
END_MAP;
```

7.3.3 NIRO Specific Restrictions

- The frame represents the transformation with respect to the absolute coordinate system.

7.3.4 Mapping to the Physical File

```
ground_occurrence =
  ENTITY_NAME "=" ground_scope_section
                 ground_keyword
             "(" ENTITY_NAME "," link_shape_type ")"
";" .

ground_scope_section =
             "&SCOPE"
             { frame_occurrence   |
               export_list_occurrence }
             "ENDSCOPE" .
```

7.4 Mechanism

7.4.1 The Original STEP Specification in EXPRESS

Refer to [6, 12/13].

```
ENTITY mechanism;
  kinematic_structure: kinematic_structure;
  base                : kinematic_link;
  actual position     : OPTIONAL axis2_placement;
  reference frame     : OPTIONAL axis2_placement;
WHERE
  (* please find additional text in [6, 12/13] *)
END_ENTITY;
```

7.4.2 The Adopted Specification for NIRO

```
MAP mechanism FROM mechanism;
WHERE
  EXISTS (actual_position);
  EXISTS (reference_frame);
END_MAP;
```

7.4.3 NIRO Specific Restrictions

- The actual position (base) frame must be given. It represents the transformation relative to the reference frame for placing the mechanism base.
- The reference frame must be specified. It is either the ground frame or an additional frame of a kinematic link in the scope of another mechanism.

7.4.4 Mapping to the Physical File

```
mechanism_occurrence =
   ENTITY_NAME "=" mechanism_scope_section
                   mechanism_keyword
               "(" ENTITY_NAME "," ENTITY_NAME ","
                   ENTITY_NAME "," ENTITY_NAME ")" ";" .

mechanism_scope_section =
                   "&SCOPE"
                   { kinematic_structure_occurrence           |
                     kinematic_link_occurrence                |
                     frame_occurrence                         |
                     index_entry_occurrence                   |
                     kinematic_joint_occurrence               |
                     kinematic_pair_occurrence                |
                     placement_occurrence                     |
                     tap_frame_occurrence                     |
                     export_list_occurrence                   |
                     link_mass_properties_occurrence          |
                     actuator_occurrence                      |
                     pair_placement_structure_occurrence }
                   "ENDSCOPE" .
```

o NOTE - The kinematic structure appears only once within a mechanism. The tool attachment frame may appear only once within a mechanism.

7.5 Kinematic Structure

7.5.1 The Original STEP Specification in EXPRESS

Refer to [5, 7/8].

```
ENTITY kinematic_structure;
  joints                : LIST [1:#] OF kinematic_joint;
```

```
contained_substructures :
                OPTIONAL LIST [1:#] OF
kinematic_substructure;
END_ENTITY;
```

7.5.2 The Adopted Specification for NIRO

```
MAP kinematic_structure FROM kinematic_structure;
WHERE
  NOT EXISTS (contained_substructures);
END_MAP;
```

7.5.3 NIRO Specific Restrictions

- NIRO does not support substructures (high level topology)

7.5.4 Mapping to the Physical File

```
kinematic_structure_occurrence =
 ENTITY_NAME "=" kinematic_structure_keyword
        "(" reference_list "," default_token ")" ";".
```

7.5.5 Usage

- The reference list contains references to the kinematic joints in
  the kinematic structure.

7.6 Kinematic Joint

7.6.1 The Original STEP Specification in EXPRESS.

Refer to [5, 8/9].

```
ENTITY kinematic_joint ;
  first_link  : kinematic_link;
  second_link : kinematic_link;
  pair        : kinematic_pair;
UNIQUE
  pair;
END_ENTITY;
```

7.6.2 The Adopted Specification for NIRO

```
MAP kinematic_joint FROM kinematic_joint;
END_MAP;
```

7.6.3 NIRO Specific Restrictions

- Only lower pairs are allowed.

7.6.4 Mapping to the Physical File

```
kinematic_joint_occurrence =
            ENTITY_NAME "=" kinematic_joint_keyword
                        "(" ENTITY_NAME "," ENTITY_NAME
                        "," ENTITY_NAME ")" ";" .
```

7.7 Kinematic Link

7.7.1 The Original STEP Specification in EXPRESS

Refer to [5, 9].

```
ENTITY kinematic_link;
  pairs            : SET [1:#] OF
pair_placement_structure;
  additional_frames :      OPTIONAL SET [1:#] OF
placement;
  geometric_representation : OPTIONAL link_shape;
DERIVE
  link_frame: frame := compute_link_frame
(kinematic_link);
END_ENTITY;
```

7.7.2 The Adopted Specification for NIRO

```
MAP kinematic_link FROM kinematic_link;
END_MAP;
```

7.7.3 NIRO Specific Restrictions

- The attributes which are related to high level topology are not present.

7.7.4 Mapping to the Physical File

```
kinematic_link_occurrence =
  ENTITY_NAME "=" kinematic_link_keyword
              "(" reference_list "," reference_list ","
              link_shape_type ")" ";" .
```

- It was formerly seen that the optional link-shape type may take the form of the default token. Similarly, when there are no additional frames for the link, the corresponding optional reference list is substituted by the default token. In this case, the kinematic_link_occurrence takes the form:

```
kinematic_link_occurrence =
  ENTITY_NAME "=" kinematic_link_keyword
              "(" reference_list "," default_token   ","
              link_shape_type ")" ";" .
```

7.7.5 Usage

- The "link frame" shall represent an implicitly defined local coordinate system that has been chosen in the sending system. This coordinate system shall be defined by the link frame attribute in absolute coordinates. It can be computed only in a complete kinematic analysis of the whole mechanism. All placement information on the link (such as pair_placement or additional_frames) shall be given relative to the link frame.
- Since the mechanism entity was adopted from the Sapporo version [6] of STEP part 105, (allowing a mechanism to be "attached" on a link of another mechanism), the computation of the "link frame" shall be relative to the absolute coordinate system rather than relative to the ground.
- e solids referenced in the link shape type are understood to be positioned in space relative to the "link frame".

7.8 Pair Placement Structure

7.8.1 The Original STEP Specification in EXPRESS

Refer to [5, 10].

```
ENTITY pair_placement_structure;
  pair       : kinematic_pair;
  placement : placement;
END_ENTITY;
```

7.8.2 The Adopted Specification for NIRO

```
MAP pair_placement_structure FROM
pair_placement_structure;
END_MAP;
```

7.8.3 NIRO Specific Restrictions

- Both the kinematic pair and the placement must be defined within the scope of the mechanism where the pair_placement_structure appears.
- Placement is relative to the "link frame" of the kinematic link.

7.8.4 Mapping to the Physical File

```
pair_placement_structure_occurrence =
    ENTITY_NAME "=" pair_placement_structure_keyword
          "(" ENTITY_NAME "," ENTITY_NAME ")" ";".
```

7.9 Placement

7.9.1 The Original STEP Specification in EXPRESS

Refer to [5, 10].

```
ENTITY placement SUPERTYPE OF (ONEOF( frame,
su_parameters ));
END_ENTITY;
```

7.9.2 The Adopted Specification for NIRO

```
MAP placement FROM placement;
SUPERTYPE OF (frame);
END_MAP;
```

7.9.3 NIRO Specific Restrictions

- A placement is always described as a frame matrix.

7.9.4 Mapping to the Physical File

```
placement_occurrence = frame_occurrence .
```

7.10 Frame

7.10.1 The Original STEP Specification in EXPRESS

Refer to [5, 11].

```
MAP frame FROM axis2_placement;
SUBTYPE OF (placement);
DERIVE
  x_axis : direction :=
      place_axis (3,axis,ref_direction) [1];
  y_axis : direction :=
      place_axis (3,axis,ref_direction) [2];
  z_axis : direction :=
      place_axis (3,axis,ref_direction) [3];
END_MAP;
```

7.10.2 The Adopted Specification for NIRO

```
MAP frame FROM frame;
END_MAP;
```

7.10.3 NIRO Specific Restrictions

- The frame entity appears once in the scope of the ground entity expressing the transformation with respect to the absolute coordinate system. Within a mechanism, it appears several times: once to express the actual position of the base relative to the reference frame (either the ground frame or an additional frame of a link of another mechanism), and the others to express the placement of joints and additional control frames on links, relative to the link_frame.

- The frame entity also appears within the scope of the tool entity: Once to express the unmounted position of the tool origin relative to the absolute coordinate system, and one or more times to define the tool_centre_point frames relative to the tool origin.

7.10.4 Mapping to the Physical File

```
frame_occurrence = axis2_placement_occurrence .
```

7.10.5 Usage

- The frame represents a rigid rotation and translation.
- The frame is realised via the axis2_placement entity. The cartesian_point and directions in the scope of axis2_placement refer to coordinate systems which are implicit in the context in which the frame is used as explained in 7.10.3.

7.11 Kinematic Pair

7.11.1 The Original STEP Specification in EXPRESS

Refer to [5, 14].

```
ENTITY kinematic_pair
SUPERTYPE OF
(ONEOF( kinematic_lower_pair, kinematic_higher_pair ));
DERIVE
  (* Please find additional text in [5, 14] *)
END_ENTITY;
```

7.11.2 The Adopted Specification for NIRO

```
MAP kinematic_pair FROM kinematic_pair;
SUPERTYPE OF (kinematic_lower_pair);
END_MAP;
```

7.11.3 NIRO Specific Restrictions

- Only kinematic lower pairs are allowed.

7.11.4 Mapping to the Physical File

```
kinematic_pair_occurrence =
kinematic_lower_pair_occurrence.
```

7.12 Kinematic Lower Pair

7.12.1 The Original STEP Specification in EXPRESS

Refer to [5, 17].

```
ENTITY kinematic_lower_pair
SUPERTYPE OF (ONEOF( revolute_pair, prismatic_pair,
                     screw_pair, cylindrical_pair,
                     spherical_pair, planar_pair)
SUBTYPE OF (kinematic_pair);
END_ENTITY;
```

7.12.2 The Adopted Specification for NIRO

```
MAP TYPE kinematic_lower_pair FROM
kinematic_lower_pair;
SUPERTYPE OF (ONEOF( revolute_pair, prismatic_pair ));
END_MAP;
```

7.12.3 NIRO Specific Restrictions

- Only revolute and prismatic pairs are allowed.

7.12.4 Mapping to the Physical File

```
kinematic_lower_pair_occurrence =
      revolute_pair_occurrence |
      prismatic_pair_occurrence.
```

7.13 Revolute Pair in EXPRESS

Refer to [5, 17].

```
7.13.1  The Original STEP Specification ENTITY
revolute_pair
SUBTYPE OF ( kinematic_lower_pair);
  actual_rotation              : kinematic_angle_value;
  lower_limit_actual_rotation : OPTIONAL
                                 kinematic_angle_value;
  upper_limit_actual_rotation : OPTIONAL
                                 kinematic_angle_value;
WHERE
  number_of_possible_motions    = 1;
  number_of_independent_motions = 1;
  actual_pair_parameter [1] =  actual_rotation;
  pair_parameter_type [1]   = 'kinematic_angle_value';
END_ENTITY;
```

7.13.2 The Adopted Specification for NIRO

```
MAP revolute_pair FROM revolute_pair;
WHERE
  EXISTS (lower_limit_actual_rotation);
  EXISTS (upper_limit_actual_rotation);
  kinematic_angle_value  = REAL;
END_MAP;
```

7.13.3 NIRO Specific Restrictions

- NIRO requires the rotation limits to be given.

7.13.4 Mapping to the Physical File

```
revolute_pair_occurrence =
        ENTITY_NAME "=" revolute_pair_keyword
                 "(" REAL "," REAL "," REAL ")" ";"
```

7.13.5 Usage

- The revolute pair shall restrict the motion between two adjacent
  links to a rotation about a common axis. To measure the angle
  of rotation a local coordinate system shall be defined in each of
  the links such that local origins and the z-axes coincide and
  their positive directions agree. The motion of the second link
  with respect to the first link shall be defined as the angle

required to rotate the x-axis of the first pair coordinate system in positive direction around the common z-axis until it matches the x-axis of the second pair coordinate system.

- The real values are in radians.

7.14 Prismatic Pair

7.14.1 The Original STEP Specification in EXPRESS

Refer to [5, 19].

```
ENTITY prismatic_pair
SUBTYPE OF (kinematic_lower_pair);
  actual_translation              :
kinematic_length_value;
  lower_limit_actual_translation : OPTIONAL

kinematic_length_value;
  upper_limit_actual_translation : OPTIONAL

kinematic_length_value;
WHERE
  number_of_possible_motions    = 1;
  number_of_independent_motions = 1;
  actual_pair_parameter [1] =  actual_translation;
  pair_parameter_type [1]   = 'kinematic_length_value';
END_ENTITY;
```

7.14.2 The Adopted Specification for NIRO

```
MAP prismatic_pair FROM prismatic_pair;
WHERE
  EXISTS (lower_limit_actual_translation);
  EXISTS (upper_limit_actual_translation);
  kinematic_length_value  = REAL;
END_MAP;
```

7.14.3 NIRO Specific Restrictions

- NIRO requires the translation limits to be given.

7.14.4 Mapping to the Physical File

```
prismatic_pair_occurrence =
    ENTITY_NAME "=" prismatic_pair_keyword
                "(" REAL "," REAL "," REAL ")" ";" .
```

7.14.5 Usage

- The prismatic pair shall restrict the motion between two adjacent links to a translation along a common axis. To measure the distance of translation a local coordinate system shall be defined on each of the links such that their corresponding coordinate axes coincide and their positive directions agree. The motion of the second link with respect to the first link shall be defined as the distance required to translate the xy-plane of the first pair coordinate system in positive direction of the common z-axis until it coincides with the xy-plane of the second pair coordinate system.
- The real values are in meters.

8. NIRO-Specific STEP Types and Entities

The types and entities in this section do not exist in the STEP documents used as reference.

They are to be mapped to the physical file as user defined entities, i.e. with an exclamation mark as the first character of the keyword, and ought to be syntactically (though not necessarily semantically) understood by postprocessors, at least to the extent that allows post-processors to recognize them and ignore them.

User defined entities must conform to the same general syntax of all DATA section entities. Post-processor translator software should be prepared to ignore those user defined entities it does not recognize. This precaution allows the introduction of new user defined entities without the need to modify existing translator software.

8.1 Aliasable_Entity

8.1.1 The Original STEP Specification in EXPRESS

- o NOTE: There are some references to aliasable entity in [4, 41].

8.1.2 The Adopted Specification for NIRO

```
TYPE aliasable_entity = SELECT (
            facetted_brep,  kinematic_joint,
            kinematic_link, kinematic_model,
            kinematic_structure,  mechanism,
            solid_instance, ground, tool, tap_frame );
END_TYPE;
```

8.2 Render_Face

8.2.1 The Adopted Specification for NIRO

```
ENTITY render_face;
   face                    : face;
   colour                  : INTEGER;
   point_direction_pair :
            SET [0:#] OF point_direction_pair;
UNIQUE
   face;
END_ENTITY;
```

8.2.2 Description

The face (defined by the face attribute) shall be assumed to be represented by a surface whose outward normal directions (away from the material) in the vertices of the face are given by the point_direction_pairs. The colour is given by the colour code, in accordance with the colour table, or, when the table is not given, with the default codes (0 to 15).

Whenever information on face rendering is specified, it shall override the colour information assigned to a solid for that face.

8.2.3 Mapping to the Physical File

```
render_face_occurrence =
            ENTITY_NAME "=" render_face_keyword
                     "(" ENTITY_NAME "," INTEGER ","
                     reference_list ")" ";"
```

8.3 Point_Direction_Pair

8.3.1 The Adopted Specification for NIRO

```
ENTITY point_direction_pair;
  point     : point;
  direction : direction;
END_ENTITY;
```

8.3.2 Description

See description of render_face.

8.3.3 Mapping to the Physical File

```
point_direction_pair_occurrence =
  ENTITY_NAME "=" point_direction_pair_keyword
             "(" ENTITY_NAME "," ENTITY_NAME ")" ";".
```

8.4 RGB_Colour_Table

8.4.1 The Adopted Specification for NIRO

```
ENTITY rgb_colour_table;
  table   : LIST [1:#] of rgb_colour;
WHERE
  WR1: SIZEOF (table) <= 256;
END_ENTITY;
```

8.4.2 Description

The rgb-colour-table, if present, appears only once in the DATA section of the STEP file. It defines the integer codes of colours (from 0 to 255) as the position (minus one) of the reference to a rgb-colour in the list of references. The rgb-colour-table may also not be present, in which case 16 default colour codes (0 to 15) may be used according to the following convention:

Colour code	colour	rgb-weights		
		RED	GREEN	BLUE

0	black	0.0000	0.0000	0.0000
1	white	0.8000	0.8000	0.8000
2	red	0.7000	0.0000	0.0000
3	green	0.0000	0.7000	0.0000
4	blue	0.0000	0.0000	0.7000
5	yellow	0.8000	0.8000	0.0000
6	brown	0.5000	0.2000	0.0000
7	grey	0.5000	0.5000	0.5000
8	bright white	1.0000	1.0000	1.0000
9	bright red	1.0000	0.0000	0.0000
10	bright green	0.0000	1.0000	0.0000
11	bright blue	0.0000	0.0000	1.0000
12	cyan	0.0000	0.7000	0.7000
13	magenta	0.7000	0.0000	0.7000
14	dark grey	0.2000	0.2000	0.2000
15	orange	1.0000	0.4000	0.0000

The colour code can be assigned to a solid via the colour_attribute, or to a face via the render_face entity.

8.4.3 Mapping to the Physical File

```
rgb_colour_table_occurrence =
        ENTITY_NAME "=" rgb_scope_section
                       rgb_colour_table_keyword
                "(" reference_list ")" ";".

rgb_scope_section = "&SCOPE"
                    { rgb_colour_occurrence }
                    "ENDSCOPE" .
```

8.5 Colour_Attribute

8.5.1 The Adopted Specification for NIRO

```
ENTITY colour_attribute;
  entity_to_be_coloured : solid_model;
  colour_code           : INTEGER;

UNIQUE

  entity_to_be_coloured;

END_ENTITY;
```

8.5.2 Description

The colour attribute entity assigns a colour code to an entire solid. If the solid is rendered with flat shading and there is rendering information for faces on it, the render_face overrides the colour attribute for those faces. The colour attribute entity may occur in the same scope as solids do, i.e. within the scope of a kinematic model or in the DATA section of the STEP file, following the usual referencing rules.

The INTEGER represents the colour code, in accordance with the colour table, or, when the table is not given, with the default codes (0 to 15).

8.5.3 Mapping to the Physical File

```
colour_attribute_occurrence =
    ENTITY_NAME "=" colour_attribute_keyword
                  "(" ENTITY_NAME "," INTEGER ")" ";" .
```

8.6 RGB_Colour

8.6.1 The Adopted Specification for NIRO

```
ENTITY rgb_colour;
   red   : REAL;
   green : REAL;
   blue  : REAL;
WHERE
   WR1 : 0.0 <= red   <= 1.0;
   WR2 : 0.0 <= green <= 1.0;
   WR3 : 0.0 <= blue  <= 1.0;
END_ENTITY;
```

8.6.2 Description

The rgb-colour entity defines a colour for the rgb-colour-table. The rgb-colour entities appear only within the scope of the rgb-colour-table. They are implicitly assigned an integer code according to the sequence in which they are referenced in the rgb-colour-table entity. In case the rgb-colour-table is not present, there are 16 default, predefined colours (see rgb_colour_table).

8.6.3 Mapping to the Physical File

```
rgb_colour_occurrence =
    ENTITY_NAME "=" rgb_colour_keyword
                "(" REAL "," REAL "," REAL ")" ";" .
```

8.7 Solid_Mass_Properties

8.7.1 The Adopted Specification for NIRO

```
ENTITY solid_mass_properties;
    solids   : link_shape;
    density  : REAL;
    volume   : REAL;
    x        : REAL;
    y        : REAL;
    z        : REAL;
    Gx       : REAL;
    Gy       : REAL;
    Gz       : REAL;
    Gyz      : REAL;
    Gxz      : REAL;
    Gxy      : REAL;
END_ENTITY;
```

8.7.2 Description

This entity expresses the mass properties for a set of solids. It gives the density in kilograms per cubic meter, the volume in cubic meters, the coordinates of the centre of mass in meters, the principal moments of inertia, and the inertia products in kilograms times square meters.

Rather than defining the centre of mass as a cartesian point, which would necessitate a scope section for this entity, it was considered advantageous to define its three coordinates at the same level as the rest of the information.

The coordinates of the centre of mass are relative to the absolute coordinate system. The inertia values are with respect to a coordinate system with origin on the centre of mass and with the same orientation as the absolute coordinate system.

The solid-mass-properties entity may occur in the same scope as the solids referenced, i.e. in the DATA section of the STEP file or within the scope of the kinematic model.

8.7.3 Mapping to the Physical File

```
solid_mass_properties_occurrence =
    ENTITY_NAME "=" solid_mass_properties_keyword
            "(" link_shape_type "," REAL "," REAL ","
            REAL "," REAL "," REAL ","
            REAL "," REAL "," REAL ","
            REAL "," REAL "," REAL ")" ";".
```

8.8 Link_Mass_Properties

8.8.1 The Adopted Specification for NIRO

```
ENTITY link_mass_properties;
  link : kinematic_link;
  mass : REAL;
  x    : REAL;
  y    : REAL;
  z    : REAL;
  Gx   : REAL;
  Gy   : REAL;
  Gz   : REAL;
  Gyz  : REAL;
  Gxz  : REAL;
  Gxy  : REAL;
UNIQUE
  link;
END_ENTITY;
```

8.8.2 Description

If, by some means (e.g. actual measurement), the mass properties of a complete link are known, these can be transmitted in the neutral file via this entity. The link-mass- properties entity is more relevant than the solid-mass-properties for the simulation of robot arm dynamics. It is more difficult to obtain automatically from an approximate model though. Rather than a set of solids, the link-mass-properties refers to a kinematic link. It gives the mass of the link in kilograms, the coordinates of the centre of mass in meters, the principal moments of inertia, and the inertia products in kilograms times square meters.

Rather than defining the centre of mass as a cartesian point, which would necessitate a scope section for this entity, it was considered advantageous to define its three coordinates at the same level as the rest of the information.

The coordinates of the centre of mass are relative to the "link frame". The inertia values are with respect to a coordinate system with origin on the centre of mass and the same orientation as the link frame.

The link-mass-properties entity may occur in the scope of the mechanism containing the referenced kinematic link.

8.8.3 Mapping to the Physical File

```
link_mass_properties_occurrence =
    ENTITY_NAME "=" link_mass_properties_keyword
                "(" ENTITY_NAME   "," REAL ","
                    REAL "," REAL "," REAL ","
                    REAL "," REAL "," REAL ","
                    REAL "," REAL "," REAL ")" ";"  .
```

8.9 Actuator

8.9.1 The Adopted Specification for NIRO

```
ENTITY actuator;
  pair  : kinematic_pair;
  motor : OPTIONAL servo_motor;
  gear  : OPTIONAL gear;
UNIQUE
  pair, motor, gear ;
WHERE
  NOT EXISTS (motor);
  NOT EXISTS (gear );
END_ENTITY;
```

8.9.2 Description

In the description of a mechanism in STEP, there is no indication on which kinematic pairs are driven, and which are driving. This entity should also associate the driving pair with a motor and a gear. However, since these entities are not defined, they are specified as non existent. This entity may occur within the scope of the mechanism in which the relevant kinematic pair is defined.

8.9.3 Mapping to the Physical File

```
actuator_occurrence =
ENTITY_NAME "=" actuator_keyword
```

```
"("  ENTITY_NAME  ","
     default_token  ","  default_token  ")"  ";".
```

8.10 Tool_Attachment_Frame

8.10.1 The Adopted Specification for NIRO

```
ENTITY tap_frame;
  tap : frame;
END_ENTITY;
```

8.10.2 Description

The tool attachment point (tap) frame is an entity that tells which one of the additional frames on the last link of a robotic mechanism is to be considered for the attachment of robotic tools. There can be at most one tap-frame in a mechanism. If the tap_frame is referenced in a mount-tool entity, it must be exported from the mechanism via an export-list entity. The tap-frame entity may occur within the scope of the mechanism that contains the kinematic link with the referenced additional frame.

8.10.3 Mapping to the Physical File

```
tap_frame_occurrence =
     ENTITY_NAME "=" tap_frame_keyword
                "(" ENTITY_NAME ")" ";" .
```

   o   The reference must be to an "additional frame".

8.11 Tool

8.11.1 The Adopted Specification for NIRO

```
ENTITY tool;
  unmounted_position       : frame;
  tool_centre_points       : SET  [1:#] OF frame;
  geometric_representation  : link_shape;
  delay_parameters         : LIST [1:#] OF REAL;
WHERE
  WR1 : SIZEOF (delay_parameters) =
        SIZEOF(tool_centre_points);
END_ENTITY;
```

8.11.2 Description

The robotic tool in STEP is represented by a set of tool-centre-point (tcp) frames. These frames express the offset and orientation of the control work points on the tool relative to the tool origin (the absolute coordinate system).

When the tool is mounted on a robot arm, the tool origin is understood to coincide with the tap-frame of the robotic mechanism. When unmounted, the tool origin is located and oriented in space according to the unmounted-position frame.

The tool entity also contains references to solids that model its geometric shape.

A list of real numbers (one per tcp-frame and sequenced correspondingly), represents overall parameters for the tool. Each real number is a delay time in seconds for the corresponding tcp to become active.

The role of such a tool modeling parameter is as follows: Whenever one of the tool centre points of the tool is activated during a simulation, the entire kinematic structure to which the tool is belongs is to remain stationary during a period of time corresponding to the delay time. Examples for such delay times are the time needed to close a gripper, or the bum time of a laser beam.

More elaborate definitions of the tool entity might contain specific tools as sub-types of the generic definition given above. These specific tools (e.g. gripper, cutter, welder, etc.) would carry all the descriptive attributes characterizing them, besides the ones inherited from the generic tool entity considered in this specification.

The tool entity may occur within the scope of the kinematic model or in the DATA section of the STEP file. When a tool is mounted , the current tcp-frame will be referenced by the mount-tool entity. This tcp-frame has to be exported from the scope of the tool via an export-list entity.

8.11.3 Mapping to the Physical File

```
tool_occurrence =
  ENTITY_NAME "=" tool_scope_section
                  tool_keyword
           "(" ENTITY_NAME "," reference_list ","
               link_shape_type "," real_list ")" ";".

tool_scope_section = "&SCOPE"
                        { frame_occurrence      |
                          export_list_occurrence }
                     "ENDSCOPE" .
```

8.12 Mount_Tool

8.12.1 The Adopted Specification for NIRO

```
ENTITY mount_tool;
   tool_to_mount      : tool ;
   current_tcp        : frame;
   attachment_frame : tap_frame;
UNIQUE
   tool_to_mount, current_tcp, attachment_frame;
END_ENTITY;
```

8.12.2 Description

This entity is intended to inform on the STEP file, that a tool is mounted on a robotic mechanism, and indicate what is the currently active tcp of the tool. This is a snapshot at the moment the STEP file is created. The mount-tool entity may occur within the scope of the kinematic model. Since there are references to scoped entities, namely the tap-frame and the active tcp, these should be exported accordingly.

8.12.3 Mapping to the Physical File

```
mount_tool_occurrence =
   ENTITY_NAME "=" mount_tool_keyword
                "(" ENTITY_NAME ","
                    ENTITY_NAME "," ENTITY_NAME ")" ";".
```

9. Keywords for the STEP Physical File

We have adopted the keyword abbreviations provided in STEP when available in the corresponding STEP documents. When not available, full entity type names were used as keywords. Keywords for user defined entities were agreed upon in NIRO project meetings. Following STEP rules, they are preceded by an exclamation mark.

```
axis2_placement_keyword        = "AXSPLZ".
cartesian_point_keyword        = "CRTPNT".
closed_shell_keyword           = "CLSSHL".
direction_keyword              = "DIRCTN".
export_list_keyword            = "EXPORT_LIST".
face_keyword                   = "FACE".
facetted_brep_keyword          = "FCTBRP".
```

```
file_description_keyword            = "FILE_DESCRIPTION".
file_identifier_keyword             = "FILE_IDENTIFIER".
ground_keyword                      = "GROUND".
index_entry_keyword                 = "INDEX_ENTRY".
kinematic_joint_keyword             = "KINEMATIC_JOINT".
kinematic_link_keyword              = "KINEMATIC_LINK".
kinematic_model_keyword             = "KINEMATIC_MODEL".
kinematic_structure_keyword         =
                                      "KINEMATIC_STRUCTURE".
mechanism_keyword                   = "MECHANISM".
pair_placement_structure_keyword =
                              "PAIR_PLACEMENT_STRUCTURE".
poly_loop_keyword                   = "PLYLOP".
prismatic_pair_keyword              = "PRISMATIC_PAIR".
revolute_pair_keyword               = "REVOLUTE_PAIR".
solid_instance_keyword              = "SOLID_INSTANCE".
transformation_keyword              = "TRNSFR".
```

9.1 NIRO Specific Entities.

```
point_direction_pair_keyword        =
                                   "!POINT_DIRECTION_PAIR".
render_face_keyword                 = "!RENDER_FACE".
rgb_colour_keyword                  = "!RGBCLR".
rgb_colour_table_keyword            = "!RGBTBL".
colour_attribute_keyword            = "!CLRATT".
solid_mass_properties_keyword       = "!S_MASS_PR".
link_mass_properties_keyword        = "!L_MASS_PR".
actuator_keyword                    = "!ACTUATOR".
tap_frame_keyword                   = "!TAP_FRM".
tool_keyword                        = "!TOOL".
mount_tool_keyword                  = "!MOUNT".
```

10. An Overview of the STEP Schema Used in NIRO

```
    DATA;

        rgb_colour_table
            rgb_colour*

        facetted_brep
            cartesian_point*
            direction*
            poly_loop
            face
            closed_shell
            point_direction_pair
            render_face

        solid_instance
            transformation
```

```
            cartesian_point*
            direction*

colour_attribute

solid_mass_properties

tool
    axis2_placement
            cartesian_point*
            direction*
    export_list

index_entry
kinematic_model

    facetted_brep
    solid_instance
    colour_attribute
    solid_mass_properties
    tool

    mount_tool

    ground
        axis2_placement
            cartesian_point*
            direction*
        export_list

    mechanism
        kinematic_pair*
        axis2_placement
            cartesian_point*
            direction*
        pair_placement_structure
        kinematic_link
        actuator
        link_mass_properties
        kinematic_structure
        kinematic_joint
        tap_frame
        export_list
        index_entry

        index_entry
    ENDSEC;
```

Entities marked with an asterisk '*' are terminal entities. That is they have no references to other entities. The indentations illustrate the scope structure of the different entities.

A graph overview of the kinematic model schema is presented in Fig. A1-1 as on the next page.

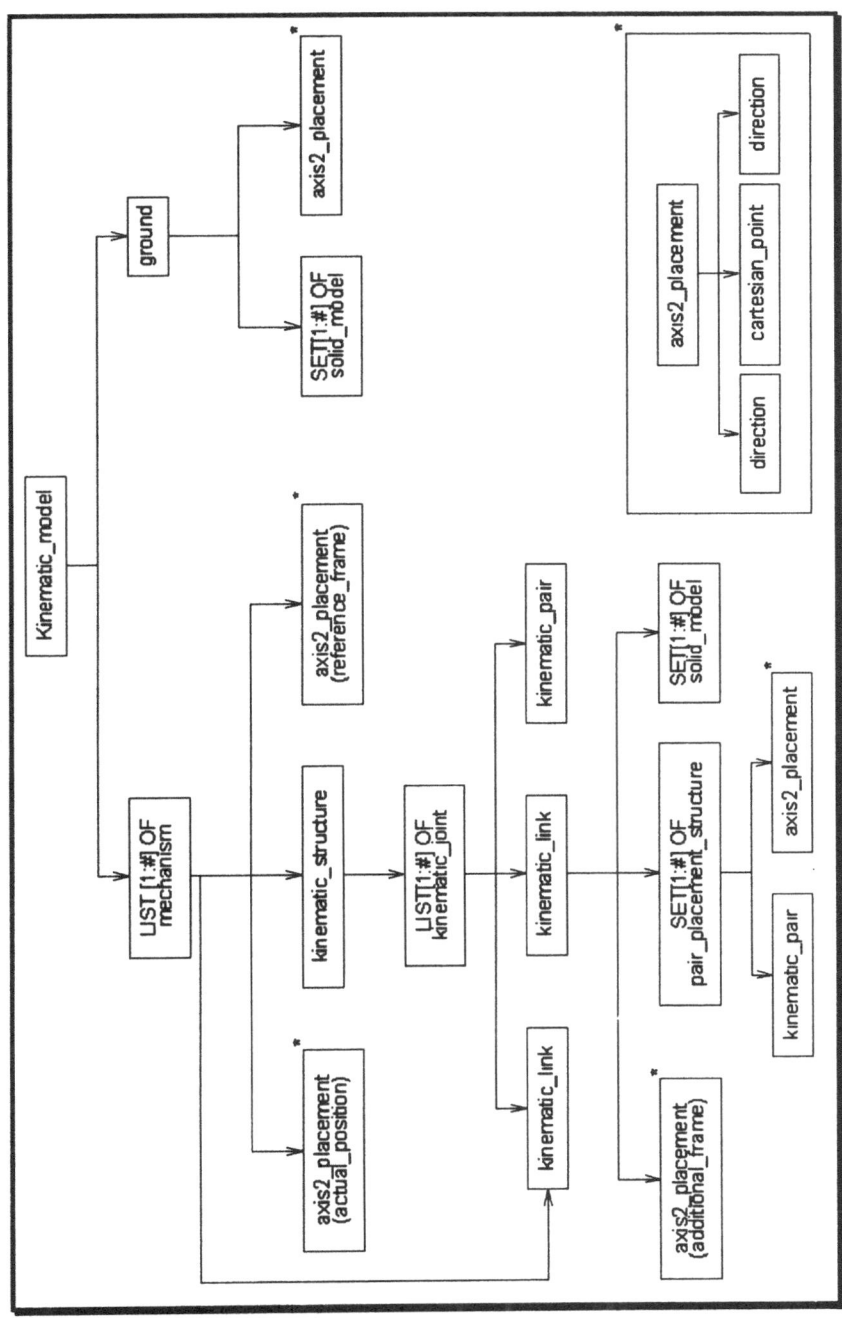

Fig. A1-1: Graph of the kinematic model structure

11. Example of a STEP Physical File (CATIA)

In order to illustrate the realization of the specification, actual STEP files generated from identical models in the CATIA and BRAVO systems are presented in this and the next section respectively. Fig. A1-2 below shows the geometry and kinematic structures.

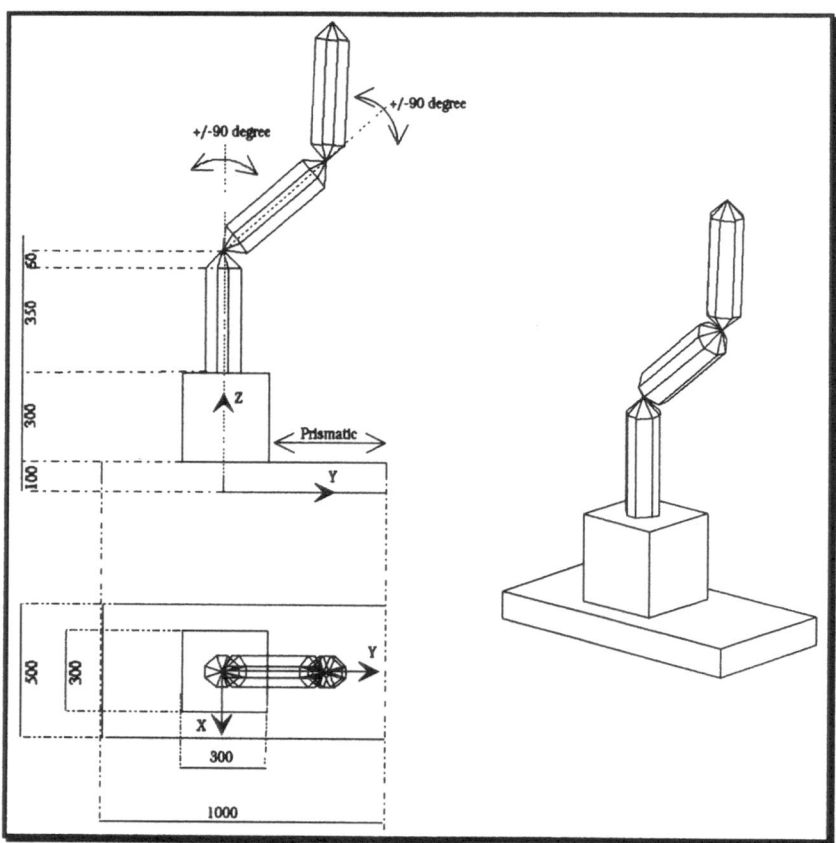

Fig. A1-2: The geometry and kinematics structure of the simple robot. Two corresponding STEP files generated from this model in the CATIA and the BRAVO systems are shown in the current and the next section respectively.

```
STEP;
HEADER;
  FILE_IDENTIFIER('spec20',
                  '1991-12-06T11:18:52',
                  ('U.I. Kroszynski and T. Soerensen'),
                  ('IfS - Control Engineering Institute',
                  'Technical University of Denmark - DTH',
                  'Building 424','DK-2800 Lyngby - DENMARK'),
                  'STEP L02V00','Preprocessor CATNIRO v2.2',
                  'CATIA (AIX) version 2.3');
  FILE_DESCRIPTION(('Example for Specs. December 91'),
                  'Implementation level NIRO spec. L02V00');
ENDSEC;
DATA;
/* polyhedron */
#1=&SCOPE
   #2=CRTPNT(+5.0000E-01,+1.0000E+00,+1.0000E-01);
   #3=CRTPNT(+0.0000E+00,+1.0000E+00,+1.0000E-01);
   #4=CRTPNT(+0.0000E+00,+0.0000E+00,+1.0000E-01);
   #5=CRTPNT(+5.0000E-01,+0.0000E+00,+1.0000E-01);
   #6=CRTPNT(+5.0000E-01,+1.0000E+00,+0.0000E+00);
   #7=CRTPNT(+0.0000E+00,+1.0000E+00,+0.0000E+00);
   #8=CRTPNT(+0.0000E+00,+0.0000E+00,+0.0000E+00);
   #9=CRTPNT(+5.0000E-01,+0.0000E+00,+0.0000E+00);
   #10=PLYLOP((#9,#6,#2,#5));
   #11=PLYLOP((#6,#7,#3,#2));
   #12=PLYLOP((#2,#3,#4,#5));
   #13=PLYLOP((#7,#8,#4,#3));
   #14=PLYLOP((#8,#9,#5,#4));
   #15=PLYLOP((#6,#9,#8,#7));
   #16=FACE($,(#10),$);
   #17=FACE($,(#11),$);
   #18=FACE($,(#12),$);
   #19=FACE($,(#13),$);
   #20=FACE($,(#14),$);
   #21=FACE($,(#15),$);
   #22=CLSSHL((#16,#17,#18,#19,#20,#21));
ENDSCOPE
FCTBRP(#22,$);
/* #23=!S_MASS_PR((#1),+1.0000E+00,+5.0000E-02,         */
/*              +2.5000E-01,+5.0000E-01,+5.0000E-02,    */
/*              +4.2083E-03,+1.0833E-03,+5.2083E-03,    */
/*              +0.0000E+00,+0.0000E+00,+0.0000E+00); */
/* polyhedron */
#24=&SCOPE
   #25=CRTPNT(+3.0000E-01,+3.0000E-01,+3.0000E-01);
   #26=CRTPNT(+0.0000E+00,+3.0000E-01,+3.0000E-01);
   #27=CRTPNT(+0.0000E+00,+0.0000E+00,+3.0000E-01);
   #28=CRTPNT(+3.0000E-01,+0.0000E+00,+3.0000E-01);
   #29=CRTPNT(+3.0000E-01,+3.0000E-01,+0.0000E+00);
   #30=CRTPNT(+0.0000E+00,+3.0000E-01,+0.0000E+00);
   #31=CRTPNT(+0.0000E+00,+0.0000E+00,+0.0000E+00);
   #32=CRTPNT(+3.0000E-01,+0.0000E+00,+0.0000E+00);
   #33=CRTPNT(+2.1000E-01,+1.5000E-01,+6.5000E-01);
   #34=CRTPNT(+1.9243E-01,+1.9243E-01,+6.5000E-01);
   #35=CRTPNT(+1.5000E-01,+2.1000E-01,+6.5000E-01);
   #36=CRTPNT(+1.0757E-01,+1.9243E-01,+6.5000E-01);
   #37=CRTPNT(+9.0000E-02,+1.5000E-01,+6.5000E-01);
   #38=CRTPNT(+1.0757E-01,+1.0757E-01,+6.5000E-01);
   #39=CRTPNT(+1.5000E-01,+9.0000E-02,+6.5000E-01);
   #40=CRTPNT(+1.9243E-01,+1.0757E-01,+6.5000E-01);
   #41=CRTPNT(+1.5000E-01,+1.5000E-01,+7.1000E-01);
```

```
#42=CRTPNT(+2.1000E-01,+1.5000E-01,+3.0000E-01);
#43=CRTPNT(+1.9243E-01,+1.0757E-01,+3.0000E-01);
#44=CRTPNT(+1.5000E-01,+9.0000E-02,+3.0000E-01);
#45=CRTPNT(+1.0757E-01,+1.0757E-01,+3.0000E-01);
#46=CRTPNT(+9.0000E-02,+1.5000E-01,+3.0000E-01);
#47=CRTPNT(+1.0757E-01,+1.9243E-01,+3.0000E-01);
#48=CRTPNT(+1.5000E-01,+2.1000E-01,+3.0000E-01);
#49=CRTPNT(+1.9243E-01,+1.9243E-01,+3.0000E-01);
#50=PLYLOP((#32,#29,#25,#28));
#51=PLYLOP((#29,#30,#26,#25));
#52=PLYLOP((#27,#28,#25,#26));
#53=PLYLOP((#44,#45,#46,#47,#48,#49,#42,#43));
#54=PLYLOP((#30,#31,#27,#26));
#55=PLYLOP((#31,#32,#28,#27));
#56=PLYLOP((#29,#32,#31,#30));
#57=PLYLOP((#34,#41,#33));
#58=PLYLOP((#35,#41,#34));
#59=PLYLOP((#36,#41,#35));
#60=PLYLOP((#37,#41,#36));
#61=PLYLOP((#38,#41,#37));
#62=PLYLOP((#39,#41,#38));
#63=PLYLOP((#40,#41,#39));
#64=PLYLOP((#33,#41,#40));
#65=PLYLOP((#33,#40,#43,#42));
#66=PLYLOP((#40,#39,#44,#43));
#67=PLYLOP((#39,#38,#45,#44));
#68=PLYLOP((#38,#37,#46,#45));
#69=PLYLOP((#37,#36,#47,#46));
#70=PLYLOP((#36,#35,#48,#47));
#71=PLYLOP((#35,#34,#49,#48));
#72=PLYLOP((#34,#33,#42,#49));
#73=FACE($,(#50),$);
#74=FACE($,(#51),$);
#75=FACE($,(#52,#53),$);
#76=FACE($,(#54),$);
#77=FACE($,(#55),$);
#78=FACE($,(#56),$);
#79=FACE($,(#57),$);
#80=FACE($,(#58),$);
#81=FACE($,(#59),$);
#82=FACE($,(#60),$);
#83=FACE($,(#61),$);
#84=FACE($,(#62),$);
#85=FACE($,(#63),$);
#86=FACE($,(#64),$);
#87=FACE($,(#65),$);
#88=FACE($,(#66),$);
#89=FACE($,(#67),$);
#90=FACE($,(#68),$);
#91=FACE($,(#69),$);
#92=FACE($,(#70),$);
#93=FACE($,(#71),$);
#94=FACE($,(#72),$);

#95=CLSSHL((#73,#74,#75,#76,#77,#78,#79,#80,#81,#82,#83,#84,#85,#86,
           #87,#88,#89,#90,#91,#92,#93,#94));
ENDSCOPE
FCTBRP(#95,$);
 * #96=!S_MASS_PR((#24),+1.0000E+00,+3.0767E-02,      */
/*              +1.5000E-01,+1.5000E-01,+1.9105E-01,  */
/*              +8.2299E-04,+8.2299E-04,+4.1099E-04,  */
/*              +0.0000E+00,+0.0000E+00,+0.0000E+00); */
```

```
/* polyhedron */
#97=&SCOPE
    #98=CRTPNT(+6.0000E-02,+0.0000E+00,+3.5000E-01);
    #99=CRTPNT(+4.2426E-02,+4.2426E-02,+3.5000E-01);
    #100=CRTPNT(+0.0000E+00,+6.0000E-02,+3.5000E-01);
    #101=CRTPNT(-4.2426E-02,+4.2426E-02,+3.5000E-01);
    #102=CRTPNT(-6.0000E-02,+0.0000E+00,+3.5000E-01);
    #103=CRTPNT(-4.2426E-02,-4.2426E-02,+3.5000E-01);
    #104=CRTPNT(+0.0000E+00,-6.0000E-02,+3.5000E-01);
    #105=CRTPNT(+4.2426E-02,-4.2426E-02,+3.5000E-01);
    #106=CRTPNT(+0.0000E+00,+0.0000E+00,+4.1000E-01);
    #107=CRTPNT(+6.0000E-02,+0.0000E+00,+0.0000E+00);
    #108=CRTPNT(+4.2426E-02,-4.2426E-02,+0.0000E+00);
    #109=CRTPNT(+0.0000E+00,-6.0000E-02,+0.0000E+00);
    #110=CRTPNT(-4.2426E-02,-4.2426E-02,+0.0000E+00);
    #111=CRTPNT(-6.0000E-02,+0.0000E+00,+0.0000E+00);
    #112=CRTPNT(-4.2426E-02,+4.2426E-02,+0.0000E+00);
    #113=CRTPNT(+0.0000E+00,+6.0000E-02,+0.0000E+00);
    #114=CRTPNT(+4.2426E-02,+4.2426E-02,+0.0000E+00);
    #115=CRTPNT(+0.0000E+00,+0.0000E+00,-6.0000E-02);
    #116=PLYLOP((#99,#106,#98));
    #117=PLYLOP((#100,#106,#99));
    #118=PLYLOP((#101,#106,#100));
    #119=PLYLOP((#102,#106,#101));
    #120=PLYLOP((#103,#106,#102));
    #121=PLYLOP((#104,#106,#103));
    #122=PLYLOP((#105,#106,#104));
    #123=PLYLOP((#98,#106,#105));
    #124=PLYLOP((#98,#105,#108,#107));
    #125=PLYLOP((#105,#104,#109,#108));
    #126=PLYLOP((#104,#103,#110,#109));
    #127=PLYLOP((#103,#102,#111,#110));
    #128=PLYLOP((#102,#101,#112,#111));
    #129=PLYLOP((#101,#100,#113,#112));
    #130=PLYLOP((#100,#99,#114,#113));
    #131=PLYLOP((#99,#98,#107,#114));
    #132=PLYLOP((#108,#115,#107));
    #133=PLYLOP((#109,#115,#108));
    #134=PLYLOP((#110,#115,#109));
    #135=PLYLOP((#111,#115,#110));
    #136=PLYLOP((#112,#115,#111));
    #137=PLYLOP((#113,#115,#112));
    #138=PLYLOP((#114,#115,#113));
    #139=PLYLOP((#107,#115,#114));
    #140=FACE($,(#116),$);
    #141=FACE($,(#117),$);
    #142=FACE($,(#118),$);
    #143=FACE($,(#119),$);
    #144=FACE($,(#120),$);
    #145=FACE($,(#121),$);
    #146=FACE($,(#122),$);
    #147=FACE($,(#123),$);
    #148=FACE($,(#124),$);
    #149=FACE($,(#125),$);
    #150=FACE($,(#126),$);
    #151=FACE($,(#127),$);
    #152=FACE($,(#128),$);
    #153=FACE($,(#129),$);
    #154=FACE($,(#130),$);
    #155=FACE($,(#131),$);
    #156=FACE($,(#132),$);
    #157=FACE($,(#133),$);
```

```
    #158=FACE($,(#134),$);
    #159=FACE($,(#135),$);
    #160=FACE($,(#136),$);
    #161=FACE($,(#137),$);
    #162=FACE($,(#138),$);
    #163=FACE($,(#139),$);

#164=CLSSHL((#140,#141,#142,#143,#144,#145,#146,#147,#148,#149,#150,

#151,#152,#153,#154,#155,#156,#157,#158,#159,#160,#161,
                #162,#163));
ENDSCOPE
FCTBRP(#164,$);
/* #165=!S_MASS_PR((#97),+1.0000E+00,+3.9711E-03,          */
/*                 +0.0000E+00,+0.0000E+00,+1.7500E-01,    */
/*                 +5.4232E-05,+5.4232E-05,+6.1855E-06,    */
/*                 +0.0000E+00,+0.0000E+00,+0.0000E+00); */
/* solid instance */
#166=&SCOPE
    #167=&SCOPE
       #168=DIRCTN(+1.0000E+00,+0.0000E+00,+0.0000E+00);
    /* #169=DIRCTN(+0.0000E+00,+1.0000E+00,+0.0000E+00); */
       #170=DIRCTN(+0.0000E+00,+0.0000E+00,+1.0000E+00);
       #171=CRTPNT(-2.5000E-01,-5.0000E-01,+0.0000E+00);
    ENDSCOPE
    TRNSFR(#168,$,#170,#171,$);
ENDSCOPE
SOLID_INSTANCE(#1,#167);
/* solid instance */
#172=&SCOPE
    #173=&SCOPE
       #174=DIRCTN(+1.0000E+00,+0.0000E+00,+0.0000E+00);
    /* #175=DIRCTN(+0.0000E+00,+1.0000E+00,+0.0000E+00); */
       #176=DIRCTN(+0.0000E+00,+0.0000E+00,+1.0000E+00);
       #177=CRTPNT(-1.5000E-01,-2.5000E-01,+1.0000E-01);
    ENDSCOPE
    TRNSFR(#174,$,#176,#177,$);
ENDSCOPE
SOLID_INSTANCE(#24,#173);
/* solid instance */
#178=&SCOPE
    #179=&SCOPE
       #180=DIRCTN(+1.0000E+00,+0.0000E+00,+0.0000E+00);
    /* #181=DIRCTN(+0.0000E+00,+7.0711E-01,-7.0711E-01); */
       #182=DIRCTN(+0.0000E+00,+7.0711E-01,+7.0711E-01);
       #183=CRTPNT(+0.0000E+00,-5.7574E-02,+8.5243E-01);
    ENDSCOPE
    TRNSFR(#180,$,#182,#183,$);
ENDSCOPE
SOLID_INSTANCE(#97,#179);
/* solid instance */
#184=&SCOPE
    #185=&SCOPE
       #186=DIRCTN(+1.0000E+00,+0.0000E+00,+0.0000E+00);
    /* #187=DIRCTN(+0.0000E+00,+1.0000E+00,+0.0000E+00); */
       #188=DIRCTN(+0.0000E+00,+0.0000E+00,+1.0000E+00);
       #189=CRTPNT(+0.0000E+00,+2.3234E-01,+1.2023E+00);
    ENDSCOPE
    TRNSFR(#186,$,#188,#189,$);
ENDSCOPE
SOLID_INSTANCE(#97,#185);
#190=!CLRATT(#166, 2);
```

```
#191=!CLRATT(#172, 3);
#192=!CLRATT(#178, 4);
#193=!CLRATT(#184, 5);
/* kinematic model */
#194=&SCOPE
   /* mechanism */
   #195=&SCOPE
      /* actual location (base) frame */
      #196=&SCOPE
         #197=DIRCTN(+1.0000E+00,+0.0000E+00,+0.0000E+00);
      /* #198=DIRCTN(+0.0000E+00,+1.0000E+00,+0.0000E+00); */
         #199=DIRCTN(+0.0000E+00,+0.0000E+00,+1.0000E+00);
         #200=CRTPNT(+0.0000E+00,+0.0000E+00,+0.0000E+00);
      ENDSCOPE
      AXSPLZ(#200,#199,#197);
      #201=PRISMATIC_PAIR(-1.0000E-01,-3.5000E-01,+3.5000E-01);
      #202=&SCOPE
         #203=DIRCTN(+0.0000E+00,+0.0000E+00,+1.0000E+00);
      /* #204=DIRCTN(+1.0000E+00,+0.0000E+00,+0.0000E+00); */
         #205=DIRCTN(+0.0000E+00,+1.0000E+00,+0.0000E+00);
         #206=CRTPNT(+0.0000E+00,+0.0000E+00,+1.0000E-01);
      ENDSCOPE
      AXSPLZ(#206,#205,#203);
      #207=PAIR_PLACEMENT_STRUCTURE(#201,#202);
      #208=&SCOPE
         #209=DIRCTN(+0.0000E+00,+0.0000E+00,+1.0000E+00);
      /* #210=DIRCTN(+1.0000E+00,+0.0000E+00,+0.0000E+00); */
         #211=DIRCTN(+0.0000E+00,+1.0000E+00,+0.0000E+00);
         #212=CRTPNT(+0.0000E+00,-1.0000E-01,+1.0000E-01);
      ENDSCOPE
      AXSPLZ(#212,#211,#209);
      #213=PAIR_PLACEMENT_STRUCTURE(#201,#208);
      #214=REVOLUTE_PAIR(+7.8540E-01,-1.5708E+00,+1.5708E+00);
      #215=&SCOPE
         #216=DIRCTN(+0.0000E+00,-1.0000E+00,+0.0000E+00);
      /* #217=DIRCTN(+0.0000E+00,+0.0000E+00,+1.0000E+00); */
         #218=DIRCTN(-1.0000E+00,+0.0000E+00,+0.0000E+00);
         #219=CRTPNT(+0.0000E+00,-1.0000E-01,+8.1000E-01);
      ENDSCOPE
      AXSPLZ(#219,#218,#216);
      #220=PAIR_PLACEMENT_STRUCTURE(#214,#215);
      #221=&SCOPE
         #222=DIRCTN(+0.0000E+00,-7.0711E-01,+7.0711E-01);
      /* #223=DIRCTN(+0.0000E+00,+7.0711E-01,+7.0711E-01); */
         #224=DIRCTN(-1.0000E+00,+0.0000E+00,+0.0000E+00);
         #225=CRTPNT(+0.0000E+00,-1.0000E-01,+8.1000E-01);
      ENDSCOPE
      AXSPLZ(#225,#224,#222);
      #226=PAIR_PLACEMENT_STRUCTURE(#214,#221);
      #227=REVOLUTE_PAIR(-7.8540E-01,-1.5708E+00,+1.5708E+00);
      #228=&SCOPE
         #229=DIRCTN(+0.0000E+00,-7.0711E-01,+7.0711E-01);
      /* #230=DIRCTN(+0.0000E+00,+7.0711E-01,+7.0711E-01); */
         #231=DIRCTN(-1.0000E+00,+0.0000E+00,+0.0000E+00);
         #232=CRTPNT(+0.0000E+00,+2.3234E-01,+1.1423E+00);
      ENDSCOPE
      AXSPLZ(#232,#231,#229);
      #233=PAIR_PLACEMENT_STRUCTURE(#227,#228);
      #234=&SCOPE
         #235=DIRCTN(+0.0000E+00,-1.0000E+00,+0.0000E+00);
      /* #236=DIRCTN(+0.0000E+00,+0.0000E+00,+1.0000E+00); */
         #237=DIRCTN(-1.0000E+00,+0.0000E+00,+0.0000E+00);
```

```
            #238=CRTPNT(+0.0000E+00,+2.3234E-01,+1.1423E+00);
         ENDSCOPE
         AXSPLZ(#238,#237,#235);
         #239=PAIR_PLACEMENT_STRUCTURE(#227,#234);
         #240=KINEMATIC_LINK((#207),$,(#166));
         #241=KINEMATIC_LINK((#213,#220),$,(#172));
         #242=KINEMATIC_LINK((#226,#233),$,(#178));
         #243=KINEMATIC_LINK((#239),$,(#184));
         #244=KINEMATIC_JOINT(#240,#241,#201);
         #245=KINEMATIC_JOINT(#241,#242,#214);
         #246=KINEMATIC_JOINT(#242,#243,#227);
         #247=KINEMATIC_STRUCTURE((#244,#245,#246),$);
      ENDSCOPE
      MECHANISM(#247,#240,#196,#249);
      /* the ground */
      #248=&SCOPE
         #249=&SCOPE
            #250=DIRCTN(+1.0000E+00,+0.0000E+00,+0.0000E+00);
         /* #251=DIRCTN(+0.0000E+00,+1.0000E+00,+0.0000E+00); */
            #252=DIRCTN(+0.0000E+00,+0.0000E+00,+1.0000E+00);
            #253=CRTPNT(+0.0000E+00,+0.0000E+00,+0.0000E+00);
         ENDSCOPE
         AXSPLZ(#253,#252,#250);
         #254=EXPORT_LIST((#249));
      ENDSCOPE
      GROUND(#249,$);
   ENDSCOPE
   KINEMATIC_MODEL(#248,(#195),$,$);
ENDSEC;
ENDSTEP;
```

12. Example of a STEP Physical File (BRAVO)

```
STEP;
HEADER;
FILE_IDENTIFIER('niro_env.stp','1991-12-05T19:14:51',
('Arnold Ludwig'),
('KfK/IRE'),
'STEP L02V00','ROB_KIS/STEP Version 1','BRAVO3/ROBOT Version 2.1');
FILE_DESCRIPTION(('NIRO_kinematics_test_model_from_Bravo3/ROBOT'),
'BREP_LEVEL 1.0');
ENDSEC;
DATA;

#100 =     /* --- FACETTED_BREP: ENTITY_NAME --- */
   &SCOPE
   #101 = CRTPNT( -1.500000E-01, -3.000000E-01,  1.500000E-01);
   #102 = CRTPNT( -1.500000E-01, -3.000000E-01, -1.500000E-01);
   #103 = CRTPNT(  1.500000E-01, -3.000000E-01, -1.500000E-01);
   #104 = CRTPNT(  1.500000E-01, -3.000000E-01,  1.500000E-01);
   #105 = CRTPNT(  1.500000E-01,  0.000000E+00,  1.500000E-01);
   #106 = CRTPNT( -1.500000E-01,  0.000000E+00,  1.500000E-01);
   #107 = CRTPNT(  1.500000E-01,  0.000000E+00, -1.500000E-01);
   #108 = CRTPNT( -1.500000E-01,  0.000000E+00, -1.500000E-01);
   #109 = CRTPNT( -6.000000E-02,  0.000000E+00,  0.000000E+00);
   #110 = CRTPNT( -4.242640E-02,  0.000000E+00, -4.242640E-02);
   #111 = CRTPNT(  0.000000E+00,  0.000000E+00, -6.000000E-02);
   #112 = CRTPNT(  4.242640E-02,  0.000000E+00, -4.242640E-02);
   #113 = CRTPNT(  6.000000E-02,  0.000000E+00,  0.000000E+00);
```

```
#114 = CRTPNT(  4.242640E-02,  0.000000E+00,  4.242640E-02);
#115 = CRTPNT(  0.000000E+00,  0.000000E+00,  6.000000E-02);
#116 = CRTPNT( -4.242640E-02,  0.000000E+00,  4.242640E-02);
#117 = CRTPNT( -4.242640E-02,  3.500000E-01,  4.242640E-02);
#118 = CRTPNT( -6.000000E-02,  3.500000E-01,  0.000000E+00);
#119 = CRTPNT( -3.398913E-19,  3.500000E-01,  6.000000E-02);
#120 = CRTPNT(  4.242640E-02,  3.500000E-01,  4.242640E-02);
#121 = CRTPNT(  6.000000E-02,  3.500000E-01,  3.398913E-19);
#122 = CRTPNT(  4.242640E-02,  3.500000E-01, -4.242640E-02);
#123 = CRTPNT(  3.398913E-19,  3.500000E-01, -6.000000E-02);
#124 = CRTPNT( -4.242640E-02,  3.500000E-01, -4.242640E-02);
#125 = CRTPNT(  0.000000E+00,  4.100000E-01,  0.000000E+00);

#126 = PLYLOP((#101, #102, #103, #104));
#127 = FACE($,(#126),$);
#128 = !RENDER_FACE(#127, 7, ());

#129 = PLYLOP((#101, #104, #105, #106));
#130 = FACE($,(#129),$);
#131 = !RENDER_FACE(#130, 7, ());

#132 = PLYLOP((#104, #103, #107, #105));
#133 = FACE($,(#132),$);
#134 = !RENDER_FACE(#133, 7, ());

#135 = PLYLOP((#103, #102, #108, #107));
#136 = FACE($,(#135),$);
#137 = !RENDER_FACE(#136, 7, ());

#138 = PLYLOP((#102, #101, #106, #108));
#139 = FACE($,(#138),$);
#140 = !RENDER_FACE(#139, 7, ());

#141 = PLYLOP((#106, #105, #107, #108));
#142 = PLYLOP((#109, #110, #111, #112, #113, #114, #115, #116));
#143 = FACE($,(#141, #142),$);
#144 = !RENDER_FACE(#143, 7, ());

#145 = PLYLOP((#109, #116, #117, #118));
#146 = FACE($,(#145),$);
#147 = !RENDER_FACE(#146, 7, ());

#148 = PLYLOP((#117, #116, #115, #119));
#149 = FACE($,(#148),$);
#150 = !RENDER_FACE(#149, 7, ());

#151 = PLYLOP((#119, #115, #114, #120));
#152 = FACE($,(#151),$);
#153 = !RENDER_FACE(#152, 7, ());

#154 = PLYLOP((#120, #114, #113, #121));
#155 = FACE($,(#154),$);
#156 = !RENDER_FACE(#155, 7, ());

#157 = PLYLOP((#121, #113, #112, #122));
#158 = FACE($,(#157),$);
#159 = !RENDER_FACE(#158, 7, ());

#160 = PLYLOP((#122, #112, #111, #123));
#161 = FACE($,(#160),$);
#162 = !RENDER_FACE(#161, 7, ());
```

```
    #163 = PLYLOP((#123, #111, #110, #124));
    #164 = FACE($,(#163),$);
    #165 = !RENDER_FACE(#164, 7, ());

    #166 = PLYLOP((#124, #110, #109, #118));
    #167 = FACE($,(#166),$);
    #168 = !RENDER_FACE(#167, 7, ());

    #169 = PLYLOP((#118, #117, #125));
    #170 = FACE($,(#169),$);
    #171 = !RENDER_FACE(#170, 7, ());

    #172 = PLYLOP((#125, #117, #119));
    #173 = FACE($,(#172),$);
    #174 = !RENDER_FACE(#173, 7, ());

    #175 = PLYLOP((#125, #119, #120));
    #176 = FACE($,(#175),$);
    #177 = !RENDER_FACE(#176, 7, ());

    #178 = PLYLOP((#125, #120, #121));
    #179 = FACE($,(#178),$);
    #180 = !RENDER_FACE(#179, 7, ());

    #181 = PLYLOP((#125, #121, #122));
    #182 = FACE($,(#181),$);
    #183 = !RENDER_FACE(#182, 7, ());

    #184 = PLYLOP((#125, #122, #123));
    #185 = FACE($,(#184),$);
    #186 = !RENDER_FACE(#185, 7, ());

    #187 = PLYLOP((#125, #123, #124));
    #188 = FACE($,(#187),$);
    #189 = !RENDER_FACE(#188, 7, ());

    #190 = PLYLOP((#125, #124, #118));
    #191 = FACE($,(#190),$);
    #192 = !RENDER_FACE(#191, 7, ());

    #193 = CLSSHL((
            #127, #130, #133, #136, #139, #143, #146, #149, #152, #155,
            #158, #161, #164, #167, #170, #173, #176, #179, #182, #185,
            #188, #191));
    ENDSCOPE

    FCTBRP(#193, $);
    #194 = INDEX_ENTRY('sled', #100);

#195 =    /* --- SOLID_INSTANCE: ENTITY_NAME --- */
    &SCOPE

#196 =    /* --- TRANSFORMATION: ENTITY_NAME --- */
    &SCOPE
    #197 = CRTPNT( -4.100000E-01,  0.000000E+00,  0.000000E+00);
    #198 = DIRCTN(  0.000000E+00,  0.000000E+00,  1.000000E+00);
    #199 = DIRCTN(  0.000000E+00, -1.000000E+00,  0.000000E+00);
    ENDSCOPE
    TRNSFR(#199, $, #198, #197, $);

    ENDSCOPE
    SOLID_INSTANCE(#100, #196);
```

```
#200 =      /* --- FACETTED_BREP: ENTITY_NAME --- */
    &SCOPE
    #201 = CRTPNT(  6.000000E-02,  3.500000E-01,  0.000000E+00);
    #202 = CRTPNT(  6.000000E-02,  0.000000E+00,  0.000000E+00);
    #203 = CRTPNT(  4.242640E-02,  0.000000E+00, -4.242640E-02);
    #204 = CRTPNT(  4.242640E-02,  3.500000E-01, -4.242640E-02);
    #205 = CRTPNT(  0.000000E+00,  0.000000E+00, -6.000000E-02);
    #206 = CRTPNT(  3.398913E-19,  3.500000E-01, -6.000000E-02);
    #207 = CRTPNT( -4.242640E-02,  0.000000E+00, -4.242640E-02);
    #208 = CRTPNT( -4.242640E-02,  3.500000E-01, -4.242640E-02);
    #209 = CRTPNT( -6.000000E-02,  0.000000E+00,  0.000000E+00);
    #210 = CRTPNT( -6.000000E-02,  3.500000E-01, -3.398913E-19);
    #211 = CRTPNT( -4.242640E-02,  0.000000E+00,  4.242640E-02);
    #212 = CRTPNT( -4.242640E-02,  3.500000E-01,  4.242640E-02);
    #213 = CRTPNT(  0.000000E+00,  0.000000E+00,  6.000000E-02);
    #214 = CRTPNT( -3.398913E-19,  3.500000E-01,  6.000000E-02);
    #215 = CRTPNT(  4.242640E-02,  0.000000E+00,  4.242640E-02);
    #216 = CRTPNT(  4.242640E-02,  3.500000E-01,  4.242640E-02);
    #217 = CRTPNT(  0.000000E+00,  4.100000E-01,  0.000000E+00);
    #218 = CRTPNT(  0.000000E+00, -6.000000E-02,  0.000000E+00);

    #219 = PLYLOP((#201, #202, #203, #204));
    #220 = FACE($,(#219),$);
    #221 = !RENDER_FACE(#220, 5, ());

    #222 = PLYLOP((#204, #203, #205, #206));
    #223 = FACE($,(#222),$);
    #224 = !RENDER_FACE(#223, 5, ());

    #225 = PLYLOP((#206, #205, #207, #208));
    #226 = FACE($,(#225),$);
    #227 = !RENDER_FACE(#226, 5, ());

    #228 = PLYLOP((#208, #207, #209, #210));
    #229 = FACE($,(#228),$);
    #230 = !RENDER_FACE(#229, 5, ());

    #231 = PLYLOP((#210, #209, #211, #212));
    #232 = FACE($,(#231),$);
    #233 = !RENDER_FACE(#232, 5, ());

    #234 = PLYLOP((#212, #211, #213, #214));
    #235 = FACE($,(#234),$);
    #236 = !RENDER_FACE(#235, 5, ());

    #237 = PLYLOP((#214, #213, #215, #216));
    #238 = FACE($,(#237),$);
    #239 = !RENDER_FACE(#238, 5, ());

    #240 = PLYLOP((#216, #215, #202, #201));
    #241 = FACE($,(#240),$);
    #242 = !RENDER_FACE(#241, 5, ());

    #243 = PLYLOP((#204, #206, #217));
    #244 = FACE($,(#243),$);
    #245 = !RENDER_FACE(#244, 5, ());

    #246 = PLYLOP((#217, #206, #208));
    #247 = FACE($,(#246),$);
    #248 = !RENDER_FACE(#247, 5, ());
```

```
#249 = PLYLOP((#217, #208, #210));
#250 = FACE($,(#249),$);
#251 = !RENDER_FACE(#250, 5, ());

#252 = PLYLOP((#217, #210, #212));
#253 = FACE($,(#252),$);
#254 = !RENDER_FACE(#253, 5, ());

#255 = PLYLOP((#217, #212, #214));
#256 = FACE($,(#255),$);
#257 = !RENDER_FACE(#256, 5, ());

#258 = PLYLOP((#217, #214, #216));
#259 = FACE($,(#258),$);
#260 = !RENDER_FACE(#259, 5, ());

#261 = PLYLOP((#217, #216, #201));
#262 = FACE($,(#261),$);
#263 = !RENDER_FACE(#262, 5, ());

#264 = PLYLOP((#217, #201, #204));
#265 = FACE($,(#264),$);
#266 = !RENDER_FACE(#265, 5, ());

#267 = PLYLOP((#203, #202, #218));
#268 = FACE($,(#267),$);
#269 = !RENDER_FACE(#268, 5, ());

#270 = PLYLOP((#218, #202, #215));
#271 = FACE($,(#270),$);
#272 = !RENDER_FACE(#271, 5, ());

#273 = PLYLOP((#218, #215, #213));
#274 = FACE($,(#273),$);
#275 = !RENDER_FACE(#274, 5, ());

#276 = PLYLOP((#218, #213, #211));
#277 = FACE($,(#276),$);
#278 = !RENDER_FACE(#277, 5, ());

#279 = PLYLOP((#218, #211, #209));
#280 = FACE($,(#279),$);
#281 = !RENDER_FACE(#280, 5, ());

#282 = PLYLOP((#218, #209, #207));
#283 = FACE($,(#282),$);
#284 = !RENDER_FACE(#283, 5, ());

#285 = PLYLOP((#218, #207, #205));
#286 = FACE($,(#285),$);
#287 = !RENDER_FACE(#286, 5, ());

#288 = PLYLOP((#218, #205, #203));
#289 = FACE($,(#288),$);
#290 = !RENDER_FACE(#289, 5, ());

#291 = CLSSHL((
       #220, #223, #226, #229, #232, #235, #238, #241, #244, #247,
       #250, #253, #256, #259, #262, #265, #268, #271, #274, #277,
       #280, #283, #286, #289));
ENDSCOPE
```

```
   FCTBRP(#291, $);
   #292 = INDEX_ENTRY('armshape', #200);

#293 =    /* --- SOLID_INSTANCE: ENTITY_NAME --- */
   &SCOPE

#294 =    /* --- TRANSFORMATION: ENTITY_NAME --- */
   &SCOPE
   #295 = CRTPNT( -4.100000E-01,  0.000000E+00,  0.000000E+00);
   #296 = DIRCTN(  0.000000E+00,  0.000000E+00,  1.000000E+00);
   #297 = DIRCTN(  0.000000E+00, -1.000000E+00,  0.000000E+00);
   ENDSCOPE
   TRNSFR(#297, $, #296, #295, $);

   ENDSCOPE
   SOLID_INSTANCE(#200, #294);

#298 =    /* --- SOLID_INSTANCE: ENTITY_NAME --- */
   &SCOPE

#299 =    /* --- TRANSFORMATION: ENTITY_NAME --- */
   &SCOPE
   #300 = CRTPNT(  0.000000E+00,  0.000000E+00,  6.000000E-02);
   #301 = DIRCTN(  0.000000E+00,  1.000000E+00,  0.000000E+00);
   #302 = DIRCTN( -1.000000E+00,  0.000000E+00,  0.000000E+00);
   ENDSCOPE
   TRNSFR(#302, $, #301, #300, $);

   ENDSCOPE
   SOLID_INSTANCE(#200, #299);


#303 =    /* --- FACETTED_BREP: ENTITY_NAME --- */
   &SCOPE
   #304 = CRTPNT(  0.000000E+00,  0.000000E+00,  0.000000E+00);
   #305 = CRTPNT(  0.000000E+00,  0.000000E+00, -5.000000E-01);
   #306 = CRTPNT(  1.000000E+00,  0.000000E+00, -5.000000E-01);
   #307 = CRTPNT(  1.000000E+00,  0.000000E+00,  0.000000E+00);
   #308 = CRTPNT(  0.000000E+00,  1.000000E-01,  0.000000E+00);
   #309 = CRTPNT(  0.000000E+00,  1.000000E-01, -5.000000E-01);
   #310 = CRTPNT(  1.000000E+00,  1.000000E-01, -5.000000E-01);
   #311 = CRTPNT(  1.000000E+00,  1.000000E-01,  0.000000E+00);

   #312 = PLYLOP((#304, #305, #306, #307));
   #313 = FACE($,(#312),$);
   #314 = !RENDER_FACE(#313, 3, ());

   #315 = PLYLOP((#305, #304, #308, #309));
   #316 = FACE($,(#315),$);
   #317 = !RENDER_FACE(#316, 3, ());

   #318 = PLYLOP((#306, #305, #309, #310));
   #319 = FACE($,(#318),$);
   #320 = !RENDER_FACE(#319, 3, ());

   #321 = PLYLOP((#307, #306, #310, #311));
   #322 = FACE($,(#321),$);
   #323 = !RENDER_FACE(#322, 3, ());

   #324 = PLYLOP((#308, #304, #307, #311));
   #325 = FACE($,(#324),$);
```

```
   #326 = !RENDER_FACE(#325, 3, ());

   #327 = PLYLOP((#309, #308, #311, #310));
   #328 = FACE($,(#327),$);
   #329 = !RENDER_FACE(#328, 3, ());

   #330 = CLSSHL((
          #313, #316, #319, #322, #325, #328));
   ENDSCOPE

   FCTBRP(#330, $);
   #331 = INDEX_ENTRY('ground', #303);

#332 =    /* --- SOLID_INSTANCE: ENTITY_NAME --- */
   &SCOPE

#333 =    /* --- TRANSFORMATION: ENTITY_NAME --- */
   &SCOPE
   #334 = CRTPNT(  0.000000E+00,   0.000000E+00,   0.000000E+00);
   #335 = DIRCTN(  0.000000E+00,  -1.000000E+00,   0.000000E+00);
   #336 = DIRCTN(  1.000000E+00,   0.000000E+00,   0.000000E+00);
   ENDSCOPE
   TRNSFR(#336, $, #335, #334, $);

   ENDSCOPE
   SOLID_INSTANCE(#303, #333);

#337 =    /* --- KINEMATIC_MODEL: ENTITY_NAME --- */
   &SCOPE

#338 =    /* --- GROUND: ENTITY_NAME --- */
   &SCOPE

#339 =    /* --- GROUND_FRAME: ENTITY_NAME --- */
   &SCOPE
   #340 = CRTPNT(  0.000000E+00,   0.000000E+00,   0.000000E+00);
   #341 = DIRCTN(  0.000000E+00,   0.000000E+00,   1.000000E+00);
   #342 = DIRCTN(  1.000000E+00,   0.000000E+00,   0.000000E+00);
   ENDSCOPE
   AXSPLZ(#340, #341, #342);
#343=EXPORT_LIST((#339));
   ENDSCOPE
   GROUND(#339, (#332) );

          /*         Begin niro_env.mpc      */
          /*  frame_id   : NIRO_KINBASE$1   */
          /*         Begin niro_mec.mpc      */
#344 =    /* --- MECHANISM: ENTITY_NAME ---  */
   &SCOPE

#345 =    /* --- Einheitsmatrix: ENTITY_NAME --- */
   &SCOPE
   #346 = CRTPNT(  0.000000E+00,   0.000000E+00,   0.000000E+00);
   #347 = DIRCTN(  0.000000E+00,   0.000000E+00,   1.000000E+00);
   #348 = DIRCTN(  1.000000E+00,   0.000000E+00,   0.000000E+00);
   ENDSCOPE
   AXSPLZ(#346, #347, #348);

          /*  frame_id   : KIN_BASE$2  */

#349 =    /* --- BASE_FRAME --- */
   &SCOPE
```

```
    #350 = CRTPNT(  0.000000E+00,  2.500000E-01,  1.000000E-01);
    #351 = DIRCTN(  1.000000E+00,  0.000000E+00,  0.000000E+00);
    #352 = DIRCTN(  0.000000E+00,  1.000000E+00,  0.000000E+00);
    ENDSCOPE
    AXSPLZ(#350, #351, #352);

            /*  frame_id  : SLED$2  */

#353 =     /* --- FRAME 1 --- */
    &SCOPE
    #354 = CRTPNT( -7.100000E-01,  4.000000E-01,  0.000000E+00);
    #355 = DIRCTN(  0.000000E+00, -1.000000E+00,  0.000000E+00);
    #356 = DIRCTN(  0.000000E+00,  0.000000E+00, -1.000000E+00);
    ENDSCOPE
    AXSPLZ(#354, #355, #356);

    #357 = PRISMATIC_PAIR( 0.000000E+00, -4.000000E-01, 6.000000E-01);

            /*  frame_id  : ARM_1$2H  */

#358 =     /* --- FRAME 2 --- */
    &SCOPE
    #359 = CRTPNT(  0.000000E+00,  0.000000E+00,  0.000000E+00);
    #360 = DIRCTN(  0.000000E+00,  0.000000E+00, -1.000000E+00);
    #361 = DIRCTN(  7.071068E-01, -7.071068E-01,  0.000000E+00);
    ENDSCOPE
    AXSPLZ(#359, #360, #361);

            /*  frame_id  : ARM_1$2  */

#362 =     /* --- FRAME 3 --- */
    &SCOPE
    #363 = CRTPNT( -4.700000E-01,  0.000000E+00,  0.000000E+00);
    #364 = DIRCTN(  0.000000E+00,  0.000000E+00,  1.000000E+00);
    #365 = DIRCTN(  1.000000E+00,  0.000000E+00,  0.000000E+00);
    ENDSCOPE
    AXSPLZ(#363, #364, #365);

    #366 = REVOLUTE_PAIR(  0.000000E+00, -2.356194E+00,  7.853982E-
01);

            /*  frame_id  : ARM_2$2  */

#367 =     /* --- FRAME 4 --- */
    &SCOPE
    #368 = CRTPNT(  0.000000E+00,  0.000000E+00,  0.000000E+00);
    #369 = DIRCTN(  0.000000E+00,  1.000000E+00,  0.000000E+00);
    #370 = DIRCTN(  7.071068E-01,  0.000000E+00,  7.071068E-01);
    ENDSCOPE
    AXSPLZ(#368, #369, #370);

    #371 = REVOLUTE_PAIR(  0.000000E+00, -7.853982E-01,
2.356194E+00);

    #372 = PAIR_PLACEMENT_STRUCTURE(#357, #345);
    #373 = KINEMATIC_LINK((#372), $, $ );

    #374 = PAIR_PLACEMENT_STRUCTURE(#357, #353);
    #375 = PAIR_PLACEMENT_STRUCTURE(#366, #358);
    #376 = KINEMATIC_LINK((#374, #375), $, (#195));
```

```
#377 = PAIR_PLACEMENT_STRUCTURE(#366, #362);
#378 = PAIR_PLACEMENT_STRUCTURE(#371, #345);
#379 = KINEMATIC_LINK((#377, #378), $, (#293));

#380 = PAIR_PLACEMENT_STRUCTURE(#371, #367);
#381 = KINEMATIC_LINK((#380), $, (#298));

#382 = KINEMATIC_JOINT(#373, #376, #357);
#383 = KINEMATIC_JOINT(#376, #379, #366);
#384 = KINEMATIC_JOINT(#379, #381, #371);

#385 = KINEMATIC_STRUCTURE((#382, #383, #384), $);
ENDSCOPE
MECHANISM(#385, #373, #349, #339);
#386 = INDEX_ENTRY('NIRO_test_mech', #344);
         /*           End niro_mec.mpc        */
         /*  frame_id   : GROUND$1  */
         /*           End niro_env.mpc        */
ENDSCOPE
KINEMATIC_MODEL(#338, (#344), $, $);
ENDSEC;
ENDSTEP;
```

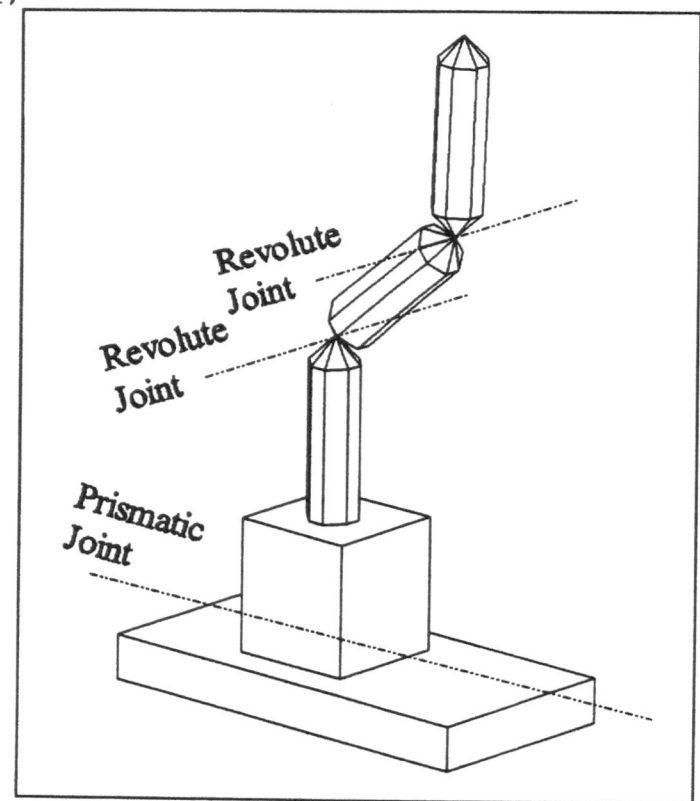

Fig. A1-3: The simple robot used in the file print-outs above.

Appendix 2

Example IRL Output for GRASP Preprocessor

This section illustrates the mapping between the GRASP robot program file and the associated preprocessed IRL program file. This program includes general program steps. It is not intended to perform any useful task, rather to demonstrate the mapping between GRASP track steps and the equivalent IRL statements. The robot program generated in GRASP is shown in Figure A2-1. The postprocessed IRL file is shown in Figure A2-2. The example contains a GRASP subtrack. Some variables/data constructs are referenced in the track which are not shown in the GRASP track file, for example, a point to point path called PTP_PATH.

Two global variables also exist which are not shown in the GRASP track file. These are :

    GB1 - input, Boolean, port 2
    GI1 - input, integer, port 1

Fig. A2-1: General steps output to GRASP track file

```
track TEST

variables
  signal  LB1 ,LB2
  integer LI1 ,LI2
  real    LR1 ,LR2
end

message 'This is an example GRASP track',

message '@IRL_OR_ICR_SPECIFIC_COMMAND ;',

pause 5,

set LI1 to 12,

park path PTP_PATH  ,

position WORKPLACE
 (shift X 333.428 Z 284.230 rotate Y 156.620 Z 90.000)
```

```
 path PTP_PATH ,

Joint_position
 values (0.0000 , 0.0000 , -26.7030 , 0.0000 , 0.0000)
 path PTP_PATH ,

if GB1 then ITS_ON else ITS_OFF ,

call SUBDUMMY,

ITS_ON : message 'High',

goto FINISH ,

ITS_OFF : message 'Low',
FINISH : message 'Wait then finish',

wait until GI1 >=LI1 ,

;


track SUBDUMMY

message 'Dummy track to be called then return',
;

Stop;
```

Fig. A2-2: General steps preprocessed to IRL file

```
PROGRAM TEST ; DECLON ;

CONST
  POSE: WORKPLACE :=
    (POSITION(0.0,0.0,0.0),
     ORIZYZ2(0.0,0.0,0.0)));

  POSE: TCPNUL :=
    (POSITION(0.0,0.0,0.0),
     ORIZYZ2(0.0,0.0,0.0)));

VAR
```

```
{ Global variables used for IO }
  INPUT   INT  : GI1 AT 2 ;
  INPUT   BOOL : GB1 AT 1 ;


{ Local variables for this track }

  INT  : LI1, LI2 ;
  REAL : LR1, I ;
  BOOL : LB1, LB2 ;

PROCEDURE SUBDUMMY () ;

BEGIN

WRITELN (SCREEN,'Dummy track to be called then return'
) ;

ENDPROC ;

{ *** Program begins here *** }

BEGIN

WRITELN (SCREEN,'This is an example GRASP track' ) ;

{ Outputting statement from message step }
IRL_OR_ICR_SPECIFIC_COMMAND ;

WAIT 5 SEC ;

LI1 := 12 ;


{ -----> Park step <----- }

$TOOL := TCPNUL ;

MOVE PTP
 ROBTARGET
  (POSE(POSITION(430.0001,0.0,430.0),
        ORIZYZ2(0.0,135.0,90.0)))
 SPEED_PTP := 100.0 ;
```

```
{ -----> Position step <----- }

MOVE PTP
 ROBTARGET
  (WORKPLACE*
   POSE(POSITION(333.428,0.0,284.23),
        ORIZYZ2(0.0,156.62,89.99999)))
 SPEED_PTP := 100.0 ;


{ -----> Joint position step <----- }

MOVE PTP
  JOINT(MAIN_JOINT(0.0,0.0,-26.703,0.0,0.0))
 SPEED_PTP := 100.0 ;

IF GB1 = TRUE THEN
  GOTO ITS_ON ;
ELSE
  GOTO ITS_OFF ;
ENDIF ;

SUBDUMMY () ;

ITS_ON : WRITELN (SCREEN,'High' );

GOTO FINISH ;

ITS_OFF : WRITELN (SCREEN,'Low' );

FINISH : WRITELN (SCREEN,'Wait then finish' );

WAIT FOR GI1>=LI1 ;

ENDPROGRAM ;
```

Appendix 3

Example ICR Output for GRASP Preprocessor

This section illustrates the mapping between the GRASP robot program file and the associated preprocessed ICR program file. This program includes general program steps. It is not intended to perform any useful task, rather to demonstrate the mapping between GRASP track steps and the equivalent ICR statements. The robot program generated in GRASP is shown in Figure A3-1. The postprocessed ICR file is shown in Figure A3-2. The example contains a GRASP subtrack. Some variables/data constructs are referenced in the track which are not shown in the GRASP track file, for example, a point to point path called PTP_PATH.

Two global variables also exist which are not shown in the GRASP track file. These are :

      GB1 - input, Boolean, port 2
      GI1 - input, integer, port 1

Fig. A3-1: General steps output to GRASP track file

```
track TEST

variables
  signal  LB1 ,LB2
  integer LI1 ,LI2
  real    LR1 ,LR2
end

message 'This is an example GRASP track',

message '@IRL_OR_ICR_SPECIFIC_COMMAND ;',

pause 5,

set LI1 to 12,

park path PTP_PATH  ,

position WORKPLACE
 (shift X 333.428 Z 284.230 rotate Y 156.620 Z 90.000)
```

```
 path PTP_PATH ,

Joint_position
 values (0.0000 , 0.0000 , -26.7030 , 0.0000 , 0.0000)
 path PTP_PATH ,

if GB1 then ITS_ON else ITS_OFF ,

call SUBDUMMY,

ITS_ON : message 'High',

goto FINISH ,

ITS_OFF : message 'Low',
FINISH : message 'Wait then finish',

wait until GI1 >=LI1 ,

;


track SUBDUMMY

message 'Dummy track to be called then return',
;

Stop;
```

Fig. A3-2: General steps preprocessed to ICR program file

```
1,PBEG,32,"TEST"
2,REMARK,0,"";
3,REMARK,0,"~~~~~~~~~~~~~~~~~~~~~~~~~~~~~~~~~~~~~~~~~~~~~~~"
;
4,SUBPBEG,"SUBDUMMY";
5,BLKBEG,1,1;
6,REMARK,0," Get call stack arguments if necessary ";
7,REMARK,0,"Dummy track to be called then return";
8,BLKEND;
9,SUBPEND;
```

```
10,REMARK,0,"~~~~~~~~~~~~~~~~~~~~~~~~~~~~~~~~~~~~~~~~~~~~~~"
;
11,REMARK,0,"";
12,REMARK,0,"";
13,REMARK,0,"~~~~~~~~~~~~~~~~~~~~~~~~~~~~~~~~~~~~~~~~~~~~~~"
;
14,SUBPBEG,"RTARGET";
15,BLKBEG,2,1;
16,PUSHI,#0;
17,PUSHI,#0;
18,PUSHI,#0;
19,PUSHI,#0;
20,PUSHR,#0.0000;
21,PUSHR,#0.0000;
22,PUSHR,#0.0000;
23,PUSHR,#0.0000;
24,PUSHR,#0.0000;
25,GENJ;
26,GENT;
27,BLKEND;
28,SUBPEND;
29,REMARK,0,"~~~~~~~~~~~~~~~~~~~~~~~~~~~~~~~~~~~~~~~~~~~~~~"
;
30,REMARK,0,"";
31,REMARK,0,"-----------START  OF  MAIN  PROGRAM--------
";
32,BLKBEG,0,0;
33,DECLVAR,0,DL_NUMBER,1,0,DL_SOURCE,1,0,DL_DATE,5;
34,DECLVAR,0,TMP_POSE1,10,0,TMP_POSE2,10,0,TMP_POSE3,10
;
35,DECLVAR,0,TMP_POSE4,10,0,TMP_JOINT,11,0,STATUS,1;
36,DECLVAR,1,LB1,3,1,LB2,3,1,LI1,1;
37,DECLVAR,1,LI2,1,1,LR1,2,1,LR2,2;
38,DLOPEN,"TESTDAT",DL_NUMBER,DL_SOURCE,DL_DATE;
39,REMARK,0," Grasp output z-y-z orientations ";
40,CHECK_ORI,2;
41,REMARK,0," Check physical and auxiliary axes ";
42,CHECK_AXES,5,0;
43,REMARK,0," Output all tcps as end effectors ";
44,EFF_DE,5,"TCPNUL",10,#(0.0,0.0,0.0,2,0.0,0.0,0.0,0.0
);
45,REMARK,0," *** CHANNEL DATA *** ";
46,REMARK,0," Port __1 , Digital , Input GB1 , Input  ;
47,PUSHI,#0
```

```
48,PUSHI,#1;
49,CNFGCHAN;
50,PUSHI,#1;
51,OPENCHAN;
52,REMARK,0," *** CHANNEL DATA *** ";
53,REMARK,0," Port __2 , Analogue , Input GI1 , Input
;
54,PUSHI,#0
55,PUSHI,#2;
56,CNFGCHAN;
57,PUSHI,#2;
58,OPENCHAN;
59,REMARK,0,"This is an example GRASP track";
60,REMARK,0,"Outputting  instructions  from  message  ...
";
61,REMARK,0,"       IRL_OR_ICR_SPECIFIC_COMMAND ";
62,IRL_OR_ICR_SPECIFIC_COMMAND;
63,PUSHI,#5;
64,FLOAT;
65,DELAY;
66,PUSHI,#12;
67,POPI,LI1;
68,REMARK,0," ----->  Park step <----- ";
69,TOOLNA,5,"TCPNUL";
70,PUSHI,#0;
71,PUSHR,#100.0;
72,W_VEL,J,1;
73,DLEIN,"TESTDAT","G1",10,TMP_POSE1,STATUS;
74,PUSHE,TMP_POSE1;
75,CALL_S,"RTARGET";
76,RMOVE,1,T,W;
77,REMARK,0," ----->  Position step <----- ";
78,DLEIN,"TESTDAT","WORKPLACE",10,TMP_POSE1,STATUS;
79,PUSHE,TMP_POSE1;
80,DLEIN,"TESTDAT","G2",10,TMP_POSE2,STATUS;
81,PUSHE,TMP_POSE2;
82,TRANSE;
83,CALL_S,"RTARGET";
84,RMOVE,1,T,W;
85,REMARK,0," ----->  Joint Position step <----- ";
86,DLEIN,"TESTDAT","G3",11,TMP_JOINT,STATUS;
87,PUSHJ,TMP_JOINT;
88,JMOVE,1,J,W;
89,PUSHI,#1;
```

```
90,DIN;
91,PUSHI,#1;
92,EQI;
93,NOTB;
94,IFSYM,ITS_ON;
95,CALL_S,"SUBDUMMY";
96,ITS_ON;
97,REMARK,0,"High";
98,GOTOSYM,FINISH;
99,ITS_OFF;
100,REMARK,0,"Low";
101,FINISH;
102,REMARK,0,"Wait then finish";
103,PUSHI,#1;
104,CLOSECHAN;
105,PUSHI,#2;
106,CLOSECHAN;
107,DLCLS,"TESTDAT",STATUS;
108,BLKEND;
109,PEND;
```

Associated data file for ICR general steps

```
1,DLHEAD,"TESTDAT",5,1,"4,12,1991,16,18,59";
2,DLDAT,"WORKPLACE",10,0.0,0.0,0.0,2,0.0,0.0,0.00.0;
3,DLDAT,"TCPNUL",10,0.0,0.0,0.0,2,0.0,0.0,0.00.0;
4,DLDAT,"G1",10,430.0001,0.0,430.0,2,0.0,135.0,90.0,0.0
;
5,DLDAT,"G2",10,333.428,0.0,284.23,2,0.0,156.62,90.0,0.
0;
6,DLDAT,"G3",11,0.0,0.0,-26.703,0.0,0.0;
7,DLEND,"TESTDAT";
```

Appendix 4

Test Example: Movement File and Related ICR Code File

The example given in this section includes a "generic" movement file, with all movement parameters defined in sequential order , and its related ICR code file. The programs included are a "summarised" version of the test programs actually run on the robot controller where a larger number of points are actually used. It reproduces a plane square with and without sensor commands.

Fig. A4-1: Generic File

```
GENERIC FILE

JMOVE 100.0 1500.0 1600.0 6 1.0 0.0 0.0 0.0 -1000.00 -
500.00 0 0 0 0 1 T W;
LMOVE 100.0 1500.0 1400.0 6 1.0 0.0 0.0 0.0 -1000.00 -
500.00 0 0 0 0 1 T W;
LMOVE -100.0 1500.0 1400.0 6 1.0 0.0 0.0 0.0 -1000.00 -
500.00 0 0 0 0 1 T W;
LMOVE -100.0 1500.0 1600.0 6 1.0 0.0 0.0 0.0 -1000.00 -
500.00 0 0 0 0 1 T N;
LMOVE 100.0 1500.0 1600.0 6 1.0 0.0 0.0 0.0 -1000.00 -
500.00 0 0 0 0 1 T N;
LMOVE 100.0 0.0 100.0 6 0.0 0.0 0.0 0.0 0.00 0.00 0 0 0
0 2 T W;
LMOVE 0.0 0.0 -400.0 6 0.0 0.0 0.0 0.0 0.00 0.00 0 0 0
0 2 T W;
LMOVE -400.0 0.0 0.0 6 0.0 0.0 0.0 0.0 0.00 0.00 0 0 0
0 2 T N;
LMOVE 0.0 0.0 400.0 6 0.0 0.0 0.0 0.0 0.00 0.00 0 0 0 0
2 T N;
LMOVE 400.0 0.0 0.0 6 0.0 0.0 0.0 0.0 0.00 0.00 0 0 0 0
2 T N;
```

Fig. A4-2: ICR Code

```
ICR CODE

1,PBEG,221,"ROBOT";

2,LABEL,"POSICION";
3,BLKBEG,0,1;
4,PUSHR,#100.000000;
5,PUSHR,#1500.000000;
6,PUSHR,#1600.000000;
7,PUSHI,#6;
8,PUSHR,#1.000000;
9,PUSHR,#0.000000;
10,PUSHR,#0.000000;
11,PUSHR,#0.000000;
12,GENERR;
13,PUSHI,#0;
14,PUSHI,#0;
15,PUSHI,#0;
16,PUSHI,#0;
17,PUSHR,#-1000.000000;
18,PUSHR,#-500.000000;
19,GEND;
20,GENT;
21,JMOVE,1,T,W;
22,PUSHR,#100.000000;
23,PUSHR,#1500.000000;
24,PUSHR,#1400.000000;
25,PUSHI,#6;
26,PUSHR,#1.000000;
27,PUSHR,#0.000000;
28,PUSHR,#0.000000;
29,PUSHR,#0.000000;
30,GENERR;
31,PUSHI,#0;
32,PUSHI,#0;
33,PUSHI,#0;
34,PUSHI,#0;
35,PUSHR,#-1000.000000;
36,PUSHR,#-500.000000;
37,GEND;
38,GENT;
39,LMOVE,1,T,W;
```

```
40,PUSHR,#-100.000000;
41,PUSHR,#1500.000000; ·
42,PUSHR,#1400.000000;
43,PUSHI,#6;
44,PUSHR,#1.000000;
45,PUSHR,#0.000000;
46,PUSHR,#0.000000;
47,PUSHR,#0.000000;
48,GENERR;
49,PUSHI,#0;
50,PUSHI,#0;
51,PUSHI,#0;
52,PUSHI,#0;
53,PUSHR,#-1000.000000;
54,PUSHR,#-500.000000;
55,GEND;
56,GENT;
57,LMOVE,1,T,W;
58,PUSHR,#-100.000000;
59,PUSHR,#1500.000000;
60,PUSHR,#1600.000000;
61,PUSHI,#6;
62,PUSHR,#1.000000;
63,PUSHR,#0.000000;
64,PUSHR,#0.000000;
65,PUSHR,#0.000000;
66,GENERR;
67,PUSHI,#0;
68,PUSHI,#0;
69,PUSHI,#0;
70,PUSHI,#0;
71,PUSHR,#-1000.000000;
72,PUSHR,#-500.000000;
73,GEND;
74,GENT;
75,LMOVE,1,T,N;
76,PUSHR,#100.000000;
77,PUSHR,#1500.000000;
78,PUSHR,#1600.000000;
79,PUSHI,#6;
80,PUSHR,#1.000000;
81,PUSHR,#0.000000;
82,PUSHR,#0.000000;
83,PUSHR,#0.000000;
```

```
84,GENERR;
85,PUSHI,#0;
86,PUSHI,#0;
87,PUSHI,#0;
88,PUSHI,#0;
89,PUSHR,#-1000.000000;
90,PUSHR,#-500.000000;
91,GEND;
92,GENT;
93,LMOVE,1,T,N;
94,PUSHR,#100.000000;
95,PUSHR,#0.000000;
96,PUSHR,#100.000000;
97,PUSHI,#6;
98,PUSHR,#0.000000;
99,PUSHR,#0.000000;
100,PUSHR,#0.000000;
101,PUSHR,#0.000000;
102,GENERR;
103,PUSHI,#0;
104,PUSHI,#0;
105,PUSHI,#0;
106,PUSHI,#0;
107,PUSHR,#0.000000;
108,PUSHR,#0.000000;
109,GEND;
110,GENT;
111,LMOVE,2,T,W;
112,PUSHR,#0.000000;
113,PUSHR,#0.000000;
114,PUSHR,#-400.000000;
115,PUSHI,#6;
116,PUSHR,#0.000000;
117,PUSHR,#0.000000;
118,PUSHR,#0.000000;
119,PUSHR,#0.000000;
120,GENERR;
121,PUSHI,#0;
122,PUSHI,#0;
123,PUSHI,#0;
124,PUSHI,#0;
125,PUSHR,#0.000000;
126,PUSHR,#0.000000;
127,GEND;
```

```
128,GENT;
129,LMOVE,2,T,W;
130,PUSHR,#-400.000000;
131,PUSHR,#0.000000;
132,PUSHR,#0.000000;
133,PUSHI,#6;
134,PUSHR,#0.000000;
135,PUSHR,#0.000000;
136,PUSHR,#0.000000;
137,PUSHR,#0.000000;
138,GENERR;
139,PUSHI,#0;
140,PUSHI,#0;
141,PUSHI,#0;
142,PUSHI,#0;
143,PUSHR,#0.000000;
144,PUSHR,#0.000000;
145,GEND;
146,GENT;
147,LMOVE,2,T,N;
148,PUSHR,#0.000000;
149,PUSHR,#0.000000;
150,PUSHR,#400.000000;
151,PUSHI,#6;
152,PUSHR,#0.000000;
153,PUSHR,#0.000000;
154,PUSHR,#0.000000;
155,PUSHR,#0.000000;
156,GENERR;
157,PUSHI,#0;
158,PUSHI,#0;
159,PUSHI,#0;
160,PUSHI,#0;
161,PUSHR,#0.000000;
162,PUSHR,#0.000000;
163,GEND;
164,GENT;
165,LMOVE,2,T,N;
166,PUSHR,#400.000000;
167,PUSHR,#0.000000;
168,PUSHR,#0.000000;
169,PUSHI,#6;
170,PUSHR,#0.000000;
171,PUSHR,#0.000000;
```

```
172,PUSHR,#0.000000;
173,PUSHR,#0.000000;
174,GENERR;
175,PUSHI,#0;
176,PUSHI,#0;
177,PUSHI,#0;
178,PUSHI,#0;
179,PUSHR,#0.000000;
180,PUSHR,#0.000000;
181,GEND;
182,GENT;
183,LMOVE,2,T,N;
184,BLKEND;
185,SUBPEND;
186,LABEL,"INICIO";
187,BLKBEG,2,1;
188,PUSHI,#1;
189,SELECTIR;
190,PUSHI,#1;
191,PUSHI,#0;
192,PUSHI,#0;
193,PUSHI,#0;
194,PUSHI,#0;
195,CNFGCHAN;
196,PUSHI,#0;
197,OPENCHAN;
198,BLKEND;
199,SUBPEND;
200,LABEL,"PARAMETROS";
201,BLKBEG,3,1;
202,PUSHI,#0;
203,PUSHR,#66.000000;
204,W_VEL,T,1;
205,LIMIT,1,0,#10.000000;
206,LIMIT,1,1,#90.000000;
207,LIMIT,2,0,#10.000000;
208,LIMIT,2,1,#90.000000;
209,PUSHI,#1;
210,PUSHI,#0;
211,W_RESDP,T,2;
212,BLKEND;
213,SUBPEND;
214,LABEL,"CIERRE";
215,BLKBEG,4,1;
```

```
216,PUSHI,#0;
217,CLOSECHAN;
218,BLKEND;
219,SUBPEND;
220,REMARK,30,"    ****    MAIN    ****  ";
221,BLKBEG,1,0;
222,DECLVAR,0,VAX,2,0,VAY,2,0,VAZ,2;
223,DECLVAR,0,Q1,2,0,Q2,2,0,Q3,2,0,Q4,2,0,TIORI,1;
224,DECLVAR,0,ROPOSI,8,0,ROORI,9,0,ROPOSE,10;
225,DECLVAR,0,EJEEX,13,0,ERRMOV,1,0,ROFIN,14;
226,CHECK_ORI,6;
227,CHECK_AXES,6,2;
228,CALL_S,"INICIO";
229,CALL_S,"PARAMETROS";
230,CALL_S,"POSICION";
231,CALL_S,"CIERRE";
232,BLKEND;
233,PEND;
```

Fig. A4-3: Generic File

```
GENERIC FILE

JMOVE 200.0 1500.0 1700.0 6 1.0 0.0 0.0 0.0 -1000.00 -
500.00 0 0 0 0 1 T W;
SENSOR 4000 3 0;
SENSOR 2500 4 0;
SENSOR 2300 3 1;
JMOVE -200.0 1500.0 1700.0 6 1.0 0.0 0.0 0.0 -1000.00 -
500.00 0 0 0 0 1 T W;
SENSOR 4000 3 0;
SENSOR 2500 4 0;
SENSOR 2300 3 1;
JMOVE -200.0 1500.0 1500.0 6 1.0 0.0 0.0 0.0 -1000.00 -
500.00 0 0 0 0 1 T N;
SENSOR 4000 3 0;
SENSOR 2500 4 0;
SENSOR 2300 3 1;
JMOVE 200.0 1500.0 1500.0 6 1.0 0.0 0.0 0.0 -1000.00 -
500.00 0 0 0 0 2 T W;
SENSOR 4000 3 0;
SENSOR 2500 4 0;
SENSOR 2300 3 1;
```

```
JMOVE 200.0 1500.0 1700.0 6 1.0 0.0 0.0 0.0 -1000.00 -
500.00 0 0 0 0 2 T N;
SENSOR 4000 3 0;
SENSOR 2500 4 0;
SENSOR 2300 3 1;
```

Fig. A4-4: ICR Code

```
ICR CODE

1,PBEG,191,"ROBOT";
2,LABEL,"POSICION";
3,BLKBEG,0,1;
4,PUSHR,#200.000000;
5,PUSHR,#1500.000000;
6,PUSHR,#1700.000000;
7,PUSHI,#6;
8,PUSHR,#1.000000;
9,PUSHR,#0.000000;
10,PUSHR,#0.000000;
11,PUSHR,#0.000000;
12,GENERR;
13,PUSHI,#0;
14,PUSHI,#0;
15,PUSHI,#0;
16,PUSHI,#0;
17,PUSHR,#-1000.000000;
18,PUSHR,#-500.000000;
19,GEND;
20,GENT;
21,JMOVE,1,T,W;
22,PUSHI,#4000;
23,PUSHI,#3;
24,PUSHI,#0;
25,STOP,A;
26,PUSHI,#2500;
27,PUSHI,#4;
28,PUSHI,#0;
29,STOP,A;
30,PUSHI,#2300;
31,PUSHI,#3;
32,PUSHI,#1;
33,STOP,A;
```

```
34,PUSHR,#-200.000000;
35,PUSHR,#1500.000000;
36,PUSHR,#1700.000000;
37,PUSHI,#6;
38,PUSHR,#1.000000;
39,PUSHR,#0.000000;
40,PUSHR,#0.000000;
41,PUSHR,#0.000000;
42,GENERR;
43,PUSHI,#0;
44,PUSHI,#0;
45,PUSHI,#0;
46,PUSHI,#0;
47,PUSHR,#-1000.000000;
48,PUSHR,#-500.000000;
49,GEND;
50,GENT;
51,JMOVE,1,T,W;
52,PUSHI,#4000;
53,PUSHI,#3;
54,PUSHI,#0;
55,STOP,A;
56,PUSHI,#2500;
57,PUSHI,#4;
58,PUSHI,#0;
59,STOP,A;
60,PUSHI,#2300;
61,PUSHI,#3;
62,PUSHI,#1;
63,STOP,A;
64,PUSHR,#-200.000000;
65,PUSHR,#1500.000000;
66,PUSHR,#1500.000000;
67,PUSHI,#6;
68,PUSHR,#1.000000;
69,PUSHR,#0.000000;
70,PUSHR,#0.000000;
71,PUSHR,#0.000000;
72,GENERR;
73,PUSHI,#0;
74,PUSHI,#0;
75,PUSHI,#0;
76,PUSHI,#0;
77,PUSHR,#-1000.000000;
```

```
78,PUSHR,#-500.000000;
79,GEND;
80,GENT;
81,JMOVE,1,T,N;
82,PUSHI,#4000;
83,PUSHI,#3;
84,PUSHI,#0;
85,STOP,A;
86,PUSHI,#2500;
87,PUSHI,#4;
88,PUSHI,#0;
89,STOP,A;
90,PUSHI,#2300;
91,PUSHI,#3;
92,PUSHI,#1;
93,STOP,A;
94,PUSHR,#200.000000;
95,PUSHR,#1500.000000;
96,PUSHR,#1500.000000;
97,PUSHI,#6;
98,PUSHR,#1.000000;
99,PUSHR,#0.000000;
100,PUSHR,#0.000000;
101,PUSHR,#0.000000;
102,GENERR;
103,PUSHI,#0;
104,PUSHI,#0;
105,PUSHI,#0;
106,PUSHI,#0;
107,PUSHR,#-1000.000000;
108,PUSHR,#-500.000000;
109,GEND;
110,GENT;
111,JMOVE,2,T,W;
112,PUSHI,#4000;
113,PUSHI,#3;
114,PUSHI,#0;
115,STOP,A;
116,PUSHI,#2500;
117,PUSHI,#4;
118,PUSHI,#0;
119,STOP,A;
120,PUSHI,#2300;
121,PUSHI,#3;
```

```
122,PUSHI,#1;
123,STOP,A;
124,PUSHR,#200.000000;
125,PUSHR,#1500.000000;
126,PUSHR,#1700.000000;
127,PUSHI,#6;
128,PUSHR,#1.000000;
129,PUSHR,#0.000000;
130,PUSHR,#0.000000;
131,PUSHR,#0.000000;
132,GENERR;
133,PUSHI,#0;
134,PUSHI,#0;
135,PUSHI,#0;
136,PUSHI,#0;
137,PUSHR,#-1000.000000;
138,PUSHR,#-500.000000;
139,GEND;
140,GENT;
141,JMOVE,2,T,N;
142,PUSHI,#4000;
143,PUSHI,#3;
144,PUSHI,#0;
145,STOP,A;
146,PUSHI,#2500;
147,PUSHI,#4;
148,PUSHI,#0;
149,STOP,A;
150,PUSHI,#2300;
151,PUSHI,#3;
152,PUSHI,#1;
153,STOP,A;
154,BLKEND;
155,SUBPEND;
156,LABEL,"INICIO";
157,BLKBEG,2,1;
158,PUSHI,#1;
159,SELECTIR;
160,PUSHI,#1;
161,PUSHI,#0;
162,PUSHI,#0;
163,PUSHI,#0;
164,PUSHI,#0;
165,CNFGCHAN;
```

```
166,PUSHI,#0;
167,OPENCHAN;
168,BLKEND;
169,SUBPEND;
170,LABEL,"PARAMETROS";
171,BLKBEG,3,1;
172,PUSHI,#0;
173,PUSHR,#66.000000;
174,W_VEL,T,1;
175,LIMIT,1,0,#10.000000;
176,LIMIT,1,1,#90.000000;
177,LIMIT,2,0,#10.000000;
178,LIMIT,2,1,#90.000000;
179,PUSHI,#1;
180,PUSHI,#0;
181,W_RESDP,T,2;
182,BLKEND;
183,SUBPEND;
184,LABEL,"CIERRE";
185,BLKBEG,4,1;
186,PUSHI,#0;
187,CLOSECHAN;
188,BLKEND;
189,SUBPEND;
190,REMARK,30,"   ****   MAIN   **** ";
191,BLKBEG,1,0;
192,DECLVAR,0,VAX,2,0,VAY,2,0,VAZ,2;
193,DECLVAR,0,Q1,2,0,Q2,2,0,Q3,2,0,Q4,2,0,TIORI,1;
194,DECLVAR,0,ROPOSI,8,0,ROORI,9,0,ROPOSE,10;
195,DECLVAR,0,EJEEX,13,0,ERRMOV,1,0,ROFIN,14;
196,CHECK_ORI,6;
197,CHECK_AXES,6,2;
198,CALL_S,"INICIO";
199,CALL_S,"PARAMETROS";
200,CALL_S,"POSICION";
201,CALL_S,"CIERRE";
202,BLKEND;
203,PEND;
```

Springer-Verlag und Umwelt

Als internationaler wissenschaftlicher Verlag sind wir uns unserer besonderen Verpflichtung der Umwelt gegenüber bewußt und beziehen umweltorientierte Grundsätze in Unternehmensentscheidungen mit ein.

Von unseren Geschäftspartnern (Druckereien, Papierfabriken, Verpackungsherstellern usw.) verlangen wir, daß sie sowohl beim Herstellungsprozeß selbst als auch beim Einsatz der zur Verwendung kommenden Materialien ökologische Gesichtspunkte berücksichtigen.

Das für dieses Buch verwendete Papier ist aus chlorfrei bzw. chlorarm hergestelltem Zellstoff gefertigt und im pH-Wert neutral.